Spatial Analysis Methods of Road Traffic Collisions

Spatial Analysis Methods of Road Traffic Collisions

Becky P.Y. Loo
Tessa Kate Anderson

CRC Press is an imprint of the
Taylor & Francis Group, an **informa** business

CRC Press
Taylor & Francis Group
6000 Broken Sound Parkway NW, Suite 300
Boca Raton, FL 33487-2742

© 2016 by Taylor & Francis Group, LLC
CRC Press is an imprint of Taylor & Francis Group, an Informa business

No claim to original U.S. Government works

Printed on acid-free paper
Version Date: 20150722

International Standard Book Number-13: 978-1-4398-7412-7 (Hardback)

This book contains information obtained from authentic and highly regarded sources. Reasonable efforts have been made to publish reliable data and information, but the author and publisher cannot assume responsibility for the validity of all materials or the consequences of their use. The authors and publishers have attempted to trace the copyright holders of all material reproduced in this publication and apologize to copyright holders if permission to publish in this form has not been obtained. If any copyright material has not been acknowledged please write and let us know so we may rectify in any future reprint.

Except as permitted under U.S. Copyright Law, no part of this book may be reprinted, reproduced, transmitted, or utilized in any form by any electronic, mechanical, or other means, now known or hereafter invented, including photocopying, microfilming, and recording, or in any information storage or retrieval system, without written permission from the publishers.

For permission to photocopy or use material electronically from this work, please access www.copyright.com (http://www.copyright.com/) or contact the Copyright Clearance Center, Inc. (CCC), 222 Rosewood Drive, Danvers, MA 01923, 978-750-8400. CCC is a not-for-profit organization that provides licenses and registration for a variety of users. For organizations that have been granted a photocopy license by the CCC, a separate system of payment has been arranged.

Trademark Notice: Product or corporate names may be trademarks or registered trademarks, and are used only for identification and explanation without intent to infringe.

Visit the Taylor & Francis Web site at
http://www.taylorandfrancis.com

and the CRC Press Web site at
http://www.crcpress.com

Printed and bound in Great Britain by
TJ International Ltd, Padstow, Cornwall

Contents

List of Figures xiii
List of Tables xv
Preface xvii
Authors xix
Abbreviations xxi

Chapter 1 Collisions as Spatial Events 1

    1.1   Introduction 1
    1.2   Distance-Based Methods 8
    1.3   Simple Means Methods 9
    1.4   Simple Variance Methods 11
    1.5   Nearest Neighbor Analysis 12
    1.6   Conclusion 16
    References 17

Chapter 2 Collision Density in Two-Dimensional Space 19

    2.1   Introduction 19
    2.2   Quadrat Methods 20
    2.3   Simple Density Functions 21
        2.3.1   Histograms 22
        2.3.2   K-Function Method 23
    2.4   Spatial Autocorrelation 23
        2.4.1   Global Order Effects 25
        2.4.2   Local Indicators of Spatial Autocorrelation (LISA) 26
    2.5   Kernel Density Estimation 27
        2.5.1   Optimum Bandwidth 29
        2.5.2   Case Study: Road Collisions in London, United Kingdom 30
    2.6   Geographically Weighted Regression 32
    2.7   Conclusion 34
    References 35

Chapter 3 Road Safety as a Public Health Issue 39

    3.1   Why Would Road Collisions Be Considered a Public Health Issue? 39
    3.2   Current Global Estimates 41
    3.3   Irtad Database Coverage and Underreporting 42
    3.4   Economic, Social, and Health Burdens 48
    3.5   Global Geography of Road Risk 50

v

	3.6	Road Safety and Development	50
	3.7	Global Statistics, Data, and Assessment	51
	3.8	Global Divide of Injury and Death, and Ultimately Burden	51
	3.9	Road Collision Costing	53
	3.10	International Road Infrastructure: A Neglected Measure?	54
	3.11	Conclusion	55
	References		57

Chapter 4 Risk and Socioeconomic Factors ... 59

4.1	Relationships and Risk		59
4.2	Socioeconomic Characteristics		60
	4.2.1	Deprivation	60
		4.2.1.1 What Is Deprivation?	60
		4.2.1.2 What Are the Influencing Factors?	60
		4.2.1.3 Child Pedestrians and Deprivation	61
		4.2.1.4 Scales of Factors Linking Deprivation, Disadvantage, and Road Collisions	63
	4.2.2	Ethnicity	65
	4.2.3	Exposure and Inequality	66
	4.2.4	Geodemographics	67
4.3	Measurement and Analysis		69
	4.3.1	Data	69
	4.3.2	Database Construction	70
	4.3.3	Methods	72
		4.3.3.1 Qualitative Data Analysis	73
		4.3.3.2 Descriptive Statistics	73
		4.3.3.3 Regression Analysis	73
		4.3.3.4 Geostatistics	74
		4.3.3.5 Typology Analysis	74
		4.3.3.6 Case Study: Geodemographics in London, United Kingdom	75
	4.3.4	Methodological Issues	77
	4.3.5	Can You Measure Risky Behavior?	77
4.4	Policy and Intervention		77
4.5	Conclusion		81
References			81

Chapter 5 Road Collisions and Risk-Taking Behaviors ... 85

5.1	Introduction	85
5.2	What Is *Risk-Taking* Behavior?	92
5.3	Measuring Risky Behavior	94
5.4	Age and Gender Differences	96
5.5	Culture and Ethnicity	98

Contents vii

	5.6	Drink-Driving 99
	5.7	Drug-Driving 100
	5.8	Conclusion 100
	References 101

Chapter 6 Road Collisions and Urban Development 107

    6.1 Urban Landscape and Road Safety 107
    6.2 Changing Urban Population and Road Collisions 108
    6.3 Urban Sprawl 109
    6.4 Effective Land Use Planning 111
        6.4.1 Pedestrian Land Use Planning 116
        6.4.2 Land Use Planning Risks 117
    6.5 Planning for Safety Awareness 121
        6.5.1 University Campuses 122
        6.5.2 Driveways 123
        6.5.3 Schools 128
    6.6 Conclusion 130
    References 130

Chapter 7 Nature of Spatial Data, Accuracy, and Validation 135

    7.1 Introduction 135
    7.2 Conceptualizing Collisions as Network Phenomena 135
    7.3 Issues Involved with Collisions-in-Networks in GIS 138
        7.3.1 Requirements of Spatial Accuracy and Precision of Collision Data 138
        7.3.2 Concept of Distance in Networks 139
    7.4 Geovalidation before Collision Analysis 139
    7.5 Case Study of Hong Kong 141
        7.5.1 Database Preparation 141
        7.5.2 Methodology 143
    7.6 Conclusion 145
    References 145

Chapter 8 Collisions in Networks 147

    8.1 Introduction 147
    8.2 MAUP in Networks 147
    8.3 Network Segmentation 149
    8.4 Basic Spatial Units in Collision Analysis 150
    8.5 Assigning Collisions to Networks 151
    8.6 Spatial Autocorrelation Analysis in Networks 151
        8.6.1 Link-Attribute Approaches 152
        8.6.2 Event-Based Approaches 155
    8.7 Conclusion 158
    References 158

Chapter 9 Cluster Identifications in Networks 161

- 9.1 Introduction 161
- 9.2 What Are Hazardous Road Locations? 161
  - 9.2.1 On the Definition of Sites 162
  - 9.2.2 Setting the Criteria 163
    - 9.2.2.1 Magic Figures 163
    - 9.2.2.2 Statistical Definitions 163
    - 9.2.2.3 Model-Based Definitions 164
- 9.3 Ranking Issues, False Positives, and False Negatives 166
- 9.4 HRL Identification Using Spatial Analysis 169
  - 9.4.1 Defining the Spatial Unit of Analysis and Calculating Collision Statistics 169
  - 9.4.2 Hot Zone Identification 169
    - 9.4.2.1 Link-Attribute Approach 170
    - 9.4.2.2 Event-Based Approach 172
- 9.5 Some Additional Methodological Remarks 173
  - 9.5.1 Study Period 173
  - 9.5.2 Degree of Injury 173
- 9.6 Conclusion 174
- References 175

Chapter 10 Exposure Factor 1: Traffic Volume 179

- 10.1 Introduction 179
- 10.2 Relationship between Traffic Flow and Collisions 179
- 10.3 Traffic Volume 181
- 10.4 Methods 185
  - 10.4.1 Simple Ratios 185
  - 10.4.2 Simple Exponents 185
  - 10.4.3 Linear Regression Models 186
  - 10.4.4 Poisson Regressions 186
  - 10.4.5 Negative Binomial Methods 190
- 10.5 Implications on Interventions 192
  - 10.5.1 Collision Count versus Collision Rate in Road Safety Analysis 192
  - 10.5.2 "Regression-to-Mean" Problems 193
- 10.6 Conclusion 193
- References 193

Chapter 11 Exposure Factor 2: Road Environment 197

- 11.1 Introduction 197
- 11.2 Relationship between Road Environment and Collisions 197
  - 11.2.1 Intersections and Mid-Block Locations 197
  - 11.2.2 Other Geometric Features 198

	11.3	Methods	198
		11.3.1 Logistical Regression	198
		11.3.2 Geographically Weighted Regression	198
		11.3.3 Empirical Bayes Methods	199
		11.3.4 Hierarchical Bayes Methods	201
	11.4	Intervention	201
	11.5	Evaluation	201
	11.6	Conclusion	204
	References		205

Chapter 12 Exposure Factor 3: Distance Traveled 207

12.1	Introduction	207
12.2	Methods	207
	12.2.1 Road Collision per Population and per Vehicle Registered	207
	12.2.2 Road Collision per Vehicle- and Passenger-km	208
	12.2.3 Time-Space Measures	208
12.3	Intervention	209
12.4	Conclusion	212
References		213

Chapter 13 Enforcement 215

13.1	Introduction	215
13.2	Managing Speeds	216
	13.2.1 Speed Limits	217
	13.2.2 Methods of Speed Enforcement	220
	13.2.2.1 Controversy	221
	13.2.2.2 Future	222
	13.2.3 Speed Cameras	223
	13.2.3.1 Background of Speed Cameras	223
	13.2.3.2 Types of Cameras	225
	13.2.3.3 Has Speed Dropped as a Result of Speed Cameras?	226
	13.2.3.4 Case Study: UK National Speed Camera Survey and Reduction of Injuries	226
	13.2.3.5 Benefits, Disadvantages, Controversies, and Effectiveness	227
13.3	Managing Drink-/Drug-Driving	231
	13.3.1 Drink-Driving	231
	13.3.2 Drug-Driving	233
13.4	Spatial Implications	235
13.5	Conclusion	235
References		236

Chapter 14 Engineering 241

- 14.1 Introduction 241
- 14.2 Location-Specific Treatments... 244
  - 14.2.1 Single Site... 244
  - 14.2.2 Mass Action.. 244
  - 14.2.3 Route Action... 244
  - 14.2.4 Area-Wide Action.. 244
- 14.3 Engineering Measures... 246
  - 14.3.1 Physical Engineering Measures 246
    - 14.3.1.1 Low Cost versus High Cost 246
    - 14.3.1.2 Roundabouts 249
  - 14.3.2 Management Measures... 249
    - 14.3.2.1 Generic Characteristics of the Road Safety Management System...................... 253
    - 14.3.2.2 Reduction and Prevention 253
  - 14.3.3 Vulnerable Road Users... 261
    - 14.3.3.1 Bicyclists... 262
    - 14.3.3.2 Pedestrians.. 263
- 14.4 Before-and-After Studies.. 264
- 14.5 Conclusion 268
- References 268

Chapter 15 Education 271

- 15.1 Introduction 271
- 15.2 Children and Youth 273
  - 15.2.1 School Education: Cycle Safety 275
  - 15.2.2 Probationary License.. 276
- 15.3 Elderly... 277
  - 15.3.1 Publicity and Campaigns 279
  - 15.3.2 Using Geodemographics to Target Road Users......... 284
- 15.4 Lost Generation 286
  - 15.4.1 Education.. 287
  - 15.4.2 Strategic Targeting 288
- 15.5 Issues of Ethnicity 288
- 15.6 Conclusion 289
- References 290

Chapter 16 Road Safety Strategy 293

- 16.1 Introduction 293
- 16.2 Traditional Approaches 293
- 16.3 Nine Components of the Road Safety Strategy 295
  - 16.3.1 Vision 295
  - 16.3.2 Objectives 295

		16.3.3	Targets 295
		16.3.4	Action Plan 296
		16.3.5	Evaluation and Monitoring 296
		16.3.6	Research and Development 296
		16.3.7	Quantitative Modeling 297
		16.3.8	Institutional Framework 297
		16.3.9	Funding 297
	16.4	Importance of Benchmarking and Incorporating Geographical Variability 297	
		16.4.1	International Best Practices 298
		16.4.2	Rural–Urban Divide 301
	16.5	Strategy in Stages 303	
		16.5.1	Short-Term Approach 303
		16.5.2	Medium-Term Approach 303
		16.5.3	Long-Term Approach 304
	16.6	Conclusion 304	
	References 305		

Appendix: STATS19 Data Record Sheets 309

Index 313

List of Figures

Figure 1.1 The integrated road safety system 5
Figure 1.2 Bar graphs showing the age of drivers and casualties involved in road collisions 11
Figure 1.3 Diagram showing patterns of dispersion to being clustered. 13
Figure 1.4 Nearest neighbor patterns. 14
Figure 2.1 Network spatial autocorrelation 25
Figure 2.2 Band 2 31
Figure 2.3 Band 4 32
Figure 3.1 Projected disability-adjusted life years (DALYs) in developing countries (children aged 5–14). 41
Figure 3.2 Collision speed–fatality relationships. 55
Figure 4.1 No-car households and child pedestrian/cyclist casualties. 61
Figure 4.2 Mosaic UK data sources 71
Figure 4.3 MAST online 79
Figure 5.1 Different kinds of risk 91
Figure 5.2 Risk thermostat 92
Figure 5.3 Relative rates of involvement in injury collisions by driver age and gender 97
Figure 6.1 Collision locations in the four urban forms 115
Figure 6.2 U.S. child fatalities in driveways. 124
Figure 6.3 Pedestrian–vehicular collisions located within quarter mile buffer of Baltimore City public schools, 2000–2002. 129
Figure 7.1 Random points in 2D and 1D space 136
Figure 7.2 (a) Road collision pattern in Hong Kong, 2008–2010. (b) Road network of Hong Kong 137
Figure 7.3 The GIS-based spatial data validation system 142
Figure 8.1 An illustration of MAUP in 2D point pattern analysis 148
Figure 8.2 A flowchart of the network dissolution procedures 150

Figure 8.3 A cross section of the kernel using the $3/\pi$ quartic function............ 156

Figure 8.4 Planar versus network *K*-function 157

Figure 9.1 A flowchart showing the steps of hot zone identification 171

Figure 10.1 Conflict points in a four-arm two-way junction.............................. 184

Figure 12.1 Comaps showing collision frequency, conditional upon six ST-slices.. 211

Figure 12.2 Comaps showing collision risk, conditional upon six ST-slices....... 212

Figure 14.1 Perspective view of *straight-through* crossroads............................ 245

Figure 14.2 Road safety management model.. 252

Figure 16.1 Evaluation of the road safety strategies for six administrations 298

Figure A.1 STATS19 vehicle records.. 310

Figure A.2 Attendant circumstances 311

Figure A.3 Casualty details 312

List of Tables

Table 1.1	A Brief Summary of the Evolution of UK Road Safety in the Twentieth Century	3
Table 1.2	Paradigms of Road Safety	4
Table 1.3	Summary of Statistics for Age of Driver and Casualty	12
Table 2.1	Variations in Search Radius and Bandwidth Using ESRI's ArcGIS Density Measure	31
Table 3.1	Leading Causes of Death in Children and Youth, Both Sexes, World, 2004	42
Table 3.2	Estimated Road Traffic Death Rate (per 100,000 Population), 2010	43
Table 3.3	Selected Data Sources about the Burden of Road Traffic Collisions in Iran, India, Mexico, and Ghana	52
Table 4.1	Child Pedestrian Injury Rates within Deprivation Deciles in London, 1999–2004	64
Table 4.2	Average Annual Pedestrian Injury Rates per 100,000 People in London, 1996–2006	66
Table 4.3	Mosaic Types and Associated Population Percentages and Index Scores for Both Casualties and Drivers	76
Table 5.1	Classification of Papers Proposed for Risk Analysis of Road Collisions	87
Table 5.2	Types of Behavioral Variables Related to Collision Risk	95
Table 7.1	Results of the Geovalidation of Traffic Collisions in Hong Kong, 2005–2010	144
Table 9.1	Problems of False Positives and False Negatives Illustrated	166
Table 10.1	Road Collision Fatality Numbers and Rates in Seven Administrations	182
Table 10.2	Estimated Parameters β_X	189
Table 10.3	Estimated β_X and β_L	190
Table 11.1	Rates (%) of Helmet Use among Motorcyclists at the Intersection with Helmet Law Enforcement Action	203
Table 11.2	Increases in the Rate (%) of Helmet Use as the Effect of the Helmet Law Enforcement, Gauged by the Naïve Before-and-After Approach and the EB Approach, Respectively	203

Table 11.3	Rates (%) of Helmet Use among Motorcyclists, Gauged by the Naïve Before-and-After Approach and the EB Approach (Method of Sample Moments)	204
Table 13.1	Factors Considered in the Setting of Speed Limits	219
Table 13.2	Number of PIC and KSI Prevented across Great Britain in Year Ending March 2004	228
Table 13.3	Controversies Associated with Speed Camera Use in Each of the Jurisdictions Grouped according to Goldenbeld's Dilemma Classifications	229
Table 14.1	Collision Situation and Engineering Remedies	243
Table 14.2	Collision Reduction Schemes in Oxfordshire, United Kingdom, 2007	246
Table 14.3	Potential Reductions (%) in Various Injury Collision Types	247
Table 14.4	List of Selected Road Engineering Safety Countermeasures	248
Table 14.5	Darwin Matrix for Traffic Calming	255
Table 14.6	Statistical Tests or Procedures for Different Designs and Criteria	266
Table 15.1	Effects of Road Safety Campaigns on Road Collisions	282

Preface

The original idea of this book was first discussed back in 2010, when we had the privileges of having lively discussions and exchange of ideas in face-to-face meetings at the University of Hong Kong. Many events took place thereafter, making heavy demand on our time and efforts. Hence, while we tried our best to spare time to write this book, it took four years to complete. Over time, many research assistants, particularly Tony Phuah and Dr. Ada Shenjun Yao, have helped in various editorial and communication matters. We are most grateful for their support. Moreover, we thankfully acknowledge the permissions from the relevant publishers/copyright holders (within parentheses) to reproduce Table 1.2 (OECD), Figure 1.1 (Victoria Transport Policy Institute), Figure 1.4 (Oxford University Press), Figure 2.1 (John Wiley & Sons), Table 3.2 (World Health Organization), Table 3.3 (Taylor & Francis Group), Figure 3.2 (Swedish National Road and Transport Research Institute), Table 4.1 (Phil Edwards, Judith Green, Ian Roberts, Chris Grundy, and Kate Lachowycz), Table 4.2 (Rebecca Steinbach, Phil Edwards, Judith Green, and Chris Grundy), Table 4.3 (Pion Ltd), Figure 4.1 (Elsevier), Figure 4.2 (Experian), Table 5.1 (Giuseppe Delfino, Corrado Rindone, Francesco Russo, Antonino Vitetta, and Association for European Transport), Table 5.2 (American Psychological Association), Figures 5.1 and 5.2 (School of Advanced Study, University of London), Figure 5.3 (Emerald Group Publishing Limited), Figure 6.1 (Marine Millot and Association for European Transport), Figure 6.2 (KidsAndCars.org), Figure 6.3 (Elsevier), Figure 7.3 (Elsevier), Figure 8.3 (Taylor & Francis Group), Figure 8.4 (Elsevier), Figure 9.1 (Taylor & Francis Group), Figure 10.1 (Elsevier), Figure 12.1 and 12.2 (Taylor & Francis Group), Table 13.2 (Royal Automobile Club Foundation for Motoring Limited), Table 14.1 (Asian Development Bank), Table 14.2 (Royal Society for the Prevention of Accidents), Table 14.5 (ARRB Group), Figure 14.1 (BMJ Publishing Group Ltd), Figure 14.2 (World Bank), Figure 16.1 (Taylor & Francis Group), and Appendix (UK Image Library of The National Archives).

On a personal front, Becky Loo thanks her husband KW Ng and three lovely children, Wilbert PS Ng, Fabian PW Ng, and Concordia PL Ng, for their love and support.

Becky P.Y. Loo
Tessa Kate Anderson

Authors

Becky P.Y. Loo is professor of geography and director of the Institute of Transport Studies at The University of Hong Kong, Pokfulam, Hong Kong. Her research interests are transportation, e-technologies, and society. In particular, she is interested in applying spatial analysis, surveys, and statistical methods in analyzing pertinent issues related to sustainable transportation. She is the founding editor-in-chief of *Travel Behaviour and Society* and associate editor of the *Journal of Transport Geography*. She is on the editorial boards of major research journals, including *Asian Geographer, Injury Epidemiology, International Journal of Shipping and Logistics, International Journal of Sustainable Transportation, Journal of Urban Technology, Transportmetrica A: Transport Science*, and *Transportation*, among others.

Tessa Kate Anderson is a researcher at the Technical University of Denmark in Copenhagen, Denmark. She has previously worked at the University of Hong Kong, the University of Queensland (Australia), and the University of Canterbury (New Zealand). She completed her PhD in 2007 at the Centre for Advanced Spatial Analysis on road accidents in London. Her research interests are transportation, road safety, and socioeconomics. In particular, she is interested in the links between socioeconomics and road safety, the effects of climate change on road safety and transport, and the application of spatial analysis to further our understanding of these issues. She has published research papers in *Accident Analysis and Prevention, Environment and Planning B*, and *Cities*.

Abbreviations

1D	One-dimensional
2D	Two-dimensional
3D	Three-dimensional
AADT	Average annual daily traffic
ACORN	A Classification of Residential Neighbourhoods
ALGSP	Arizona Local Government Safety Project
ANOVA	Analysis of variance
ANPR	Automatic number plate recognition
BAC	Blood alcohol concentration
BAME	Black and minority ethnic
BSU	Basic spatial unit
CARRS-Q	Centre for Accident Research and Road Safety–Queensland
CI	Confidence interval
CL	Confidence level
CN	Critical number
COM	Component object model
CR	Critical collision rate
DALY	Disability-adjusted life year
DDS	Discrete density surface
DETR	Department for the Environment, Transport and the Regions
DfT	Department for Transport
DTLR	Department for Transport, Local Government and the Regions
DUID	Driving under the influence of drugs
DUMAS	Developing Urban Management and Safety
DWI	Driving while impaired
EB	Empirical Bayes
ECMT	European Conference of Ministers of Transport
ED	Euclidian distance
EKC	Environmental Kuznets curve
ESDA	Exploratory spatial data analysis
ESRI	Environmental Systems Research Institute
EU	European Union
FYRR	First year rate of return
GAM	Geographical analysis machine
GB	Great Britain
GDP	Gross domestic product
GIS	Geographical information system
GIS-T	Geographical information system–transportation
GLM	Generalized linear model
GLS	Graduated license system
GNP	Gross national product
GPS	Global positioning system

GRSP	Global Road Safety Partnership
GWR	Geographically weighted regression
HASS	Home Accident Surveillance System
HRL	Hazardous road location
ICD-10	International Classification of Diseases, 10th Revision
IIHS	Insurance Institute for Highway Safety
IMD	Index of Multiple Deprivation
IOM	Institute of Medicine
IRTAD	International Road Traffic and Accident Database
ITS	Intelligent transport system
KDE	Kernel density estimation
KSI	Killed and seriously injured
LAAU	London Accident Analysis Unit
LASS	Leisure Accident Surveillance System
LHA	Local Health Authority
LISA	Local Indicators of Spatial Association
MAST	Market analysis and segmentation tools
MAUP	Modifiable areal unit problem
MDGs	Millennium development goals
MISE	Mean integrated squared error
MVC	Motor vehicle collision
NB	Negative binomial
NCHRP	National Cooperative Highway Research Program
NHS	National Health Service
NHTSA	National Highway Transportation Safety Administration
NIMBY	Not in my back yard
NNI	Nearest Neighbor Index
NRSI	Neighbourhood Road Safety Initiative
OECD	Organisation for Economic Co-operation and Development
Ofsted	Office for Standards in Education, Children's Services and Skills
OLS	Ordinary least squares
ONS	Office for National Statistics
PCR	Potential for collision reduction
PIC	Personal injury collision
POP	Population density
PIL	Priority Investigation Location
PPA	Potential path area
PPT	Potential path tree
RAC	Royal Automobile Club
RBT	Random breath testing
RoSPA	Royal Society for the Prevention of Collisions
RP	Reference point
RSA	Road Safety Analysis Limited
RTI	Road traffic injuries
SEM	Spatial error model
SLM	Spatial lag model

Abbreviations

SOA	Super output area
SQL	Structured query language
SR	Simple ranking
SRTS	Safe routes to school
STP	Space-time path
TCS	Travel Characteristics Survey
TIGER	Topologically integrated geographic encoding and referencing
TL	Threshold level
TRADS	Traffic Road Accident Database System
TRB	Transportation Research Board
TRIS	Transportation Research Information Services
TRL	Transport Research Laboratory
UK	United Kingdom
UN	United Nations
US	United States
USA	United States of America
VDM	Value difference metric
VKT	Vehicle-kilometers traveled
VMT	Vehicle-miles traveled
WHO	World Health Organization
ZIP	Zero-inflated Poisson

1 Collisions as Spatial Events

1.1 INTRODUCTION

Throughout the world hundreds of millions of motor vehicles mix with billions of people. A pedestrian crossing a busy street tries to make eye contact with the approaching motorist. Will he slow down? The motorist tries to divine the intentions of the pedestrian. Will he give way? Shall I? Shan't I? Will he? Won't he? As the distance between them closes, signals implicit and explicit pass between them at the speed of light. Risk is perceived as risk acted upon. It changes in the twinkling of an eye as eye lights upon it.

Adams (1995)

The analysis of road traffic collisions is not easy, due to their complexity. The American Automobile Association estimates that road traffic collisions claim a life every 13 min in the United States, and the World Health Organization (WHO) estimates 1.18 million people were killed in 2002 in road collisions, which was 2.1% of the global mortality (Peden et al. 2004). Road traffic collisions have been considered by the WHO to be the leading injury-related cause of death among people aged 15–44. Road traffic collisions have formed part of our everyday lives. Every person is at risk. Even if one is not a vehicle driver, one is likely to be a pedestrian, a passenger, and/or a cyclist, and at some point every person is subject to using the road network and, therefore, be at risk of being involved in a road traffic collision.

There are two main approaches to road safety and road collision reduction. The first of these is preventing the collision itself. The second approach to road safety can be determined by the need to reduce the damage that occurs in a collision. Critics have labeled this approach "safe collision," and some argue that this approach has been overemphasized by government policies and traffic safety agencies alike (Gladwell 2001). However, the backbone of any collision analysis is the datum and its quality. There has been an increasing interest over the recent years on the management, collection, and analysis of data related to road collisions.

It has often been said by road safety professionals that data, together with their analysis, are the cornerstone of all road safety activities. Good-quality data are ultimately essential for the diagnosis of the road collision problem and the reduction or management of road collisions. It is important to identify what categories of road users are involved in collisions, what maneuvers and behavior patterns lead to collisions, and under what conditions collisions occur, in order to define appropriate safety measures. The analysis of road collisions varies considerably, and there are neither bespoke universal guidelines of how road traffic collisions should be analyzed nor best practice guides on prediction and prevention for practitioners and academics alike. For instance, within London, although all boroughs are managed and funded by the London Accident Analysis Unit (LAAU), it is the individual boroughs that are responsible for their own area and subsequent analysis and preventative measures.

It is worth noting at this point that there is some division within the literature concerning the definitions of "collision," "crash," "incident," and "accident." In this book, the term "collision" will be used because it is important to acknowledge that a vast majority of "road collisions" are in fact not "accidents." The word "incident" does not properly portray the notion that an injury has occurred. A road traffic collision can be defined as "the product of an unwelcome interaction between two or more moving objects, or a fixed and a moving object" (Whitelegg 1987, 162). "Collision" is also preferred to "crash," because the latter often suggests sudden damage or even destruction on violent impact that may not be true for many road traffic collisions. Road safety relates to many other fields of activity including education, driver training, publicity campaigns, police enforcement, road traffic policing, the court system, national health services, and vehicle engineering.

The field of transportation has come to embrace geographical information systems (GIS) as a key technology to support its research and operational need. The acronym GIS-T (geographical information system–transportation) is often employed to refer to the application and adaptation of GIS to research, planning, and management in transportation. GIS-T covers a broad arena of disciplines of which road traffic collision detection is just one theme. Other themes within the discipline of GIS-T include in-vehicle navigation systems and global positioning systems (GPS). Initially, the use of GIS in transportation was only restricted to query simple collision questions, such as depicting the relative incidence of collisions in wet weather or the adequacy of street lighting, or to flag high absolute or relative incidences of collisions (Anderson 2002; Loo and Yao 2012). Recently, there has been increased acknowledgment that there is a requirement to go beyond these simple questions and to extend the analysis. It has been widely claimed by academics and the police that knowing where road collisions occur will lead to better road policing, education, engineering, and awareness.

There have been a number of developments in the road safety domain that shape the current research and policy-driven initiatives today. Therefore, it is useful to reflect on the most important advancements within road safety, as summarized in Table 1.1.

Table 1.1 highlights the recent nature of road safety. Road safety research can be argued to be only in its infancy with scope for more robust research, which will fundamentally address the nature of the geography of road collisions and how they interrelate to their environment and not just the road environment. Another, perhaps more helpful, way of approaching the evolution of road safety is to segregate the various trends of approach to road safety and the analysis of collisions. Table 1.2 illustrates the shift in paradigms over the past century.

In order to understand the development of road safety research, it is important to know how the scientific view has changed during the short history of systematic road safety research. It is possible to distinguish four phases of scientific views or paradigms that overlap and interact in a complex way (Table 1.2):

TABLE 1.1
A Brief Summary of the Evolution of UK Road Safety in the Twentieth Century

Year	Road Safety Milestone
1896	First road death recorded, Bridget Driscoll killed by a horse drawn carriage
1899	First fatal road collision involving a motor vehicle
1903	Speed limit increased to 20 mph
1919	Ministry of Transport set up
1930	Minimum driving age introduced
1930	Road Traffic Act set different speed limits for different vehicles
1934	Compulsory driving test introduced
1941	Royal Society for the Prevention of Accidents (RoSPA) set up
1957	First motorway opened
1965	50 mph limit on certain roads in order to reduce collisions
1975	First roundabout in Croydon
1978	First "drink drive" campaign
1983	Seat belts compulsory
1990	Department of Transport set up
1990s	"Kill your speed" campaign set up
1991	20 mph zones in urban areas
1992	Speed cameras made permanent
2000	Ten year plan outlined in "Tomorrow's roads: safer for everyone"
2003	Congestion charging introduced in Central London

Source: Data from Cummins, G., The history of road safety, 2003, http://www.driveandstayalive.com/Info%20Section/history/history.htm.

1. Control of the automobiles was seen as the problem. There was limited research but more of a description of what was happening. This phase coincided with the rise of the automobiles from the beginning of the twentieth century to 1935.
2. Control of traffic situations was seen to be the problem. The countermeasures and the research were centered on the classical three "Es" approach of engineering, education, and enforcement. This is when systematic road safety research was born and when a number of new disciplines came into road safety research. This occurred from 1935 to approximately 1970.
3. Management of the traffic system was seen to be a problem. In this systems approach, mathematical models for the description and prediction of traffic collisions were developed. This phase occurred from 1970 to approximately 1985.
4. Management of the transport system as a whole was seen as the problem. The scope is widened from just focusing on the road itself. This is the current trend of road safety thinking.

TABLE 1.2
Paradigms of Road Safety

Evolution of Road Safety Paradigms

Aspects	Paradigm I	Paradigm II	Paradigm III	Paradigm IV
Decennia of dominating position	1900–1925/35	1925/35–1965/70	1965/70–1980/85	1980/85–present
Description	Control of motorized carriage	Mastering traffic situations	Managing traffic system	Managing transport system
Main disciplines involved	Law enforcement	Car and road engineering, psychology	Traffic engineering, advanced statistics	Advanced technology, systems analysis, sociology, communications
Terms used about unwanted events	Collision	Accident	Crash, casualty	Suffering, costs
Premise concerning unsafety	Transitional problem, passing stage of maladjustment	Individual problem, inadequate moral and skills	Defective traffic system	Risk exposure
Data ideals in research	Basic statistics, answers on "what"	Causes of accidents; "why"	Cost-benefit ratio of means; "how"	Multidimensional
Organizational form of safety work	Separate efforts on trial and error basis	Coordinated efforts on voluntary basis	Programmed efforts, authorized politically	Decentralization, local management
Typical countermeasures	Vehicle codes and inspection, school patrols	The three E's doctrine, screening of accident prone drivers	Combined samples of measures for diminishing risks	Networking and pricing
Effects	Gradual increase in traffic risks and health risks	Rapid increase of health risk with decreasing traffic risk	Successive cycles of decrease of health risks and traffic risks	Continuous reduction of serious road accidents

Source: OECD, Road Transport Research, *Models in Road Safety*, OECD, Paris, France, 1997.

Collisions as Spatial Events

Road safety research has been studied from top-down (aggregate level) and bottom-up (individual level) approaches. The ultimate purpose of road safety research is to find and implement countermeasure strategies and countermeasure actions that effectively reduce the road safety problems identified. Researchers have, however, mainly focused their interest and efforts on the main road collision variables and to some extent on countermeasure effectiveness. They have rarely extended their interest and efforts to the next stage—how to implement the theoretical and empirical knowledge acquired concerning main road collision variables and effective remedial measures. Figure 1.1 represents the management of road safety and how engineering and behavioral factors are integrated together with the aim to reduce or manage road collisions. This diagram seeks to act as a general model for road safety and how it can be approached and ultimately managed in countries such as the United States, Europe, and some Asian countries. As road traffic collisions involve roads, motor vehicles, and also human beings, the geographical analysis needs to address issues covering road engineering, signage, vehicle design, education of road users, and enforcement of traffic safety measures on a holistic basis. Figure 1.1 illustrates the relationships among various actors. The two major categories are engineering and behavior changes. The former involves safer vehicles and roadways. Wilson and Lipinski (2004), for example, describe many of the engineering strategies for improving traffic safety. The latter include mobility management (changes in travel mode, route, destination, frequency, and speed), more cautious driving, and actions by vehicle occupants such as using seat belts, child restraints, and helmets.

This section seeks to outline and interpret the relative importance and benefits of using GIS for road safety and more directly road collision analysis. What follows includes methods that are unique to GIS and how they integrate into the methods

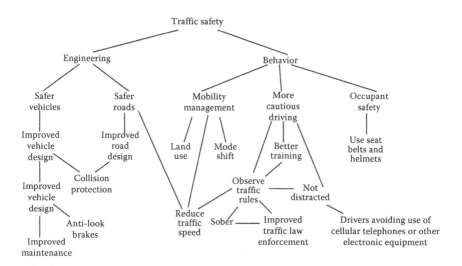

FIGURE 1.1 The integrated road safety system. (Courtesy of Victoria Transport Policy Institute, Traffic Safety Strategies, Online TDM Encyclopedia, http://www.vtpi.org/tdm/tdm86.htm, accessed September 2, 2005.)

outlined in this book. GIS has been employed to relate, organize, and analyze road traffic collisions worldwide. It is clear, however, that GIS cannot replace the need for local analysts to interpret the results and recommend improvements. A recent study by Loo et al. (2013) illustrates the process and benefits of multidisciplinary efforts in improving road safety using GIS and local engineering measures. This section seeks to outline an approach that underpins the benefits of using GIS for road collision data, as opposed to using merely the data in a statistical package or onsite identification of road collision causes.

It therefore raises the question: What additional benefits can GIS provide that do not already exist in terms of road safety analysis provision? This question is important for this book as it captures a question asked by many local government agencies and police who use software to analyze road collisions that is not conventionally classified as "GIS."

One of the most common uses of GIS in road safety is to visually digest a large amount of information quickly, for example, showing a map of high frequency road collision locations. A study in North Carolina used a "sliding scale" whereby a segment of a specific length along a road was dynamically moved until that segment met a threshold such as a minimum number of collisions or collisions of a particular severity. In this case, the threshold can be varied. The task of studying road collisions in a GIS may be represented as a spatial analysis problem. There can be two key benefits that can be deemed from a visual representation of collision locations:

1. An understanding of any clustering of high collision locations.
2. Visual patterns may be used to discern geographic and spatial relationships based on selected variables such as drivers' age. However, this can be narrowed down to a specific query by looking, for example, at those collisions that occurred Friday and Saturday evenings between 9 and 6 p.m. that involve male drivers under the age of 24; therefore, certain types of problem can be identified.

Austin (1995), however, states that two other types of inquiry can make better use of GIS. The first of these is error checking, where the features in the database can be compared to the features of the collisions coded by the attending police officer (e.g., differences in speed limit from database to incident report). The second aspect of GIS usage is to identify collision regions or zones as opposed to identifying specific intersections or segments. This allows the analyst to categorize areas by land use and compare how they affect the number and spatial layout of collisions. An example application would be analyzing child pedestrian safety on the route to school. An in-depth analysis could be made of the neighborhood and an evaluation of routes to school and their relative safety.

Most road traffic collisions may be considered to be random events that depend on time and location. Thus the annual traffic collision count at a particular location will vary from year to year, and for a particular year, the annual traffic collision count will vary from location to location. This means that road collision counts are subject to both temporal and spatial variations. Some of the collisions may not be completely random, in that the temporal and spatial variations in their occurrence can be explained in part by variations in systematic factors involved in collision occurrence.

Collisions are rare events and generally not uniformly or equally distributed over the road system. They are often clustered at sites, along routes, or within areas. The basis for a strategic approach to road traffic collision reduction by specific engineering measures is to develop a framework within which priorities may be set for implementing measures identified through collision reduction analysis techniques.

Old approaches to road collision research emphasized the concept of problem solving in road safety, but it is better to recognize that road safety activities do not *solve* problems. For instance, when a safer road design is implemented, hopefully the number of collisions or their seriousness will go down, but they will not disappear. It is more correct to see road safety as an area where the implementation of correct policies, programs, and measures will reduce collision numbers or consequences, but it will not be solved.

This realization is important because it changes the focus from a problem that will go away if we devote enough resources to it, to a situation requiring ongoing management. This management in turn requires the development of scientifically based techniques, which will enable us to predict with confidence that safety resources are well spent and likely to be effective. Some of the major challenges to road collision spatial analysis are outlined in the following:

- Tailoring data management, analysis, and especially visualization of results to the requirements of the user.
  The range of agencies requiring information is broad and purposes to which the information is applied vary. At a national level, these include government, academics and researchers, organizations working in the field, the private sector, and the media. At a regional scale, interested agencies include local health authorities (LHAs) for health promotion, planning and preventative work, voluntary sector agencies, interagency groupings, and local practitioners.
- Collating information on the geographic distribution of potential socioeconomic, demographic, behavioral, and environmental risk factors for comparison with collision distribution.
  It has been suggested that there are gaps in the data available concerning causal factors (behavioral and environmental), which lie behind the occurrence of collisions as well as information linking collisions with socioeconomic profiles and characteristics. Furthermore, local studies have shown wide differences between collisions occurring in different districts, which can be explained geographically, environmentally, and socially.
- Integration of different base denominator data.
  Problems in analysis have been caused by incompatibility of coding systems, use of different populations and denominators, and lack of temporal continuity. Standardization is required to gain a better picture of trends in injuries (numbers of injuries per unit population) and type of collisions (e.g., number of road collisions per unit traffic volume).

- Temporal analysis for intervention management.
  In road collision analysis, it is considered essential to review trends and plot changes over time. This is required to examine whether intervention measures are successful to manage resources for future prevention schemes.
- Modeling.
  Estimates of the overall picture can often only be made by extrapolating the findings of local studies.

1.2 DISTANCE-BASED METHODS

With regards to distance-based methods and road collision analysis, there are a number of different types of distance methods depending on the data available and the type of outcome needed. The different types are outlined as follows:

- Home to collision location (Euclidian)
- Home to collision location (network)
- Work to collision location (Euclidian)
- Work to collision location (network)
- Journey time as distance
- Distance between specific collisions (Euclidian and network)
- Distance between spatial clusters of collisions (Euclidian and network)

The methodology for measuring the distance between home location and road collision is usually the Euclidian approach. This technique depends on the type of data being analyzed. For example, if measured by police collision data alone, there is often no mandatory requirement for home address to be recorded of the road users. However, if hospital data are used, there is usually a requirement for home address to be recorded as part of the admissions process. The Euclidian approach refers to the straight-line distance between two points and it can be calculated in a GIS using standard SQL functions or using Pythagoras theorem in spreadsheet software such as Excel.

We can disseminate the distance measurement types as follows:

- *Manhattan or Euclidian distance*—This method calculates the shortest distance between two points using either horizontal or vertical directions. This can be calculated in a GIS or software package.
- *Street route/network distance*—This calculates the shortest path following the street network from the driver/casualty location to the road collision location. This process involves specialist street routing data and often other specialist software.
- *Journey time distance*—This is the measurement of the time it takes to travel a distance. This is a more complex task as it has to take into account traffic, speed limits, and mode of travel. In the road safety literature, this is often referred to as exposure (in terms of the number of kilometers or the length of journey time a driver or passenger travels).

Collisions as Spatial Events

For the Euclidian distance, if we have two locations whose coordinates are (x_1, y_1) and (x_2, y_2), the Euclidian distance between them is

$$d_{1,2} = \sqrt{(x_1 - x_2)^2 + (y_1 - y_2)^2} \qquad (1.1)$$

Following the notion of Gatrell (1983), it is possible to reconfigure the coordinates of the two locations as (x_{11}, x_{12}), (x_{21}, x_{22}) so that the first subscript refers to the location and the second subscript refers to the coordinate. According to Fotheringham et al. (2000), the Euclidian distance between two locations i and j with coordinates (x_{i1}, x_{i2}) and (x_{j1}, x_{j2}) can be written as

$$d_E(i,j) = \left[\sum_{k=1}^{2} (x_{ik} - x_{jk})^2 \right]^{1/2} \qquad (1.2)$$

The first point to make is that the Euclidian or Manhattan techniques do not take into account physical road barriers such as railways, rivers, lakes, buildings, open space, and any other area that is not accessible. This often then assumes that the network method is more accurate. Whilst this can be argued to be true, what often is missing in terms of data is the information of the journey taken. People are not necessarily traveling from home, and there are many different locations that the road users could be coming from (work, school, recreation, shopping, etc.). Without knowing the exact trip parameters, it can be difficult to accurately assess distance.

In a recent study (Siddiqui 2009), the postcode of the driver/road user was known from the collision data. They took the centroid of the postcode (as a grid coordinate) and did a straight-line Euclidian distance calculation to the road collision location. They found that over 60% of all the road collisions occurred between 2 and 10 miles of the home location.

1.3 SIMPLE MEANS METHODS

Although this book is largely concerned with the spatial elements and analysis of road collisions, descriptive statistics can often give a good understanding of the scale and distribution of point data. Descriptive statistics are especially useful when comparing two sets of point data. In road safety analysis, we often see these descriptive statistics presented as report-based evidence and research rather than detailed journal research. Descriptive statistics should, however, be treated with caution, especially when dealing with complex datasets such as road collisions. However, they do offer useful insight into the overall nature of road collisions and the dataset. This section is going to focus on some examples of descriptive statistics and preliminary thoughts on road safety databases.

The application of road collision and other relevant datasets (e.g., traffic flow, demographic, land infrastructure, and road use) are critical for a better understanding and management of road safety in general. Road safety analysis is often framed by the use of the road collision dataset, either a sampled specific dataset or police-recorded collision data.

Examples of the types of statistics would include distribution, central tendency, and dispersion. Often it is only possible to use descriptive statistics and simple means methods on one variable at a time. Descriptive statistical analysis should be performed at the beginning of any type of analysis. It is always essential to look at the data before any models or hypotheses are formally fitted.

Taking the London dataset as an example, for each injury road collision known to have occurred in their areas, the police authorities complete a statistical return (which is called a "Stats 19" return), which provides details of the collision circumstances, separate information for each vehicle that was involved in the collision, and separate information for each person who was injured in the collision. Therefore, the data are disaggregated into three tables according to the STATS19 records (see Appendix for STATS19 data record sheets). Most of the variables are categorically coded and the codes are shown in Appendix. These tables are segmented into *Attendant Circumstances*, *Casualty Details*, and *Vehicle Details*. These can be summarized as follows:

1. *Attendant circumstances*: This section of the data records the general circumstances for the collision. There is one row in the dataset for each injury collision recorded. Data included in this table would include, for example, geographical reference (eastings and northings), time, date, collision description, number of injuries, general level of severity, and so on.
2. *Casualty details*: This dataset includes information specifically on the injuries of the collision—their age, gender, severity, whether they were a pedestrian, cyclist, or car occupant, and so on. There is one row for every casualty recorded.
3. *Vehicle details*: This table contains information regarding the type of vehicle(s) involved, information about the driver(s), age and gender, vehicle speed, and whether the collision caused any injury.

Figure 1.2 shows a contextual summary of the age and gender disaggregation of the collision data. The age of the driver follows a coherent pattern and in line with the idea of a large proportion of inexperienced drivers on the roads from the age of 17 upward. The 17-and-under age group will be largely associated with pedal cyclists or scooters/mopeds. There is a small peak in the age of injuries at 9–11 years old, which has been identified as a high-risk age group compared to their population in the United Kingdom. The driver and casualty age data (Figure 1.2) also suggest rounding errors at the local peaks of 25, 30, 35, 40, and so on.

Table 1.3 illustrates information regarding the age of both driver and casualty. It is clear from this information that the mean age is similar for both variables and other statistics including the percentiles and range. These tables also report the number of missing variables in the data and one can clearly see that the frequency of missing variables for driver's age is higher than for casualty's age. Another feature, which these tables do not show but is clearly visible from the visual representations of the data, is the high proportion of child and elderly injuries.

Collisions as Spatial Events 11

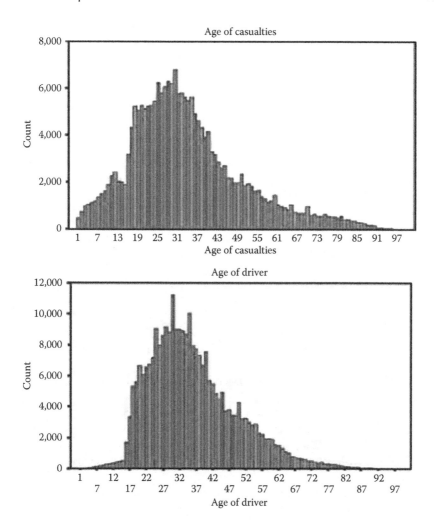

FIGURE 1.2 Bar graphs showing the age of drivers and casualties involved in road collisions.

1.4 SIMPLE VARIANCE METHODS

The analysis of variance (ANOVA) is essentially a collection of statistical methods in which the observed variance is determined. In its simplest form, ANOVA provides a statistical test of whether or not the means of several groups are all equal, and therefore generalizes *t*-test to more than two groups. It is not the focus of this book to go into theoretical detail about variance methods; however, this section will introduce the readers to simple examples used for analyzing road collisions. Methods of variance are often used as the basis for formulating more complex data models. ANOVA statistical tests are conducted to see if there are any significant differences in the data. One of the more common variance tests is the *t*-test. Often independent variables are used (such as speed limits, weather, seat belt use, etc.) in order to test

TABLE 1.3
Summary of Statistics for Age of Driver and Casualty

	Age of Casualty	Age of Driver
N—Valid	221,789	314,436
N—Missing	2,693	10,620
Mean	32.12	31.53
Median	30.00	32.00
Mode	0	0
Standard deviation	17.96	17.20
Variance	322.52	295.85
Range	99	99
Minimum	0	0
Maximum	99	99
25th Percentile	21.00	23.00
50th Percentile	30.00	32.00
75th Percentile	41.00	42.00

the determinants for the cause of road collisions. The t-test allows the user to analyze the data and determine (using a significance level) whether the null hypothesis can be accepted or rejected.

The main testing that most practitioners/researchers will encounter is in

- The comparison of collision frequencies where a Chi-squared test may be used, or paired t-test if the distribution of collisions is assessed as coming from a normal distribution (the Fisher Exact Test can be used instead of a Chi-squared test when any value in the cells of a 2×2 comparison matrix falls below 10)
- The comparison of collision rates using a paired t-test
- The comparison of proportions using a Z-test

In a statistical analysis of collision reductions, the "95% confidence level" is typically used, although in some circumstances it is acceptable to use a 90% level (meaning that there is a 1-in-10 chance of the outcome occurring purely by chance).

1.5 NEAREST NEIGHBOR ANALYSIS

The nearest neighbor index is a distance statistic for point pattern datasets, which makes it useful for road collision data. It gives the analyst an indication of the degree of clustering of the points. It is used primarily as a form of exploratory data analysis. A nearest neighbor analysis compares the characteristics of an observed set of distances between pairs of closest points with distances that would be expected if points were randomly placed. Many of the recent studies in the road safety literature have focused on using nearest neighbor on a network, which will be discussed in

Collisions as Spatial Events

later chapters. It has also been used extensively in measuring crime patterns (Levine 2007). During the analysis, the distance from each point to its nearest neighbor is calculated. This value gets added to a running total of all minimum distances, and once every point has been examined, the sum is divided by the number of points. This then produces what we call a "mean minimum distance" or "nearest neighbor distance." The equation looks like this:

$$\bar{d} = \sum_{i=0}^{n} \frac{d_{ij}}{n} \quad (1.3)$$

where
$\bar{d}$ is the mean nearest neighbor distance
d_{ij} is the distance between the point i and its nearest neighbor j
n is the number of points in the dataset

Figure 1.3 shows three different types of spatial pattern. A clustered pattern is often the most common found in road collision data. Road collisions are often a result of dangerous road or driving at a particular area. Often, collisions are not randomly distributed (sometimes they will be, but you will often find groupings of clusters in the dataset). If a collision pattern is more spread out, it exhibits the second type of spatial pattern, that is, a random distribution. Although there may be some local clusters in this type of pattern, the overall pattern of road collisions is spread across the study area without any apparent pattern. In other words, the road collision has an equal chance to be anywhere in the study area. The third type of pattern is a uniform one, which is rarely seen in road collision research. This occurs when points are spaced roughly the same distance apart.

$$R_n = 2\bar{D}\sqrt{\left(\frac{n}{a}\right)} \quad (1.4)$$

where
R_n is the nearest neighbor index
$\bar{D}$ is the average distance between each point and its nearest neighbor
n is the number of points under study
a is the size of the area under study

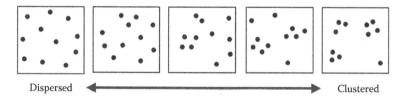

FIGURE 1.3 Diagram showing patterns of dispersion to being clustered.

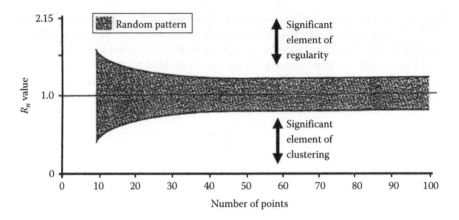

FIGURE 1.4 Nearest neighbor patterns. (Reprinted from Nagle, G. and Witherick, M., *Skills and Techniques for Geography A-Level*, Stanley Thornes (Publishers) Ltd., Cheltenham, U.K., p. 28, 1998, by permission of Oxford University Press.)

$$\bar{D} = \sum \frac{d}{n} \qquad (1.5)$$

where d is the distance between each point and its nearest neighbor. The formula produced by the nearest neighbor analysis produces a figure, expressed as R_n (the nearest neighbor index), which measures the extent to which the pattern is clustered, random, or regular (Figure 1.4).

- *Clustered*: $R_n = 0$—All the dots are close to the same point.
- *Random*: $R_n = 1.0$—There is no pattern.
- *Regular*: $R_n = 2.15$—There is a perfectly uniform pattern where each dot is equidistant from its neighbor.

One of the major drawbacks of nearest neighbor analysis is that it only analyzes the location of the points not the attributes.

The nearest neighbor analysis is a classification method in which the class of an unknown record is assigned after comparisons between the unknown record and all known records (training data) in data repository are made. The degree of similarity between different records is determined by a function called the distance function. Nukoolkit and Chen (2001) used two different distance functions—Euclidian distance (ED) and value difference metric (VDM) distance both combined with k-mode clustering in predicting whether a car collision will have either an injury or a noninjury outcome using a subset of year 2000 Alabama interstate alcohol-related collisions. The prediction errors of 33% and 45% were observed using ED and VDM methods, respectively. The study further proposed an improved technique that combines the distance function with decision tree clustering, which reduced the prediction error to 19%. The existence of variables that vary in form and magnitude makes it difficult to establish the distance function. While some variables are continuous, others are discrete. In addition, even within the continuous and discrete variable groups, the range of magnitudes and the number of categories differ from variable to variable. This lessens the appropriateness of the nearest neighbor technique in collision prediction.

Nearest neighbor methods make use of precise information on the locations of collisions and avoid the arbitrary choices (e.g., quadrat size, shape, and location) associated with quadrat methods. Given the disadvantages of the quadrat approach, the nearest neighbor approach is often preferred for analyzing collision spatial distributions. Moreover, both distances and directions to nearest neighbor should be analyzed to assist the detection of clustering at sites or along routes.

There are a number of different approaches that can be made. These tests can be used on a single road collision distribution to explore the concept of spatial randomness. In addition, these tests can be used to compare the general spatial randomness of one type of road collision with another (e.g., pedestrian and bicycle collisions) or from one time period to another. We will have a look at these in more detail (as well as nearest neighbor on a network in later chapters); there are a number of good texts that refer in more detail to the concepts on nearest neighbor such as Ripley (1991), Diggle (2003), and Bailey and Gatrell (1995).

Spatial dependence in a single road collision pattern is investigated by examining the observed distribution of nearest neighbor measures and comparing the mean across the dataset with an expected, theoretical distribution that would occur if the points were dispersed in a random manner. The random distribution is a function of the size of the study area and the number of point

$$\bar{\delta} = \frac{1}{2\sqrt{A/n}} \tag{1.6}$$

where
$\bar{\delta}$ is the expected mean distance between nearest neighbors
A/n is the point density, expressed as the area of the study region divided by the number of points

The calculation of the nearest neighbor index is a simple ratio of the two calculations:

$$R(\text{NNI}) = \frac{\bar{d}}{\bar{\delta}} \tag{1.7}$$

where $R(\text{NNI})$ is the nearest neighbor index expressed as a ratio. Often, researchers will be able to get the specialized software (ArcGIS or MAAP) to calculate these values. Most software packages will calculate a statistical significance for the nearest neighbor index. The problem with statistical measures of nearest neighbor significance is the difficulty in correcting for "edge effects." Few study areas in road safety are perfectly rectangular in shape, and therefore, the estimates of the mean nearest neighbor distance are often larger in reality. This is because points that lie close to the study boundary are excluded from the possibility of having a nearest neighbor just the other side of the boundary.

Some events (collisions) or points may be closer to the boundary of the region than to their nearest neighbors within the region. If the nearest neighbor is taken to be the closest event within the region, nearest neighbor distances will be greater for sampled events or points near the boundary of the region than for events or points near the center of the region. There will be a bias in the nearest neighbor statistics, unless a correction for the edge effect is made. There are three general approaches for correcting for edge effects (Cressie 1992):

1. Construction of a "guard area" inside the perimeter of the region, with no events being selected from within the guard area
2. Assuming that the region is the center plot of a 3×3 grid of plots identical to the region (i.e., the region is surrounded by eight identical regions, or the spatial distribution is on a toroid)
3. Obtaining empirical, finite-sample corrections for statistics or indices

The third approach has a major drawback, in that the corrections relate to specific situations and are not applicable generally. The disadvantage of the first method is the exclusion of points and events within the guard area (i.e., not all the data are used). When the region is rectangular, the second method is very easy to implement, and it means that all the data can be analyzed. If there is a strong linear pattern of events within the region (e.g., collisions are strongly clustered along a line), the second approach will result in some reduction in the strength of the linear clustering effect, due to the discontinuity at the boundary. In this case, the first method may well be better overall. If the region is irregular (as will generally be the case if the region is a city, county, state, or province), the second method is impractical because of the gaps between the region and the surrounding identical regions. In this case, the guard area approach would be the best.

1.6 CONCLUSION

This chapter introduces the readers to some practical and essential methods and tools for analyzing road collisions. We know that road collisions do not occur randomly, and there is a level of spatial dependency involved in the processes and events leading to a road collision. As the title of this book suggests, we are primarily interested in the spatial patterns and processes of road collisions. In this chapter, we see road collisions being dealt with as two-dimensional (2D) point patterns, where the data are only locations of a set of point objects. This represents the simplest possible spatial data. This 2D point pattern analysis arguably forms the backbone of any spatial analysis of road collisions, which makes this chapter important. Point pattern can be very complex to analyze, and in this chapter we outlined what is meant by a point pattern and how it can be analyzed on a 2D plane, or in other words, distance-based methods. In applied geography and GIS, point patterns are fairly common. Generally speaking, we are concerned with road collisions as point patterns and whether or not there is some sort of concentration of events, or clustering. It is also important to point out that we should not ignore areas where there are no collisions at all or where the pattern displays no particular clustering. A point pattern consists of a number of events in a study region, and there are a number of other requirements (O'Sullivan and Unwin 2003):

- The pattern of road collisions should be mapped on the plane, and they should have a longitude and latitude and be projected appropriately.
- The study area should be defined appropriately. There should be clear boundaries of the study area. This chapter and the next chapter deal with non-network analysis of road collisions, whilst later on the subject of network analysis is approached.
- The pattern should not be a sample, but all the relevant entities should be included in the study region.
- There should be one-to-one correspondence between events in the pattern.
- Event location must be proper; there should, for example, not be centroids of areal units.

There are two main spatial processes that are fundamental to the understanding of road collisions: spatial dependence and spatial heterogeneity.

An early collision analysis that includes the spatial component was published by Levine et al. (1995). Collisions were geocoded to the nearest intersection or ramp, and then different spatial statistics, including mean center, standard distance deviation based on great circle distance, the standard deviational ellipse (first and second principal component), and the nearest neighbor index (based on the x and y coordinates of the road collisions), were calculated. For the purposes of research or policy-making analysis of a single intervention, an impact study may be conducted. Usually, these studies are of limited scope dealing with a small portion of the road network, a limited time period, and a small sample of road users. Often, they presume all other variables remain constant. This is what lies at the heart of the complexity of analyzing road collisions in space: it is often impossible to control for all variables that might be present. It is possible, however, to analyze simultaneously a number of intervention variables in order to identify those most crucial in reducing level/severity of road collisions. Generally speaking, the field of road collision analysis is in its early stages. When compared to similar statistical fields such as health and crime, the analysis of road collisions is fairly recent. What issues road safety analysts face now are also related to the shift in a Western-dominated world of road collisions occurring on fairly uniform roads and in fairly uniform situations, compared to an increasing number of road collisions occurring in fast urbanizing and modernizing countries such as China. We will discuss the nature of this in more detail in the following chapters.

REFERENCES

Adams, J. 1995. *Risk*. London, U.K.: Routledge.
Anderson, T. 2002. Spatial analysis of road accidents in Warwickshire. Bachelor dissertation, University College London, London, U.K.
Austin, K. 1995. The identification of mistakes in road accident records: Part 1, Locational variables. *Accident Analysis & Prevention* 27 (2): 261–276.

Bailey, T. C. and A. C. Gatrell. 1995. *Interactive Spatial Data Analysis*. Essex, U.K.: Longman Scientific & Technical.

Cressie, N. 1992. Statistics for spatial data. *Terra Nova* 4 (5): 613–617.

Cummins, G. 2003. The history of road safety. http://www.driveandstayalive.com/Info%20 Section/history/history.htm.

Diggle, P. 2003. *Statistical Analysis of Spatial Point Patterns*. London, U.K.: Arnold.

Fotheringham, A. S., M. E. Charlton, and C. Brunsdon. 2000. *Quantitative Geography: Perspectives on Spatial Data Analysis*. Thousand Oaks, CA: Sage.

Gatrell, A. C. 1983. *Distance and Space: A Geographical Perspective*. Oxford, U.K.: Clarendon Press.

Gladwell, M. 2001. Wrong turn: How the fight to make America's roadways safer went off course. *New Yorker*, June 11, 2001, pp. 50–61.

Levine, N. 2007. *CrimeStat III: A Spatial Statistics Program for the Analysis of Crime Incident Locations (Version 3.1)*. Houston, TX: Ned Levine & Associates.

Levine, N., K. E. Kim, and L. H. Nitz. 1995. Daily fluctuations in Honolulu motor vehicle accidents. *Accident Analysis & Prevention* 27 (6): 785–796.

Loo, B. P. Y., C. B. Chow, M. Leung, T. H. J. Kwong, S. F. A. Lai, and Y. H. Chau. 2013. Multi-disciplinary efforts toward sustained road safety benefits: Integrating place-based and people-based safety analyses. *Injury Prevention* 19: 58–63.

Loo, B. P. Y. and S. Yao. 2012. Geographical information systems. In *Injury Research: Theories, Methods, and Approaches*, eds. G. Li and S. P. Baker, pp. 447–463. New York: Springer.

Nagle, G. and M. Witherick. 1998. *Skills and Techniques for Geography A-Level*. Cheltenham, U.K.: Stanley Thornes (Publishers) Ltd.

Nukoolkit, C. and H. Chen. 2001. *A Data Transformation Technique for Car Injury Prediction*. Tuscaloosa, AL: Department of Computer Science, University of Alabama.

OECD, Road Transport Research. 1997. *Models in Road Safety*. Paris, France: OECD.

O'Sullivan, D. and D. J. Unwin. 2003. *Geographic Information Analysis*. Hoboken, NJ: John Wiley & Sons.

Peden, M., R. Scurfield, D. Sleet, D. Mohan, A. Adnan, E. Jarawan, and C. Mathers, eds. 2004. World report on road traffic injury prevention. World Health Organization, Geneva, Switzerland.

Ripley, B. D. 1991. *Statistical Inference for Spatial Processes*. Cambridge, U.K.: Cambridge University Press.

Siddiqui, C. K. A. 2009. Macroscopic traffic safety analysis based on trip generation characteristics. MSc dissertation, University of Central Florida, Orlando, FL.

Whitelegg, J. 1987. A geography of road traffic accidents. *Transactions of the Institute of British Geographers* 12: 161–176.

Wilson, E. M. and M. E. Lipinski. 2004. Road safety audits. NCHRP Synthesis of Highway Practice 336. Transportation Research Board, Washington, DC. http://trb.org/publications/nchrp/nchrp_syn_336.pdf (accessed September 2, 2005).

2 Collision Density in Two-Dimensional Space

2.1 INTRODUCTION

The current practice for the analysis of spatial distributions of road collisions relies almost entirely upon visually examining a map showing the location of collisions, superimposed upon the road network. The assessment process is subjective and relies heavily on exercising judgment in order to decide whether there is a distinct pattern and what it is. Over the years, there has been an increasing move to study road collisions as cluster events. This in itself has brought a number of advantages and subsequent challenges to spatial analysis. Let us, first of all, take the example of crime. Crime does not occur randomly. There is a clear spatial dependence; and spatial heterogeneity exists when analyzing crime. Crime is affected by the surrounding area, and Tobler's first law of geography can be applied here: "Everything is related to everything else, but near things are more related than distant things" (Tobler 1970, 236). Crimes, unlike road collisions, are not constrained to the road network, and therefore the clustering of events arguably is somewhat different.

One of the first questions we often ask when looking at road collision data is "where are the hazardous road locations (HRLs)?" Engineering, public policy makers, road collision investigators all want to know where locations with higher densities of road collisions are. An HRL can be defined as a geographical area where the number of road collisions exceeds the average. The literature and research use a number of different terminologies for this higher-than-average "cluster." Examples include "black spot," "hazardous road locations," and "dangerous sites." However, for the purposes of this book, we will use the term "hazardous road locations" to refer to collision clusters in general. The collision literature provides no universally accepted definition of an HRL. Often, HRLs are relative to the area under study. In other words, an HRL represents an area of high-density road collisions relative to the overall distribution of collisions across the region of interest. Of course, these HRLs can be of varying scales, shapes, and of different interests (such as bicycle-only HRLs). Hauer (1997) describes how some researchers rank locations according to collision rate (this is usually collisions per vehicle-kilometer), while other researchers use collision frequencies (collision per road kilometer). Another dimension for contestation is that rank may be determined by the magnitude of either rate or frequency, or by the amount by which the rate or frequency exceeds what is "normal" or "expected" at a specified range of sites. Yet, there is no universal or standard threshold that can be used to determine the number or type of road collisions deemed as "normal" in a particular area.

HRL analysis for road collisions can generally be split into three distinct phases. First, the HRLs need to be identified. Second, the locations need to be rank ordered;

this could include severity (slight, serious, or fatal). Essentially, this second stage is what we call profiling the HRLs, and several techniques have been developed to tackle this challenge. The final stage involves selecting HRLs for further or initial road safety treatment (which will be based on the profile in stage two). This final stage often involves a policy decision and can depend on many factors, for example, financial obligations, cost–benefit analysis, and multiple or single location analysis. This chapter focuses specifically on the first stage, namely density cluster functions. It describes the methods of cluster analysis using a strictly non-network-based approach. These methods focus on density. Although the importance of the road network is not forgotten, it will be discussed later in the book. The previous chapter has already outlined the statistical, nonspatial element to analyzing road collisions with respect to explanatory variables. However, in order to understand the density and clustering more accurately, it is important to take space into context. The second-order methods (density-based) outlined in the next few sections generally explain collisions as being a consequence of shared common characteristics in the surrounding area.

2.2 QUADRAT METHODS

Much information is lost when calculating a single summary statistic for an overall intensity, and there is strong sensitivity to the dependence on the shape and size of a study area. One option around this problem is to record the number of events in the pattern into cells or quadrats. If one has a map of road collisions (as point events) in an area, the area can be divided into a regular grid pattern of contiguous rectangular quadrats. The number of collisions in each quadrat can then be counted. If the spatial process is completely spatially random, the counts are expected to follow a Poisson distribution. Following this, the counts can then be tested for randomness using either the Chi-squared or Kolmogorov–Smirnov tests. Alternatively, one can calculate various indices for measuring departure from complete spatial randomness, but the interpretation of some of these indices requires the quadrats to be either large or small compared to the area of clusters, which might be impractical if one does not have good previous information on the area of clusters (such as when conducting exploratory data analysis, Cressie (1992)). Nicholson (1998) outlines three major disadvantages of the quadrat method. First, the results depend upon arbitrary decisions relating to size, shape, and location of quadrats. Second, the reduction of complex point distributions to quadrat counts involves a considerable loss of spatial information, as there is no consideration of the relative position of collisions within quadrats. Third, patterns can exist at different scales and the quadrat approach can measure at one scale only (determined by the size of the quadrats). Quadrat counting can be done either as an exhaustive census of quadrats that completely fill the study area with no overlaps or by randomly placing quadrats across the study region. In road collision analysis, this can be achieved by placing the quadrats over areas such as junctions or where a large number of road collisions occur.

Whichever approach is used, the outcome will be a list of quadrat counts recording the number of events that occur in each quadrat. Then, these can be compiled into a list of frequency tables. In geography and road safety analysis, the exhaustive census-based approach is more commonly used. It is also important to note that the choice of orientation and origin of the quadrat will affect the observed frequency distribution. When dealing with the analysis of road collisions, there is rarely a defined boundary from which to start. Often, the study area boundary will be complex and undefined. Large quadrats will produce very coarse patterns; however, if the quadrat size is reduced, many cells will contain no events.

2.3 SIMPLE DENSITY FUNCTIONS

Density analysis takes known quantities of some phenomena and spreads it across the landscape based on the quantity that is measured at each location and the spatial relationship of the locations of the measured quantities. Density maps are predominantly created from point data, and a circular search area is applied to each cell in the output raster. The search area determines the distance to search for points or to spread the values out around the points to calculate a density value for each cell in the output raster. The density surface then shows where point or line features are concentrated.

Density is a fundamental measure of the concentration of something in a defined space-time region. These regions may be regular or irregular. Density can be measured at different scales and the output of a density calculation can be at any resolution. Partitioning space into regular hexagons or squares and measuring the density in each discrete spatiotemporal unit produces a discrete density surface (DDS). Mapping and eyeballing DDS can often lead to the identification of concentrations obscured in point maps. Yet again though, the visual analysis has limitations. These limitations relate to scale and resolution of the geographical partitioning. The limitation and the importance of these factors become more obvious as two or more such surfaces are compared.

The basic simple statistical method here detects if the road collision density is abnormally high. It is therefore based on comparing collision density in a studied route subsection with a so-called reference density. The user selects the reference density. It can be a network density, for example, the overall density of the studied route or the density of uniform subsections. This method requires division of the network into uniform subsections (from traffic and road type standpoints).

Point density calculates the density of point features around each output raster cell. Conceptually, a neighborhood is defined around each raster cell center, and the number of points that fall within the neighborhood is totaled and divided by the area of the neighborhood. If a population field setting other than none is used, the population field's value (the item value) determines the number of times to count the point. Therefore, an item value of 3 would cause the point to be counted as 3 points. The values can be integers or floating points. If an area unit is selected, the calculated density for the cell is multiplied by the appropriate factor before it is written to the output raster. For example, if the input ground units are meters, comparing a unit scale factor of meters to kilometers will result in multiplying the output values by 1,000,000 (1,000×1,000). The population field could be used to weigh some points more heavily than others, depending on their meaning, or to allow one point to represent several observations. For example, one address might represent a road collision of differing severity. Increasing the radius will not change the calculated density values much. Although more points will fall inside the larger neighborhood, this number will be divided by a larger area when calculating the density. The main effect of a larger radius is that density is calculated considering a larger number of points within a larger raster cell. This results in a more generalized output raster.

Simple density functions are so called because essentially it is a simple HRL analysis that it does not compare the result with a "random" result. A "simple" type of density analysis (as found in software packages such as ArcGIS) calculates the density of point features around each output grid cell. Units of density are points per unit of area. The number of points that fall within the search radius is totaled and then divided by the search radius.

2.3.1 Histograms

The histogram is an exploratory spatial data analysis (ESDA) tool that provides a univariate (one-variable) description of the road collision data. The tool displays the frequency distribution for the dataset of interest and calculates summary statistics. The three main important features of a distribution can be summarized by statistics that describe its location, spread, and shape:

1. *Measures of location*: Measures of location provide you with an idea of where the center and other parts of the distribution lie. The mean is the arithmetic average of the data. The mean provides a measure of the center of the distribution. The median value corresponds to a cumulative proportion of 0.5. If the data were arranged in an ascending order, 50% of the values would lie below the median, and 50% of the values would lie above the median. The median provides another measure of the center of the distribution. The first and third quartiles correspond to the cumulative proportion of 0.25 and 0.75, respectively. They are special cases of quantiles. The quantiles are calculated as follows:

$$\text{Quantile} = \frac{i - 0.5}{N} \qquad (2.1)$$

    where
    i is the ith rank of the ordered data values
    N is the number of values in the sample.

2. *Measures of spread*: The spread of points around the mean value is another characteristic of the displayed frequency distribution. The variance of the data is the average squared deviation of all values from the mean. The units are the square of the units of the original measurements and, because it involves squared differences, the calculated variance is sensitive to unusually high or low values. The standard deviation is the square root of the variance. It describes the spread of the data about the mean in the same unit as the original measurements. The smaller the variance and standard deviation, the tighter the cluster of measurements about the mean value.

3. *Measures of shape*: The frequency distribution is also characterized by its shape. The coefficient of skewness is a measure of the symmetry of a distribution. For symmetric distributions, the coefficient of skewness is zero. If a distribution has a long right tail of large values, it is positively skewed. If it has a long left tail of small values, it is negatively skewed. The mean is larger than the median for positively skewed distributions and vice versa for negatively skewed distributions.

2.3.2 K-Function Method

Jones et al. (1996) presents a method applicable to the examination of spatial point patterns of disease—the calculation of K-functions. The technique is used to determine the degree of clustering exhibited by the residuals from a spatially referenced logit model constructed to ascertain the factors influencing the likelihood of death in a road traffic collision. The aim of the study was to investigate the importance of various factors, in particular the role of ambulance response times, in determining the likelihood of survival for each casualty involved. The K-function is defined as the expected number of further points within a distance, say S, of an arbitrary point divided by the overall intensity of points (Jones et al. 1996). The analysis involves splitting collisions into those with a fatality and those without, and calculating and comparing their K-functions. The K-function depends on both the density of points in a region around an arbitrary point and another counter that totals the number of points in the region within a specific distance. Estimates of the K-function should be determined for a range of values of S. The separate estimates produced for each of the two point patterns can be examined to see if either exhibits comparative clustering (by dividing K-function estimates to produce a measure of fatalities over nonfatalities for all distances S at each tested point). A pattern of clustering can then be depicted by plotting a map. It is suggested that significance testing can be performed using either Monte Carlo simulation or using more systematic combinatorics. In the discussion of Jones et al. (1996, 884), it is claimed that the study to which the K-function method was applied "found an apparent localized clustering of unexpected fatalities." Yet, what Jones et al. (1996) have found is that somewhere in Norfolk there is localized clustering in a fixed time period of road collision fatalities, given the clustering of nonfatal road collisions and taking into account the mode of transport. Sadly, there is no map to identify where in space this localized clustering is. Somehow, a distance of about 2 km was found to be the best distance over which to look for clustering. There is also a suggestion that the cause of the fatalities and clustering "may be associated with dense urban areas... [and] the highly localized scale of the clustering..." indicates that clustering varies within urban areas. However, no rigorous statistical tests were conducted to establish the statistical significance of the clusters at the local scale, let alone the causes behind.

2.4 SPATIAL AUTOCORRELATION

The methods and procedures of spatial analysis can take many different guises, most obviously based upon mapping using GIS. A road collision in this context is often seen as a point or occurrence on a map. However, in many modeling and prediction techniques, traffic collisions tend to be aggregated either to links on the road network or to administrative units. This then leads to a fundamental problem experienced by many geographers, which is determining, or working with, the most appropriate size and shape of the spatial units used for analysis, since this may heavily influence the visual message of mapping and the outcome of statistical tests. In this context, Thomas (1996) explores aspects of the modifiable areal unit problem (MAUP) with the objective of finding the optimum length of road segment for road collision analysis in Belgium. Both Yule and Kendall (1950) and Openshaw and Taylor (1979) recognized the size and scale problem, which is present in many geographical studies with spatial datasets. Generally, when increasing the level of granularity of the analysis, for example, from London boroughs to census output areas for London (in terms of road collision count data summaries), the correlation between output areas will become weaker. This introduces another issue surrounding road safety analysis, that is, spatial autocorrelation. Spatial autocorrelation refers to the extent to which the value of a variable at a given location influences values of that variable at contiguous locations (Cliff and Ord 1973; Goodchild 1986; Griffith 1987; Odland 1988). For example, this could be the influence of a variable on grids that lie next to or close to each other. The notion of spatial autocorrelation rests again on the premise of Tobler's first law of geography (1970). Spatial autocorrelation simultaneously deals with both the attributes of the spatial data and the spatial feature as a location. By the same token, it deals with the location of the collision itself and the attributes of the collision such as time or severity. Figure 2.1 shows collision statistics on road links on the Ontario provincial highway network. Low spatial autocorrelation in these statistics would imply local causal factors such as HRLs, whereas strong positive autocorrelation would imply a more regional scale of variation pointing to causal factors such as lifestyles, and rural and urban land uses.

One of the key aims of the spatial analysis of road collisions is to identify HRLs. In order to identify the most suitable method, there are a number of considerations to be taken into account such as the scale at which an HRL is deemed manifest, the basic spatial unit (for example a section of road), and the boundary problems associated with any determination of a basic spatial unit. Flahaut et al. (2003) argue that the spatial aspects of road collisions have often been neglected in the literature and that several basic methodological aspects such as the definition of an HRL are still under-researched. It is evident from the literature (see Thomas 1996; Flahaut et al. 2003), however, that spatial concentrations of road collisions suggest there is a spatial dependence and interaction between contiguous collision locations.

When more than one HRL is identified, it is often assumed that these high concentrations of collisions are spatially independent of each other (albeit that the collisions in the HRL are spatially dependent). However, these HRLs may also be related spatially. The analysis of road collisions can be aggregated into basic spatial units of different sizes (generated from kernel density estimations). This is known as an aggregation problem and is part of the MAUP. There are two main choices with the first being road network segments and the other being grid cells. Both share aggregation drawbacks. The road network neglects information regarding to schools, shopping centers, and land use. In comparison, grid cells can include socioeconomic data

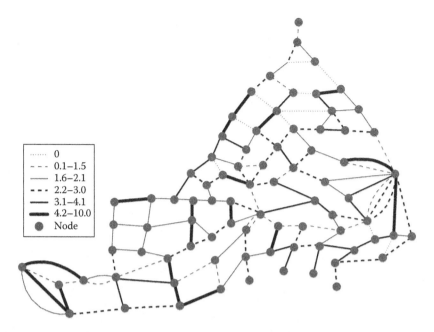

FIGURE 2.1 Network spatial autocorrelation. (Longley, P.A., Goodchild, M.F., Maguire, D.J., and Rhind, D.W.: *Geographical Information Systems and Science*. p. 98, Figure 4.11(B). 2005. Copyright John Wiley & Sons Ltd. Reproduced with permission.)

not recorded on the road network as well as the total length of road network within the grid cell(s). Many of these socioeconomic variables are not related to the road network environment.

2.4.1 Global Order Effects

The Getis and Ord's G_i and G_i^* statistics are viable options for road collision investigation. Research by Kingham et al. (2011) and Khan et al. (2006) have highlighted the use of the statistics when analyzing road collisions in relation to the school run and weather, respectively. Generally speaking, the G_i and G_i^* statistics can be categorized as examples of ESDA techniques, alongside many of the techniques mentioned in this chapter. Lack of spatial independence in geographic data has given rise to spatial statistical techniques to measure spatial autocorrelation in data, which can be incorporated in modeling procedures to eliminate errors and account for spatial dependencies. Generally, two kinds of ESDA techniques are employed to explore the underlying structure of spatial data. The first are graphical methods, such as histograms, scatter plots, box plots, and so on. The second are statistical methods that are used to describe characteristics of data distribution, quantification of spatial autocorrelation, and detection of spatial patterns of regularity.

For collisions that are influenced by weather or some other spatial phenomena, the assumption of spatial independence is violated. Although road collisions are random events, it is important to analyze the spatial heterogeneity/homogeneity of the data spread in space, especially when analyzing them from a geographical context, to make correct assumptions about the nature of the data and the analysis conducted. Road collisions can be analyzed from different spatial contexts to establish spatial associations. The measurements of these spatial dependencies or spatial autocorrelation through ESDA techniques integrated with GIS can help in analyzing spatial patterns and clusters in collision data as well as help in better modeling procedures and error estimates.

The local measures of spatial association can quantify spatial autocorrelation at a local scale that may be masked by global measures. Both distance statistics, that is, Getis–Ord's G_i statistics and local Moran's I or LISA (local indicators of spatial association) proposed by Anselin, are well-known types of local measures of spatial association.

$G_i(d)$ and $G_i^*(d)$ statistics are described by Ord and Getis. They indicate the extent to which a location is surrounded by a cluster of high or low values. This statistic shows areas where higher-than-average values tend to be found near each other or where lower-than-average values tend to be found near each other. The $G_i(d)$ statistic excludes the value at the given location in which the spatial autocorrelation is being measured while $G_i^*(d)$ includes the value at that location. The standardized G calculates a single Z-score value for each location in the study area. Positive value indicates clustering of high-attribute value locations and negative value indicates clustering of low-attribute value locations. The more positive or negative the value, the more significant the results are.

2.4.2 Local Indicators of Spatial Autocorrelation (LISA)

It is important for one to distinguish between global and local spatial autocorrelation. The global spatial autocorrelation measures investigate globally if locations that belong to the study area are spatially correlated. They give an idea about the study area as a whole. If there is no global spatial autocorrelation detected, it may still happen that parts of the study area actually are exhibiting a spatial autocorrelation. If global spatial autocorrelation is present, the local indices will be useful to point at the contribution of smaller parts of the area under investigation. These local indices are considered to be local indicators of spatial association (LISA), if they meet two conditions:

1. It needs to measure the extent of spatial autocorrelation around a particular observation, and this is for each observation in the dataset.
2. The sum of the local indices needs to be proportional to the global measure of spatial association.

In recent years, many local statistical processes have been developed for road collision analysis. These LISA statistics include local Moran's I, local Geary's c, and the Getis–Ord's G_i and G_i^* statistics. The local version of Moran's I can be written as follows:

$$I_i = \frac{(X_i - \bar{X})}{\sum (X_i - \bar{X})^2} \sum_j W_{ij}(X_j - \bar{X}) \qquad (2.2)$$

where

I_i is the I value on location i
W_{ij} is the weight describing the spatial relationship between location i and j
X is the attribute value
$\bar{X}$ is the global mean value calculated as the average of all data

These local spatial autocorrelation statistics can offer insights into the location, scale, shape, and extent of local clusters or HRLs within a study area. In the road safety literature, there has been a tendency to use spatial data analysis techniques next to statistical (Poisson) regression models to determine locations with a high number of collisions. This enables the typical statistical model to account for the spatial character of a location.

2.5 KERNEL DENSITY ESTIMATION

Although there has been some use of kernel density estimation (Sabel et al. 2005) for analyzing road collision point data, there has been limited explanation of the method in the road safety literature. For example, the optimum bandwidth discussion for collision data has been neglected. Its use in recent years in social science has focused on representing the density or volume of crimes distributed across the study area (Chainey and Ratcliffe 2005; van Eck et al. 2006). Chainey and Ratcliffe (2005) suggest that what is still missing from the generic kernel density estimation is a method that can statistically define those areas that are HRLs. In road safety research, kernel density estimation is an interpolation technique, which is a method for generalizing collision locations (points) to an entire area (Silverman 1986; Bailey and Gatrell 1995). In short, whereas spatial distribution and HRL techniques provide comprehensive statistical summaries for the collision data, interpolation techniques generalize the collisions over the study region. There are many interpolation techniques such as Kriging and local regression models.

By interpreting the collision point data in the form of a density surface, a number of decisions have to be made to facilitate appropriate and robust surfaces and ultimately the results. The literature is characterized by many similar academic studies on road collision patterns using the original data points of the collision locations and representing this "population" as the original points using symbols and possibly varying colors for the types of severity of type of vehicle involved in the collision. These are known as dot density maps. However, what these maps fail to show is the spread of risk that a collision generates. There has been limited work that has focused on the density maps in terms of bandwidth choice and kernel values. The advantages of these surface representations particularly of road collisions are that they can provide a more realistic continuous model of HRL patterns, reflecting the changes in density that are often difficult to represent using geographically constrained boundary-based models such as the transport network or census tracts.

There are many advantages to the use of kernel density estimation (KDE) as opposed to statistical and clustering techniques such as K-means. The main advantage for this particular method lies in determining the spread of risk of a collision. This spread of risk can be defined as the area around a defined cluster in which there is a higher likelihood for a collision to occur based on spatial dependence. This degree of risk would not be measured using the clustering techniques. Instead, kernel density estimation involves placing a symmetrical surface over each point and then evaluating the distance from the point to a reference location based on a mathematical function and then summing the value for all the surfaces for that reference location. This procedure is repeated for successive points. This method therefore allows us to place a kernel over each observation, and summing these individual kernels gives us the density estimate for the distribution of collision points (Fotheringham et al. 2000).

The concept of this method originated in the 1950s as an alternative method for the density of a histogram. This concept was first applied to univariate data. Used in a geographical context, it is applied to multivariate data in order to appreciate the spatial distribution and intensity of the points. Its application to collision analysis will be based on using the x, y coordinates for the location and obtaining the density from the count data. The KDE equation can be written as (Fotheringham et al. 2000)

$$\hat{f}(u,v) = \frac{1}{nh^2} \sum_{i=1}^{n} K\left(\frac{d_i}{h}\right) \qquad (2.3)$$

where
$f(u, v)$ is the density estimate at the location (u, v)
n is the number of observations
h is the bandwidth or kernel size
K is the kernel function
d_i is the distance between the location (u, v) and the location of the ith observation

The effect of placing these humps or kernels over the points is to create a smooth and continuous surface. When computing this method, there are many decisions that need to be made regarding the kernel shape, bandwidth, and cell size.

Using the kernel density function allows a surface to be created, which will visualize the locations of area (or cells) with high (and low) numbers of collisions. Using a spatial density measure for this purpose is more accurate than using count data across space. This is important, as this methodology seeks to determine risk levels not only at specific collision points but also in the neighborhood of these points. The count data are used in the initial stages to establish the density, which can reflect a spread of risk that may or may not occur around the collision.

The method is known as kernel density estimation, because around each point (at which the indicator is observed) a circular area (the kernel) of defined bandwidth is created. This takes the value of the indicator at that point spread into it according to some appropriate function. Summing all of these values at all places, including

Collision Density in Two-Dimensional Space

those at which no incidences of the indicator variable were recorded, gives a surface of density estimates. Density can be measured by two methods: simple and kernel. The simple method divides the entire study area to predetermined number of cells and draws a circular neighborhood around each cell to calculate the individual cell density values, which is the ratio of number of features that fall within the search area to the size of the area. Radius of the circular neighborhood affects the resulting density map. If the radius is increased, there is a possibility that the circular neighborhood would include more feature points that result in a smoother density surface.

The kernel density estimation method uses a mathematically complex way to estimate the density compared to the simple method. The kernel method divides the entire study area into predetermined number of cells. Rather than considering a circular neighborhood around each cell (the simple method), the kernel method draws a circular neighborhood around each feature point (the collision) and then a mathematical equation is applied that goes from 1 at the position of the feature point to 0 at the neighborhood boundary. The chosen radius of the circular neighborhood affects the resulting density map. If the radius is increased, all other things being equal, the kernel becomes flatter. This kernel function is applied to each collision point, and individual cell density value is the sum of the overlapping kernel values over that cell divided by the area of the search radius. A smoother-looking density surface is created by kernel density calculations than the simple density calculations.

2.5.1 Optimum Bandwidth

The bandwidth is the search radius within which intensity values for each point are calculated. Points are weighted, where collisions closer to the kernel center contribute a higher value to the cell's intensity value of the cell (Ratcliffe 1999). The choice of bandwidth will affect the outcome of the spatial clusters of road collisions identified. The bandwidth could cover an area the size of a census output area or the size of a street, and this in turn will affect the location and size of the HRLs. The larger the bandwidth, the larger the HRLs will be. Some degree of aggregation and smoothing is required if we are to identify urban HRLs using a more local small-area approach.

Arguably the most important criterion for determining the most appropriate density surface is the bandwidth (Brunsdon 1991, 1995; Bailey and Gatrell 1995; Fotheringham et al. 2000). Depending on the dataset and the scale of the dataset, a number of methods to work out the optimum size have been suggested in the literature. The first being Environmental Systems Research Institute (ESRI) product defaults, whereby the minimum dimension (x or y) of the extent of the input theme is divided by 30, that is, $\min(x, y)/30$. Bailey and Gatrell (1995) suggest a bandwidth defined as 0.68 times the number of points (n) raised to the -0.2 power, or $0.68(n)^{-0.2}$. This can be adjusted depending on the areal extent by multiplying it by the square root of the size of the study area. The problem with both of these methods for estimating bandwidth is that neither takes into account the spatial distribution of the points. Bailey and Gatrell's (1995) estimate is based on point density, but this is limited. Large sample sizes will result in small bandwidths, while small sample sizes will result in large bandwidths. No consideration is given to the relative spacing of the points. The arbitrary nature of the coefficient and power is also problematic. A very large number of combinations would yield similar results. A more practical approach to selecting a bandwidth would take into consideration the relative distribution of points across the study area. One way to achieve this is to base the bandwidth on average distances among points.

In addition, some have argued that the bandwidth should be no larger than the finest resolution and others have argued for a variation on random nearest neighbor distances (Chainey and Ratcliffe 2005). Others have argued for particular sizes, two methods for which have been outlined in the previous paragraph. Generally, a narrower bandwidth interval leads to a finer mesh density estimate, whereas a larger bandwidth interval will lead to a less clear pattern of variability and, therefore, less variability between areas. While smaller bandwidths show greater differentiation between areas, one has to keep in mind the statistical precision of the estimates. For example, if a sample size is not very large, a smaller bandwidth will lead to greater imprecision in the estimates. On the other hand, if the sample size is large, a finer density estimate can be produced. This has entailed detailed experimentation with varying bandwidths.

Finding the optimum parameters (bandwidth and cell size) for kernel density estimation when analyzing road collision density specifically is not an easy process, as there are no strict statistical guidelines that can be followed. The limited range of studies that have documented parameters for road collision density measurements means that the process of deciding the bandwidth and grid cell size is somewhat subjective. In retrospect, even if previous research had suggested viable parameters, it is evident that the area being measured in each study will vary. For example, Flahaut's (2004) density measures are of one road in Belgium (N29). He uses kernel density estimation to estimate the density along this busy and dangerous road in Belgium. This particular method has subsequently been applied to road collision analysis in New Zealand (Sabel et al. 2005). These studies have very different study area sizes.

2.5.2 Case Study: Road Collisions in London, United Kingdom

Using the road collision data in London, Table 2.1 shows the varying parameters that have been tested and used for KDE. Figures 2.2 and 2.3 show the subsequent images in ArcGIS. The figures in the table need to be interpreted with caution, as a large majority of the negative (zero) readings, where collisions do not occur, fall outside of the study area, which is the boundary of the London boroughs where no data have been recorded for the study (the density function creates a square over the study area whereby the majority of negative readings have to be disregarded). Therefore, it was important to create quite a fine mesh (in terms of cell size) over the data to have a better understanding of the areas where no collisions occur, because if the cell size is too large, the cells where there are no collisions would be overlooked, as the majority of the cells would contain at least one road collision because of the large number of collisions being analyzed.

Collision Density in Two-Dimensional Space

TABLE 2.1
Variations in Search Radius and Bandwidth Using ESRI's ArcGIS Density Measure

Bands	Search Radius	Cell Size	m/km	Total No. Cells	No.+Cells	No. − Cells	%+	%−
Band 1	750	250	m	48,510	27,884	20,666	57	43
Band 2	1000	500	m	12,180	7,335	4,845	60	40
Band 3	500	100	m	303,450	160,459	142,991	53	47
Band 4	100	20	m	7,586,777	1,946,099	5,640,978	25	75
Band 5	500	200	m	76,007	40,123	35,884	53	47
Band 6	500	250	m	48,510	25,671	22,839	53	47
Band 7	600	300	m	33,775	18,551	15,224	53	47
Band 8	500	225	m	60,138	31,717	28,421	53	47
Band 9	200	200	m	76,007	30,500	45,507	40	60
Band 10	400	100	m	241,120	103,414	137,706	43	67
Band 11	200	100	m	303,450	121,915	181,535	40	50

FIGURE 2.2 Band 2.

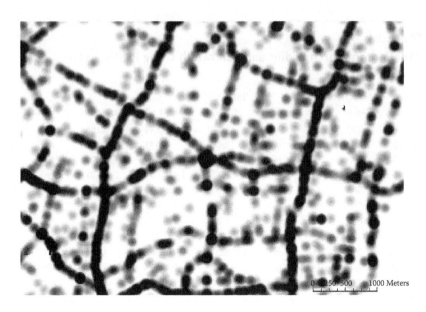

FIGURE 2.3 Band 4.

2.6 GEOGRAPHICALLY WEIGHTED REGRESSION

In recent years, spatial analysis has undergone many transitions (Páez et al. 2002). One of these changes was the general movement from an initial focus on testing for spatial pattern using spatial autocorrelation techniques (Cliff and Ord 1981) to modeling spatial patterns by means of regression models with spatial components (Griffith 1987; Anselin 1988; Haining 1990; Cressie 1992). These testing and modeling methods, however, are characterized by their global theme and determining the study area as a discrete single entity. This means that the complex spatial patterns were often overlooked (Páez et al. 2002). This drawback was somewhat rectified in the recent trend to determine the local opposite to the global. This can be observed in work by Getis and Ord and their $G_i(d)$ statistics of local spatial autocorrelation (Getis and Ord 1992; Ord and Getis 1995, 2002). Also Anselin's (1995) LISA statistics are counterparts to the global autocorrelation statistics. Specifically, these statistics are used in order to detect spatial outliers, in particular HRLs or a location where the spatial dependency between the points may be intense.

In a modeling environment, local parameter estimates are increasingly determined by essentially fitting a surface to the data, with its shape depending directly upon the complexity of the expansion parameters (Fotheringham et al. 1998). It has been argued that this surface analysis might overlook complex spatial variation. To address this issue, Fotheringham, Brunsdon, and Charlton (Brunsdon et al. 1996; Fotheringham et al. 1997, 1998) have proposed a method called geographically weighted regression (GWR), which allows investigation of whether any relationships that exist are stable over space, or whether they change to reflect the characteristics of different localities in the study area. Basically, GWR is a locally weighted regression method that works by assigning a weight to each observation i depending on the distance from a specific geographical location called x. The weighting system is based on the concept of distance decay, which uses a kernel function. A typical regression model would look like this:

$$Y = \beta_0 + \beta_1 x_1 + \beta_2 x_2 + \varepsilon \tag{2.4}$$

where
y is the dependent variable
x_1 and x_2 are the independent variables
β_0, β_1, and β_2 are the parameters to be estimated
ε is the basic random error term assumed to be normally distributed

This basic model assumes that observations are independent from one another. However, with geographical data, for example, casualty and driver location data from road collisions, this may not be the case. It also assumes that the structure of the model remains constant over space whereby GWR enables the parameter estimates to vary locally. The GWR equation can be seen as follows whereby the locations within the study area are incorporated:

$$y(u,v) = \beta_0(u,v) + \beta_1(u,v)x_1 + \beta_2(u,v)x_2 + \varepsilon(u,v) \tag{2.5}$$

This model can be fitted by least squares to a given estimate of the parameters at the location (u, v) which, in this case, are the easting and northing of the casualty and drivers home address, and there a predicted value can be achieved. This is done by, what has been briefly mentioned previously, a geographical weighting process. In short, the weights are chosen in such a way that those observations near a point in space are given a much heavier weight than those further away. The parameter estimates can be subsequently mapped.

The GWR software enables three choices of the types of model that can be used (Gaussian, logistic, and Poisson). The most common type chosen for road collisions is the Poisson option, because count data is more commonly used. Using a Poisson model for the data requires an offset variable. For example, if one is analyzing the number of burglaries in a specific area, the offset variable can be the total number of households within that area. The Poisson regression model (nonspatial) has been used frequently when analyzing road collision data (see Maher and Summersgill 1996; Mountain et al. 1996; Greibe 2003; Eisenberg 2004; MacNab 2004; Noland and Quddus 2004; Wood 2005).

One of the features of road collision data, albeit actual collision data or driver and casualty data, is that it is count data and therefore not normally distributed. Noland and Quddus (2004) explore the relationship between road injuries and the spatial variables associated with a given area. Results show (for the whole of the United Kingdom) that lower population densities have fewer injuries. They also find that areas with a higher index of multiple deprivation are associated with increased slight and serious injuries. These results are consistent with findings from Abdalla et al. (1997), Chichester et al. (1998), and Beattie et al. (2001), who also found a positive correlation between area deprivation and road traffic injuries. Other studies used Poisson distribution for the frequency of collisions in a given period of time at one specific site. Jovanis and Chang (1986) used a Poisson model to relate collision frequency to kilometrage traveled and environmental variables. Comparative studies by, for example, Joshua and Garber (1990) and Miaou and Lum (1993) have outlined the advantages of using the Poisson model over the standard regression model.

2.7 CONCLUSION

Spatial interpolation has been argued to be just intelligent guesswork. However, many years of research and successful examples have increasingly led people to think otherwise. Spatial interpolation is the procedure of estimating the value of properties at unsampled sites within the area covered by existing observations. In the previous chapter, we discussed the classic and conventional indicators that spatial interpolation techniques used, namely, Euclidian distances for space characterization. However, pure Euclidian distance has become an arbitrary measure for analyzing road collisions, particularly when focusing on density and clustering methods in the urban environment. Much of the research surrounding road collision data now focuses on prediction and evaluation. There has been little work on the subject of HRL definitions. In recent years, there has been a renewed interest in this delineation of HRLs due to the awareness of spatial interaction existing between contiguous collision locations, and therefore suggesting a spatial dependence between individual occurrences (Flahaut et al. 2003). This chapter seeks to address the challenges of this deficiency of clarity surrounding the question of how to quantify and define an HRL, focusing on determining collision density in the 2D space.

Traditionally, mapping using data point features has been carried out through representing populations as individual point objects using symbols or colors to differentiate between the data values held by the point. More recently, there has been a move toward the use of surfaces to model the likely distribution of the original points in a more useful and visually understandable format (following Haggett et al. 1977; Martin 1996; Martin et al. 2000). The conclusions from these papers that outline the advantages of using a surface-based approach indicate that surface representations can usefully summarize the distributions of point-referenced events when point locations are known.

The aim of this chapter is to outline and demonstrate the importance of using density measures rather than count measures of collision data. A density is the amount of something per unit area, expressed relative to some meaningful base category, for example, the total population per unit area within an output area or borough. Density values can be defined for all points, provided that one specifies what neighborhood or region around a point should be used to summarize the data. To create comparable densities at different locations, a comparable distance-weighting function around each point has to be chosen. Density-based approaches to analyze point patterns can be characterized in terms of first-order effects. The major drawback with all density measures is the sensitivity to the study area. This is a generic problem and can be problematic when attempting to calculate a local density.

The conceptual basis to the first section of this chapter lies in the premise of the spread of risk that a collision generates over an area. In many circumstances, it has been acceptable to define a cluster of collision points in an area or on a line, but the risk of a collision occurring again will likely spread beyond the boundaries of the historical collision cluster. If the cluster of collisions is small and spiky, the risk surrounding the cluster will be smaller. However, if the collision cluster is quite flat but covers a wider area, the risk surrounding the cluster will be larger but less intense than the cluster with the collisions closer together (Chainey and Ratcliffe 2005).

The road safety literature provides no universally accepted definition of HRLs. Identification of HRLs in the literature has often been seen as arising from the awareness of the spatial interaction existing between collision locations. The existence of these HRLs reveals concentrations and hence suggests a degree of spatial dependence between the point locations. These spatial concentrations could arise because of one or several causes of collision.

Thomas (1996) argued that the most appropriate level of spatial aggregation for road collision analysis is the road section in terms of a predetermined segment of road; however, in most studies this length is not controlled or even justified. There is no clear indication in the literature as to what would be the most meaningful length for a "dangerous" section of road. However, this does call into question whether there should be an optimal length of road section to analyze as HRLs and dangerous "segments" would vary in size and length depending on the characteristics of the individual HRL, therefore suggesting that to determine an overall length to be used for each dangerous "segment" would potentially miss out individual collisions that may nevertheless share spatial dependence with other collisions in the HRL. Two methods are compared, namely the calculation of local spatial autocorrelation measures and kernel density estimation. Both methods graduate levels of local danger and generate smoothing of the empirical process, and although each of the methods starts from different conceptual backgrounds, both provide quite similar results under a specific choice of parameters.

REFERENCES

Abdalla, I. M., R. Raeside, D. Barker, and D. R. D. McGuigan. 1997. An investigation into the relationships between area social characteristics and road accident injuries. *Accident Analysis & Prevention* 29 (5): 583–593.

Anselin, L. 1988. *Spatial Econometrics: Methods and Models.* Dordrecht, the Netherlands: Kluwer Academic Publishers.

Anselin, L. 1995. Local indicators of spatial association—LISA. *Geographical Analysis* 27 (2): 93–115.

Bailey, T. C. and A. C. Gatrell. 1995. *Interactive Spatial Data Analysis.* Essex, U.K.: Longman Scientific & Technical.

Beattie, T. F., D. R. Gorman, and J. J. Walker. 2001. The association between deprivation levels, attendance rate and triage category of children attending a children's accident and emergency department. *Emergency Medicine Journal* 18 (2): 110–111.

Brunsdon, C. 1991. Estimating probability surfaces in GIS: An adaptive technique. In *Proceedings of the Second European Conference on Geographical Information Systems*, eds. J. Harts, H. F. Ottens, and H. J. Scholten, pp. 155–163. Utrecht, the Netherlands: EGIS Foundation.

Brunsdon, C. 1995. Estimating probability surfaces for geographical point data: An adaptive kernel algorithm. *Computers & Geosciences* 21 (7): 877–894.

Brunsdon, C., A. S. Fotheringham, and M. E. Charlton. 1996. Geographically weighted regression: A method for exploring spatial nonstationarity. *Geographical Analysis* 28 (4): 281–298.

Chainey, S. and J. Ratcliffe. 2005. *GIS and Crime Mapping*. Chichester, U.K.: John Wiley & Sons.

Chichester, B. M., J. A. Gregan, D. P. Anderson, and J. M. Kerr. 1998. Associations between road traffic accidents and socio-economic deprivation on Scotland's west coast. *Scottish Medical Journal* 43 (5): 135–138.

Cliff, A. D. and J. K. Ord. 1973. *Spatial Autocorrelation*. London, U.K.: Pion.

Cliff, A. D. and J. K. Ord. 1981. *Spatial Processes: Models & Applications*. London, U.K.: Pion.

Cressie, N. 1992. Statistics for spatial data. *Terra Nova* 4 (5): 613–617.

Eisenberg, D. 2004. The mixed effects of precipitation on traffic crashes. *Accident Analysis & Prevention* 36 (4): 637–647.

Flahaut, B. 2004. Impact of infrastructure and local environment on road unsafety: Logistic modeling with spatial autocorrelation. *Accident Analysis & Prevention* 36 (6): 1055–1066.

Flahaut, B., M. Mouchart, E. S. Martin, and I. Thomas. 2003. The local spatial autocorrelation and the kernel method for identifying black zones: A comparative approach. *Accident Analysis & Prevention* 35 (6): 991–1004.

Fotheringham, A. S., M. E. Charlton, and C. Brunsdon. 1997. Measuring spatial variations in relationships with geographically weighted regression. In *Recent Developments in Spatial Analysis*, eds. M. M. Fischer and A. Getis. London, U.K.: Springer.

Fotheringham, A. S., M. E. Charlton, and C. Brunsdon. 1998. Geographically weighted regression: A natural evolution of the expansion method for spatial data analysis. *Environment and Planning A* 30 (11): 1905–1927.

Fotheringham, A. S., M. E. Charlton, and C. Brunsdon. 2000. *Quantitative Geography: Perspectives on Spatial Data Analysis*. Thousand Oaks, CA: Sage.

Getis, A. and J. K. Ord. 1992. The analysis of spatial association by use of distance statistics. *Geographical Analysis* 24 (3): 189–206.

Goodchild, M. F. 1986. *Spatial Autocorrelation*, Vol. 47: *Concepts and Techniques in Modern Geography*. Norwich, U.K.: Geo Books.

Greibe, P. 2003. Accident prediction models for urban roads. *Accident Analysis & Prevention* 35 (2): 273–285.

Griffith, D. A. 1987. *Spatial Autocorrelation: A Primer*. Washington, DC: Association of American Geographers, Resource Publications in Geography.

Haggett, P., A. D. Cliff, and A. Fry. 1977. *Location Analysis in Human Geography*, 2nd edn. New York: Wiley.

Haining, R. 1990. *Spatial Data Analysis in the Social and Environment Sciences*. Cambridge, U.K.: Cambridge University Press.

Hauer, E. 1997. *Observational Before-After Studies in Road Safety: Estimating the Effect of Highway and Traffic Engineering Measures on Road Safety*. Oxford, U.K.: Pergamon.

Jones, A. P., I. H. Langford, and G. Bentham. 1996. The application of K-function analysis to the geographical distribution of road traffic accident outcomes in Norfolk, England. *Social Science & Medicine* 42 (6): 879–885.

Joshua, S. C. and N. J. Garber. 1990. Estimating truck accident rate and involvements using linear and Poisson regression models. *Transportation Planning and Technology* 15 (1): 41–58.

Jovanis, P. P. and H. L. Chang. 1986. Modeling the relationship of accidents to miles traveled. *Transportation Research Record* 1068: 42–51.

Khan, G., X. Qin, and D. A. Noyce. 2006. Spatial analysis of weather crash patterns in Wisconsin. Paper presented at the *85 Annual Meeting of the Transportation Research Board*, Washington, DC.

Kingham, S., C. E. Sabel, and P. Bartie. 2011. The impact of the "school run" on road traffic accidents: A spatio-temporal analysis. *Journal of Transport Geography* 19 (4): 705–711.

Longley, P. A., M. F. Goodchild, D. J. Maguire, and D. W. Rhind. 2005. *Geographical Information Systems and Science*. Chichester, U.K.: John Wiley & Sons.

MacNab, Y. C. 2004. Bayesian spatial and ecological models for small-area accident and injury analysis. *Accident Analysis & Prevention* 36 (6): 1019–1028.

Maher, M. J. and I. Summersgill. 1996. A comprehensive methodology for the fitting of predictive accident models. *Accident Analysis & Prevention* 28 (3): 281–296.

Martin, D. 1996. An assessment of surface and zonal models of population. *International Journal of Geographical Information Systems* 10 (8): 973–989.

Martin, D., N. J. Tate, and M. Langford. 2000. Refining population surface models: Experiments with Northern Ireland census data. *Transactions in GIS* 4 (4): 343–360.

Miaou, S. P. and H. Lum. 1993. A statistical evaluation of the effects of highway geometric design on truck accident involvements. Oak Ridge National Laboratory, Oak Ridge, TN.

Mountain, L., B. Fawaz, and D. Jarrett. 1996. Accident prediction models for roads with minor junctions. *Accident Analysis & Prevention* 28 (6): 695–707.

Nicholson, A. 1998. Analysis of spatial distributions of accidents. *Safety Science* 31 (1): 71–91.

Noland, R. B. and M. A. Quddus. 2004. A spatially disaggregate analysis of road injuries in England. *Accident Analysis & Prevention* 36 (6): 973–984.

Odland, J. 1988. *Spatial Autocorrelation*. Newbury Park, CA: Sage Publications.

Openshaw, S. and P. J. Taylor. 1979. A million or so correlation coefficients: Three experiments on the modifiable areal unit problem. In *Statistical Applications in the Spatial Sciences*, ed. N. Wrigley. London, U.K.: Pion.

Ord, J. K. and A. Getis. 1995. Local spatial autocorrelation statistics: Distributional issues and an application. *Geographical Analysis* 27 (4): 286–306.

Ord, J. K. and A. Getis. 2002. Testing for local spatial autocorrelation in the presence of global autocorrelation. *Journal of Regional Science* 41 (3): 411–432.

Páez, A., T. Uchida, and K. Miyamoto. 2002. A general framework for estimation and inference of geographically weighted regression models: 1. Location-specific kernel bandwidths and a test for locational heterogeneity. *Environment and Planning A* 34 (4): 733–754.

Ratcliffe, J. H. 1999. Spatial pattern analysis machine version 1.2 users guide. http://web.archive.org/web/20071029014716/http://www.jratcliffe.net/ware/spam1.htm (accessed October 29, 2007).

Sabel, C. E., S. Kingham, A. Nicholson, and P. Bartie. 2005. Road traffic accident simulation modelling—A kernel estimation approach. Paper presented at the *17th Annual Colloquium of the Spatial Information Research Centre*, University of Otago, Dunedin, New Zealand.

Silverman, B. W. 1986. *Density Estimation for Statistics and Data Analysis*, 1st edn. London, U.K.: Chapman & Hall.

Thomas, I. 1996. Spatial data aggregation: Exploratory analysis of road accidents. *Accident Analysis & Prevention* 28 (2): 251–264.

Tobler, W. R. 1970. A computer movie simulating urban growth in the Detroit region. *Economic Geography* 46: 234–240.

van Eck, N. J., F. Frasincar, and J. van den Berg. 2006. Visualizing concept associations using concept density maps. In *10th International Conference on Information Visualization (IV 2006)*, London, U.K., pp. 270–275.

Wood, G. R. Confidence and prediction intervals for generalised linear accident models. 2005. *Accident Analysis & Prevention* 37 (2): 267–273.

Yule, G. U. and M. G. Kendall. 1950. *An Introduction to the Theory of Statistics*, 14th edn. London, U.K.: Griffin and Co.

3 Road Safety as a Public Health Issue

3.1 WHY WOULD ROAD COLLISIONS BE CONSIDERED A PUBLIC HEALTH ISSUE?

Road safety management does not make international headlines. The subject is absent from the agendas of global summits on poverty reduction, public health, engineering, and often transportation. Yet few issues merit more urgent attention. Road traffic deaths and injuries represent a global epidemic, and the costs of that epidemic are borne overwhelmingly by the world's poorest countries and people. When it comes to death and injury, no war or humanitarian disaster rivals the impact of road injuries. Few killer diseases pose an equivalent level of risk. Apart from the devastating human consequences, road traffic injuries are holding back progress in economic growth, poverty reduction, health, and education. With projections pointing to an increase in fatalities and injuries on the roads of the world's poorest nations, society needs to address the culture of neglect that pushes road safety to the margins of transport and development policy.

Public health is the science and practice of protecting and improving the health of *communities* through education; promotion of healthy lifestyles; and research on disease control, health promotion, and injury prevention (Sleet et al. 2007). Sleet et al. (2007) outlines that the three core functions of public health are consistent with the efforts to reduce motor vehicle injury:

1. Monitor and evaluate the health needs of communities.
2. Promote healthy practices and behaviors in populations.
3. Identify and eliminate environmental hazards to assure that populations remain healthy.

Motor vehicle injuries remain an important public health problem (Institute of Medicine [IOM] 1999). As the WHO indicates, road safety should be viewed as a shared responsibility and not the exclusive purview of a single agency (Peden et al. 2004). Traffic collisions affect not only transportation systems but also economic systems, health systems, jobs, families, and civil society. A culture of safety implies a systematic commitment by institutions, agencies, organizations, and individuals to recognize and address the unacceptable road toll and apply the best prevention strategies known to reduce it.

Transport infrastructure may seem far removed from human development concerns, but it is one of the building blocks for progress toward the Millennium

Development Goals (MDGs) of the United Nations (UN). MDGs are eight international development goals that all the 193 UN member states have agreed to achieve by 2015. Although road safety is not *one of the eight* MDGs, it arguably focuses within the context of many of the goals. Road safety therefore is another building block, or it should be. Death and injury on the world's roads is arguably the single most neglected human development challenge. The vocabulary of the road traffic injury epidemic helps to explain the neglect. While child deaths from, say, malaria are viewed as avoidable tragedies that can be stopped through government action, road traffic deaths and injuries are widely perceived as "accidents"—unpredictable events happening on a random basis to people who have the misfortune to be in the wrong place at the wrong time.

The vocabulary is out of step with reality. Road traffic fatalities and injuries are accidents only in the narrow technical sense that they are not intended outcomes. They are eminently predictable, and we know in advance the profile of the victims. Of the 3500 people who will die on the world's roads today, around 3000 will live in a developing country, and at least half will be a pedestrian or vulnerable road user who is not driving a car (Figure 3.1 and Table 3.1). Far from being the consequence of forces beyond human control, road traffic death and disability are in large also consequences of government action and inaction.

In 2002, the WHO (World Health Organization) estimated that 1.2 million people were killed and approximately 50 million injured in road collisions worldwide, costing the global communities an estimated $518 billion. In 2004, this figure had risen to 1.3 million people killed by road collisions worldwide. The International Federation of Red Cross and Red Crescent Societies have described the situation as "a worsening global disaster destroying lives and livelihoods, hampering development and leaving millions in greater vulnerability" (International Federation of Red Cross and Red Crescent Societies 1998). Without appropriate action, road traffic injuries are predicted to escalate from the ninth leading contributor to the global burden of disease in 1990 to the third by 2020. In 2004, the WHO acknowledged the growing number of deaths and injuries associated with road collisions and designated the World Health Day to *road safety*. The outcome was a comprehensive report alongside which the World Bank released a corresponding report and the UN General Assembly urged all countries to address the devastating nature of road collisions as a matter of urgency. The outcome of these actions was a working group of 42 agencies committed to reducing this preventable health burden. The agencies pledged to work within a common framework focusing on joint activities and projects including data collection and research, technical support, advocacy, policy, and financial support.

In October 2005, as a result of the initiatives, the UN General Assembly passed a new resolution on road safety asking for increased interagency working and commending the WHO on its efforts for advocacy. The assembly asked WHO to jointly organize the first UN Global Road Safety week in April 2007 and to recognize the third Sunday in November as the World Day of Remembrance for Road Traffic Victims. These global initiatives have led the academic community in recent years to take road collisions not as an engineering or transport problem but as a preventable global disease that should be managed in the global context.

Road Safety as a Public Health Issue

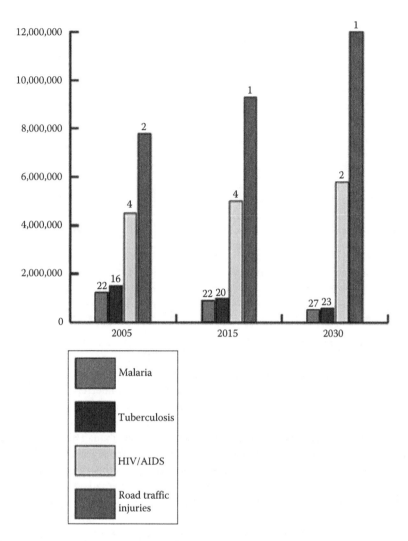

FIGURE 3.1 Projected disability-adjusted life years (DALYs) in developing countries (children aged 5–14). (Data from Mathers, C.D. and Loncar, D., Updated projections of global mortality and burden of disease, 2002–2030: Data sources, methods and results, Projected DALYs for 2005, 2015, and 2030 by country income group under the baseline scenario, World Health Organization, Geneva, Switzerland, October 2005, http://www.who.int/healthinfo/global_burden_disease/projections2002/en/.)

3.2 CURRENT GLOBAL ESTIMATES

In 2002, road collision injuries were the 11th leading cause of death in the world. The lowest rates were recorded in high-income European and Western countries; however, the highest rates were found in some African and eastern European countries.

TABLE 3.1
Leading Causes of Death in Children and Youth, Both Sexes, World, 2004

Rank	5–14 Years	15–29 Years	Total
1	Lower respiratory infections	**Road traffic injuries**	Ischemic heart disease
2	**Road traffic injuries**	HIV/AIDS	Cerebrovascular (stroke) disease
3	Malaria	Tuberculosis	Lower respiratory diseases
4	Drownings	Violence	Perinatal causes
5	Meningitis	Self-inflicted injuries	Chronic obstructive pulmonary disease
6	Diarrheal disease	Lower respiratory infections	Diarrheal disease
7	HIV/AIDS	Drownings	HIV/AIDS
8	Tuberculosis	Fires	Tuberculosis
9	Protein–energy malnutrition	War and conflict	Trachea, bronchus, lung cancers
10	Fires	Maternal hemorrhage	**Road traffic injuries**

Source: Data from World Health Organization, *The Global Burden of Disease: 2004 Update*, WHO, Geneva, Switzerland, 2008.

Table 3.2 illustrates countries and their estimated road traffic death rate (per 100,000 people). This gives an interpretation of the current global trends, with African countries having some of the highest death rates. Moreover, it is important to remember that these countries are likely to have the highest rates of underreporting as well.

The major issues associated with calculating these estimates are incomplete data, especially from the less developed countries. It was estimated that for the 2002 estimates, 35 countries out of the total of 110 used unreliable data. The point therein is that the current estimates of global fatalities might be grossly underestimated due to incomplete data. What is also important to remember is that often less developed countries such as China can have a high fatality rate due to road collisions; however, often this is offset by the size of the population, or there might be a low number of motor vehicles per head, all of which can contribute to skewed global fatality data.

There are three primary measures for comparing multinational road collision and fatality data: (1) deaths per 100,000 population or per capita rate, (2) deaths in relation to overall distance traveled (such as the vehicle-miles traveled in the United States), and (3) deaths in relation to the number of registered motor vehicles in the country. All three measures should be considered when comparing disparate countries, but using just one of these methods is generally acceptable when comparing countries of similar status (e.g., *highly motorized countries*, developed nations, and third world countries).

3.3 IRTAD DATABASE COVERAGE AND UNDERREPORTING

In many countries that belong to the Organisation for Economic Co-operation and Development (OECD), the number of road fatalities has been slowly reducing since the peak in the 1970s. The current number of road fatalities in some of the

TABLE 3.2
Estimated Road Traffic Death Rate (per 100,000 Population), 2010

Country	Estimated Road Traffic Death Rate (per 100,000 Population)
Cook Islands	9.9
Egypt	13.2
Afghanistan	19.8
Iraq	31.5
Angola	23.1
Niger	23.7
United Arab Emirates	12.7
Gambia	18.8
Iran (Islamic Republic of)	34.1
Mauritania	28.0
Ethiopia	17.6
Mozambique	18.5
Sudan	25.1
Tunisia	18.8
Guinea-Bissau	31.2
Kenya	20.9
Chad	29.7
United Republic of Tanzania	22.7
Jordan	22.9
Botswana	20.8
Madagascar	18.4
South Africa	31.9
Sao Tome and Principe	20.6
Liberia	19.0
Syrian Arab Republic	22.9
Senegal	19.5
Nigeria	33.7
Central African Republic	14.6
Democratic Republic of the Congo	20.9
Mali	23.1
Rwanda	19.9
Benin	23.9
Burkina Faso	27.7
Kazakhstan	21.9
Comoros	21.8
Ghana	22.2
Yemen	23.7
Saudi Arabia	24.8
Congo	17.1
Namibia	25.0
Lebanon	22.3

(*Continued*)

TABLE 3.2 (*Continued*)
Estimated Road Traffic Death Rate (per 100,000 Population), 2010

Country	Estimated Road Traffic Death Rate (per 100,000 Population)
Morocco	18.0
Sierra Leone	22.6
Cameroon	20.1
Togo	17.2
Zimbabwe	14.6
Lesotho	28.4
Swaziland	23.4
Malawi	19.5
Zambia	23.8
Pakistan	17.4
Russian Federation	18.6
Cape Verde	22.4
Uganda	28.9
Qatar	14.0
Malaysia	25.0
Burundi	21.3
Myanmar	15.0
Kyrgyzstan	19.2
Lithuania	11.1
Venezuela (Bolivarian Republic of)	37.2
Peru	15.9
Ukraine	13.5
Oman	30.4
Mexico	14.7
Montenegro	15.0
Philippines	9.1
Guyana	27.8
Paraguay	21.4
Thailand	38.1
Mongolia	17.8
Vanuatu	16.3
Seychelles	15.0
Brazil	22.5
Lao People's Democratic Republic	20.4
Maldives	1.9
Suriname	19.6
Latvia	10.8
Saint Lucia	14.9
Dominican Republic	41.7
Kuwait	16.5

(*Continued*)

TABLE 3.2 (*Continued*)
Estimated Road Traffic Death Rate (per 100,000 Population), 2010

Country	Estimated Road Traffic Death Rate (per 100,000 Population)
Solomon Islands	14.7
Georgia	15.7
India	18.9
Bolivia (Plurinational State of)	19.2
China	20.5
Indonesia	17.7
Timor-Leste	19.5
Viet Nam	24.7
Belarus	14.4
Belize	16.4
Trinidad and Tobago	16.7
Costa Rica	12.7
Nepal	16.0
Republic of Moldova	13.9
Slovakia	9.4
Greece	12.2
Palau	14.7
Estonia	6.5
Guatemala	6.7
Poland	11.8
Slovenia	7.2
Bahamas	13.7
Bhutan	13.2
Micronesia (Federated States of)	1.8
Nicaragua	18.8
Papua New Guinea	13.0
Tajikistan	18.1
Albania	12.7
Armenia	18.1
United States of America	11.4
Brunei Darussalam	6.8
Argentina	12.6
Chile	12.3
Croatia	10.4
Honduras	18.8
Sri Lanka	13.7
Turkey	12.0
Bulgaria	10.4
Azerbaijan	13.1
Republic of Korea	14.1
Samoa	16.4

(*Continued*)

TABLE 3.2 (*Continued*)
Estimated Road Traffic Death Rate (per 100,000 Population), 2010

Country	Estimated Road Traffic Death Rate (per 100,000 Population)
Panama	14.1
Romania	11.1
Bangladesh	11.6
El Salvador	21.9
Hungary	9.1
Jamaica	11.6
Barbados	7.3
Bahrain	10.5
Cambodia	17.2
Czech Republic	7.6
Colombia	15.6
Ecuador	27.0
Mauritius	12.2
Bosnia and Herzegovina	15.6
Cyprus	7.6
Portugal	11.8
Belgium	8.1
New Zealand	9.1
Iceland	2.8
Serbia	8.3
Uzbekistan	11.3
Italy	7.2
Spain	5.4
Canada	6.8
Cuba	7.8
Ireland	4.7
Austria	6.6
Australia	6.1
France	6.4
Kiribati	6.0
Finland	5.1
Fiji	6.3
Tonga	5.8
The former Yugoslav Republic of Macedonia	7.9
Saint Vincent and the Grenadines	4.6
Germany	4.7
Israel	4.7
United Kingdom	3.7
Sweden	3.0
Japan	5.2
Norway	4.3

(*Continued*)

TABLE 3.2 (*Continued*)
Estimated Road Traffic Death Rate (per 100,000 Population), 2010

Country	Estimated Road Traffic Death Rate (per 100,000 Population)
Switzerland	4.3
Netherlands	3.9
Singapore	5.1
Uruguay	21.5
Malta	3.8
San Marino	0.0
Marshall Islands	7.4

Source: Reprinted from World Health Organization, *Road Traffic Deaths: Data by Country*, Global Health Observatory Data Repository, http://apps.who.int/gho/data/node.main.A997?lang=en, accessed October 29, 2014. With permission.

OECD countries is approximately 50% less than their peak value. Against this background, it may be easy to assume that road collisions are gradually becoming less of a problem in the world as a whole. However, this assumption would be wrong. The reality is that the overall number of road fatalities is still increasing every year. By 2020, the WHO predicts road collisions to be the sixth leading cause of death worldwide.

The most international database and information on global road collisions is the International Traffic Safety Data and Analysis Group International Road Traffic and Accident Database (IRTAD), which was established in 1988 by the OECD. It was created to serve as a mechanism for providing an aggregated database in which international road collision and victim data as well as exposure data could be collected on a continuous basis. IRTAD is both a working group and a database. The IRTAD database includes collision and traffic data and other safety indicators for 29 countries. The International Traffic Safety Data and Analysis Group (known as the IRTAD Group) is an ongoing working group of the Joint Transport Research of the OECD and the International Transport Forum. It is composed of road safety experts and statisticians from safety research institutes, national road and transport administrations, international organizations, universities, automobilist associations, motorcar industry, and so on. Its main objectives are to contribute to international cooperation on road collision data and its analysis.

The database includes more than 500 data items aggregated by country and year (from 1970) and shows up-to-date collision and exposure data, including (International Transport Forum 2011) the following:

- Injury collisions classified by road network
- Road deaths by road usage and age, by gender and age, or by road network
- Car fatalities by driver/passengers and by age
- Hospitalized road users by road usage, age bands, or road network

- Collision involvement by road user type and associated victim data
- Risk indicators: fatalities, hospitalized, or injury collisions related to population or kilometrage figures
- Monthly road collision data (three key indicators)
- Population figures by age bands
- Vehicle population by vehicle types
- Network length classified by road network
- Kilometrage classified by road network or vehicles
- Passenger kilometrage by transport mode
- Seat belt–wearing rates of car drivers by road network
- Area of state

The IRTAD database covers the following countries: Australia, Austria, Belgium, Canada, Czech Republic, Denmark, Finland, France, Germany, Greece, Hungary, Iceland, Ireland, Israel, Italy, Japan, Korea, Luxembourg, the Netherlands, New Zealand, Norway, Poland, Portugal, Slovenia, Spain, Sweden, Switzerland, United Kingdom, and the United States. It is obvious from this list of countries that a vast majority of countries are missing, and the countries that are missing are often the ones with the most serious problem with road collisions.

It is well known that the reporting of road collisions in official statistics can be incomplete and biased. Incomplete or inaccurate road collision data are part of the larger problem concerning the availability of accurate information about road collisions in general. The first major source of error on global statistics is lack of reporting to the police. Many collisions go unreported to the police due to lack of injury, deniability, and crimes. Although at a countrywide and regional level, a lot of these unreported police data can be supplemented by insurance data (whereby the insurance companies will hold accurate information of any claims made for road collision damage to a vehicle that will not necessarily include injury). However, this information is hard and almost impossible to collect at the global scale. The sheer number of insurance companies and the data privacy of clients will mean that there will be a high number of road collisions globally, which are not reported because they resulted in no or limited injury. It is known from a large number of studies summarized by Elvik and Mysen (1999), and Loo and Tsui (2007) that the reporting of reportable injury collisions in official statistics is very incomplete. A large number of important human factors relating to the road collision are not recorded (Elvik and Mysen 1999). Finally, there are errors or missing information in some of the recorded data elements of a road collision.

3.4 ECONOMIC, SOCIAL, AND HEALTH BURDENS

Apart from the humanitarian aspect of reducing road collisions, especially in developing countries, there is an increasing need to reduce road collisions from an economic standpoint as well. Road collisions consume large amounts of financial resources. However, one must bear in mind that there are many problems in developing countries that demand a share of the funding. Difficult decisions often have to be made on the amount of resources a country can devote to road safety and preventing road collisions. In order to assist this decision-making process, it is essential that a method is devised to determine the cost of road collisions and the value of preventing them (in economic terms).

Peden et al. (2004) estimated global fatalities due to road collision cost up to 3% of the global gross domestic product (GDP) every year. The global economic cost of motor vehicle collisions was estimated at $518 billion per year in 2003, with $100 billion of that occurring in developing countries. The Center for Disease Control and Prevention in the United States estimated an economic cost of $230 billion in 2000. The first requirement for costing is at the level of national resource planning to ensure that road safety is ranked equitably in terms of investment in its improvement. Fairly broad estimates are usually sufficient for this purpose but must be compatible with competing sectors. A second need for road collision cost figures is to ensure that the best use is made of any investment and that the best (and most appropriate) safety improvements are introduced in terms of the benefits that they will generate in relation to the cost of their implementation. Failure to associate specific costs with road collisions will result in the use of widely varying criteria in the choice of measures and the assessment of projects that affect road safety. As a consequence, it is extremely unlikely that the pattern of expenditure on road safety will, in any sense, be *optimal* in terms of equity. If safety benefits are ignored in transport planning, there will inevitably be associated underinvestment in road safety.

Road traffic injuries place an enormous strain on already overstretched health systems. The systems are in effect hemorrhaging resources as finance, equipment, and skilled staff are diverted to treat the victims of road traffic injuries. For instance, road traffic injury patient represent 45%–60% of all admissions to surgical wards in Kenya (Odero et al. 2003). Studies in India show that road injuries account for 10%–30% of hospital admissions (Gururaj 2008). And one hospital in Uganda reports spending around U.S. $399 per patient treating road traffic injuries (Watkins and Sridhar 2009). This is in a country with national spending of U.S. $20 per person only. The experience of poor communities in coping with medical catastrophes is very different from that experienced by economically well-off communities. The special problems faced by poor families can include inappropriate or absence of treatment leading to complications and longer treatment time; reallocation of labor of family members and reduced productivity of whole family; permanent loss of job for the victim even if he/she survives; loss of land, personal savings, and household goods; poor health and educational attainment of surviving members; and dissolution or reconstitution of household. None of these issues are officially documented, and the economic calculation for estimating the true cost of road collisions in poor societies is impossible. The knock-on effects of someone in a poor family being affected by a road collision, whether it is death or injury, are huge. The division of labor within the family will change, often affecting people's earnings; children may miss school, and older family members will not be able to look after children or infants. The impact of this is reduced schooling, decreased income, less able to manage the home, and overall added pressure.

3.5 GLOBAL GEOGRAPHY OF ROAD RISK

As a whole, it is acknowledged that Australia, New Zealand, and Europe have among the most favorable road safety records with the traffic risk being lower than in any other parts of the world (Loo et al. 2005). North, Central, South America, and Eastern Europe have higher traffic risk. In Central and South America, the health risk is low, and this relates to the fact that the level of motorization is relatively low. However, the standard deviations of both health risk and traffic risk are high compared to the average risk.

Road collisions are a burden not just to the developed countries but also to the developing countries. Africa as a continent has some of the highest death rates associated with road collisions in the world. One of the key issues is lack of accurate data; however, the data that are available already highlights cause for concern. Some of the key causes of road collisions in Africa (see Jacobs et al. 2000) include poorly built roads, aged vehicles, tax regulations, and a culture that has less regard to human risk. While currently Southeast Asia has the highest proportion of global road fatalities (one-third of the 1.4 million occurring every year), the road traffic injury mortality rate is the highest in Africa (28.3 per 100,000 population, when corrected for underreporting). Developing countries account for approximately 85% of all road traffic deaths in the world; the increased number of vehicles per inhabitant will result in an 80% rise in injury mortality rates between 2000 and 2020. In Africa, it was estimated that 59,000 people lost their lives in road collisions in 1990, and this figure will increase to 144,000 by 2020. This 144% increase is significantly worrying. In contrast, countries in the developed world have experienced a decreasing trend since the 1960s. Due to the traditional misconception that road traffic injuries were inevitable, random, and unpredictable events, the international community's response to this worldwide public health crisis came relatively late. The number of vehicles per inhabitant is still low in Africa: less than one licensed vehicle per 100 inhabitants in low-income Africa versus 60 in high-income countries. Car ownership growth leads to increased road traffic in developing countries. This explains, for example, the reported 400% increase in road deaths in Nigeria between the 1960s and the 1980s. Available historical data from developed countries show that it is only when a development threshold is achieved that the road mortality starts to decrease (Vasconcellos 1999; Kopits and Cropper 2005; Bishai et al. 2006). This is often called the environmental Kuznets curve. Such a threshold is far from being reached in sub-Saharan Africa. Indeed, in South Africa, the most developed African country, there were already 17 licensed vehicles per 100 inhabitants in 2005, and no decline in road traffic deaths has been observed so far.

3.6 ROAD SAFETY AND DEVELOPMENT

As recently advocated by Khayesi and Peden (2005), road safety in Africa is "part of the broader development process." The situation is particularly worrying in this continent because of the combination of conflicting road users, poor vehicle condition, underdeveloped infrastructure, lack of risk awareness, and ineffective enforcement jeopardized by corruption or bribery. The road transport system is the dominant

Road Safety as a Public Health Issue 51

form of land transportation and carries more than 95% of passenger traffic. This sector is often prioritized in donor development plans in countries such as Cameroon, Ghana, Gabon, and Senegal, to cite only a few African countries receiving European Union development aid. Road transportation is essential to access markets and services, and to unlock agricultural potential, which will lead to improved incomes in rural areas.

Table 3.3 summarizes different countries' methods of collecting fatal and nonfatal data from road traffic collisions. Collection methods and procedures vary greatly from country to country and need to be accounted for when analyzing global data.

3.7 GLOBAL STATISTICS, DATA, AND ASSESSMENT

Previous reviews of global fatalities undertaken by Transport Research Laboratory (TRL) in the United Kingdom, the World Bank, and others have produced a wide range of estimates. While problems of data reliability and underreporting have been regularly acknowledged, traditional reliance has been on the use of officially published statistics based on police reports. In estimating causes of death and disability, the WHO used a different method, based on registered deaths and health sector data that produced higher estimates than those using official police statistics. For example, the WHO estimated a million deaths worldwide in 1990, while the TRL values were of the order of half of this.

In keeping with the traditional approach used by transport specialists in compiling road collision statistics, the starting point to study is the official fatality figure reported by countries. Using these values to obtain an accurate estimate of the current global fatality situation requires several factors to be taken into account as follows:

- Updating the fatality figure from the latest year (usually 1995 or 1996) to 1999.
- Estimating for those countries where fatality data were not obtained.
- Underreporting due to both recording deficiencies and nonreporting to the police.

The general problem of underreporting includes both recording deficiencies, *under recording* where injuries are reported to the police but are not included in the published statistics and nonreporting where the police are not notified of road injuries. To highlight the extent of underreporting, the problems of recording deficiencies and nonreporting have been discussed separately in this book.

3.8 GLOBAL DIVIDE OF INJURY AND DEATH, AND ULTIMATELY BURDEN

The developing world, with regard to countries such as China, India, Thailand, Vietnam, and Malaysia, has experienced rapid urban growth in recent years. Cities in

TABLE 3.3
Selected Data Sources about the Burden of Road Traffic Collisions in Iran, India, Mexico, and Ghana

Country	Deaths	Non-fatal Injuries
Iran	*National death registration system*: Covers 29 provinces (i.e., all except Tehran); ICD-10 (International Classification of Diseases, 10th revision) derivative causes of death. *National forensic medicine system*: Estimates available for all provinces.	*Hospital data sample*: Data collected from all hospitals in 12 provinces (outpatient for 4 days, and hospital admissions for 4 weeks), followed back to household post discharge. *Demographic and Health Survey (DHS)*: Approx. 110,000 households, included questions about road traffic injury involvement and care.
India	*National Sample Registration System*: Nationally representative sample of deaths in India causes evaluated by verbal autopsy. *National Medical Certification of Cause of Death (MCCD) System*: Cause of death for reporting hospital in urban areas; covers approx. 500,000 deaths from all causes annually.	*World Health Survey (WHS)*: Representative sample with questions about road traffic injuries and care; conducted in six states *Survey—New Delhi*: 5,412 households, all injury causes. *Survey—Bangalore*: 20,000 households, stratified by urban/rural and socioeconomic status. *Survey—near New Delhi*: Morbidity patterns in 9 villages, 25,000 households, monitored for 1 year. *Hospital—Hyderabad*: Five hospitals, approx. 800 victims, followed back to household post-discharge.
Mexico	*National death registration system*: ICD-10 coded cause of death, estimated to be near complete.	*SAEH—Ministry of Health national hospital discharge database*: Covers all Ministry of Health hospitals, approx. 115,000 unintentional injury hospital admissions. *Instituto Mexicano del Seguro Social (IMSS) national hospital discharge database*: Approx. 175,000 injury hospital admissions; external causes not recorded. *World Health Survey*: Representative sample with questions about road traffic injuries and care. *Encuesta nacional de Salud y Nutricion (ENSANUT) national health survey*: Sample size 54,068 individuals, included questions on RTI (road traffic injuries) involvement and care.
Ghana	*Mortuary data—Kumasi*: Data collected from 1996 to 1999. *Demographic Surveillance System (DSS) Sites at Navrongo*: Verbal autopsy based cause of deaths.	*World Health Survey*: Representative sample with questions about road traffic injuries and care. *Survey—Kumasi (urban) + Brong Ahafo region (rural)*: Sample of approx. 21,000 individuals. *Hospital records—Accra*: Reporting hospitals.

Source: Reprinted from Bhalla, K. et al., *Int. J. Injury Control Saf. Promot.*, 16(4), 243, 2009. With permission from Taylor & Francis Ltd.

these countries have strived to be developed quickly and efficiently. According to official statistics in China, over 73,500 people died in 2008 as a result of road collisions. China is now the world's second largest automobile market in the world, which corresponded to its road collision statistics. The development of China as a country, both income and urban growth, has meant a large increase in car ownership and usage, especially among the middle- and high-income classes. However, it is the low- and middle-income classes that are most affected, and where loss of life can be detrimental to the victim's immediate family. In China, road collisions are the leading cause of death for 15–45-year olds. They are the second cause of premature death, and this causes an acute drain on productivity due to short- and long-term disabilities. The most worrying aspect of road deaths in China is the high number of pedestrian victims (approximately 25%), closely followed by motorcyclists (23%). These statistics give us an insight into the type of road collision that is occurring, that is, vehicle–pedestrian collisions, in urban areas. Although a large proportion of collisions occur in urban areas, still over half of the road collisions occur in rural areas. Farmers and workers are the people most likely to be injured in these areas, and they are less likely to be able to afford to go to the hospital or doctors. There is lack of signs and proper infrastructure for traffic safety on many roads. Most of the road collisions occurred on roads that lack traffic management. A study by Loo et al. (2011) shows that while urban road collisions are larger in number and higher in density, rural road collisions are often more deadly. There is a lack of urban planning toward road safety especially pedestrians and cyclists.

Essentially, a lot of the use of the road environment can be argued to be cultural. In China, for example, there are a lot of migrants in large cities. These migrants have often come from rural areas in a different province. The road environment and how they interacted in that road environment would be very different from that of the larger cities in China. There are many cultural and behavioral reasons especially in developing countries for road collisions to be so common. Some of these include migrants having different ideas of risk and mortality in the road environment, they are not being used to busy road traffic in large cities, and their assumption that vehicles will stop/avoid them and that they have right-of-way. Chen et al. (2012) explore some reasons for the different cultures of migrants in affecting their risk of involving in nonmotorized traffic collisions. Yet, in general, the reasons for road collisions in low- to middle-income countries are far more complex than in high-income countries. Traffic in low-income countries has a much more diverse number of vulnerable road users.

3.9 ROAD COLLISION COSTING

Road collision costing is an important global element of road collisions themselves. Road collisions have been shown to cost annually between 1% and 3% of GDP in developing countries. The gross national product is often more readily available than the GDP figure although it is usually slightly higher than the GDP. Knowledge of road collision costs allows safety impacts to be economically justified. Road safety measures have been frequently ignored or downplayed in cost–benefit analysis on the grounds that the associated costs and benefits are too intangible. Where road safety is included in a cost–benefit analysis of road improvements, it is often factored only on a subjective basis and so does not get applied in a consistent manner required for project comparisons. In road collision costing, there are generally two elements. They are casualty-related costs (such as injury, pain, grief, and lost output) and road-collision related costs (such as property damage and administration). The cost of road collisions is the sum of these two elements.

Lost output is an important concept in road collision costing. It refers to the contribution that a road collision victim was expected to make with future earnings weighted to present value. One of the major issues is that often especially in developing countries, road collisions are more likely to affect men between 15 and 45 years, which is the prime working age and most productive to society. Lost output for serious and slight injuries is the daily earning rate multiplied by the number of days off work. This is usually derived from hospital and victim surveys.

Vehicle damage costs relate to the property that was damaged in the road collision. Insurance claims are the traditional source for vehicle damage costs, but the low rate of insurance coverage in many developing countries raises questions as to how representative collision claims are. Medical costs are a particularly difficult element of road collision costing. They rarely account for more than 5% of all road collision costings. Few governments/officials are able to estimate the cost of patients per night in hospital as well as outpatient costs. In developing countries, medical costs do not reflect the reality of the situation, as scarce resources limit the hospital beds and medical services available. The medical costs alone do not necessarily reflect the actual opportunity costs.

Apart from the humanitarian aspect of reducing road deaths and injuries in developing countries, a strong case can be made for reducing road collision deaths on economic grounds alone, as they consume massive financial resources that the countries can ill afford to lose. That said, it must of course be borne in mind that in developing and emerging nations, road safety is but one of the many problems demanding its share of funding and other resources. Even within the transport and highway sector, hard decisions have to be taken in the country on the resources to be devoted to road safety. As a consequence, it is extremely unlikely that the expenditure on road safety will, in any sense, be *optimal* in terms of equity. In particular, if safety benefits are ignored in transport planning, there will be underinvestment in road safety.

3.10 INTERNATIONAL ROAD INFRASTRUCTURE: A NEGLECTED MEASURE?

The "Safe System" views the road transport system holistically by seeking to manage the interaction between road users, roads and roadsides, travel speeds, and vehicles. It aims to reduce the likelihood that collisions occur and minimize the severity of those that do happen. Central to the Safe System approach is the recognition that human beings make mistakes and are fragile. As Figure 3.2 demonstrates, impacts at what might be considered reasonable speeds can significantly increase the risk of death and serious injury. The Vision Zero philosophy adopted by the Swedish Government (Johansson 2006) illustrates many of the principles required of the Safe System. Vision Zero provides a viable policy framework for sustainable safety whose basic principles can be applied in any country, at any stage of development.

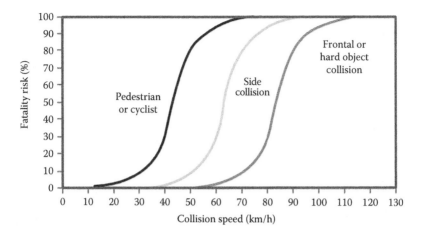

FIGURE 3.2 Collision speed–fatality relationships. (Reprinted from Wramborg, P., A new approach to a safe and sustainable traffic planning and street design for urban areas, in: *Road Safety on Four Continents Conference Proceedings*, Warsaw, Poland, 2005. With permission.)

Elements of this approach may appear utopian, but the approach lays the principles for the management of kinetic energy, the fundamental part of injury reduction.

Within this energy system, the imbalance of *kinetic mass* is such that pedestrians of 80 kg traveling at 5 km/h cannot harm a driver and 1500 kg car traveling at 90 km/h. The onus of responsibility is therefore on the driver to avoid causing injury. Sweden has demonstrated the crucial role that infrastructure can play in creating a safe and efficient road network. By developing roads that are inherently safe (e.g., using safety barriers to mitigate the risk of head-on and run-off-road collisions), Sweden has been able to increase safely the speed limits on many of its major roads. In fact, many of Sweden's safest roads are also those where speeds are the highest (Johansson 2006). Recent work (Turner et al. 2009) promotes greater use of what have been termed "primary" road safety treatments. These are treatments more likely to eliminate death and serious injury than produce only mild reductions. Examples include barriers to prevent run-off-road and head-on collisions, properly designed roundabouts at junctions, and raised platforms at junctions or locations where pedestrians cross. *Supporting* treatments such as signing and line marking plus many others may reduce collisions, but not as effectively as Safe System levels require, and generally have only limited impact on severity outcomes.

3.11 CONCLUSION

The greatest successes in public health have resulted from cultural change (Ward and Warren 2007). For example, smoking was once considered harmless and part of a healthy and active lifestyle. In the 1930s, cigarette advertisements in the United States often showcased physicians and athletes as spokespersons. With mounting scientific evidence on the hazards of smoking and a shift from emphasizing dangers to the smoker to dangers to the nonsmokers, the public began to view smoking negatively, and the health culture was permanently changed. Likewise, creating a safety culture will require a shift in how we think about traffic hazards, personal risky behaviors, and the value of prevention. Following Sleet et al. (2007), public health can contribute to this shift by the following:

- Including road safety in health promotion and disease prevention activities
- Incorporating safety culture into health education activities for adolescents so that they associate safety with all aspects of life
- Requiring safety impact assessments similar to environmental impact assessments (i.e., before new roads are built)
- Using public health tools to help the transportation sector in conducting safety audits to identify unsafe roads and intersections
- Incorporating safety and mobility into healthy aging, for example, by focusing on the mobility needs of older adults, especially as they relinquish their driving privileges
- Applying modern evaluation techniques to measure the impact of road safety programs and injury prevention interventions
- Measuring health-care costs and public health consequences of traffic injuries
- Assisting states and communities with local injury data collection and traffic injury surveillance systems
- Reducing health disparities by assuring equal access to community preventive services such as child safety seats, bicycle helmets, and neighborhood sidewalks for poor or underserved populations
- Strengthening pre-hospital and hospital care for trauma victims by supporting comprehensive trauma care systems, nationwide

It is clear from the data that the most consistent road injury information is derived from high-income countries and focuses on the benefits for vehicle occupants. However, this group forms a small proportion of road users at the global level. There must be prioritization toward the data collection, analysis, and implementation in middle- and lower-income countries. One of the remaining obstacles is the public's misconception that injuries are accidents that occur by chance. It has been difficult to summon popular sentiment for motor vehicle injury because there is no single cause or cure. It is not widely recognized as a public health problem, and most people consider injury the result of an uncontrollable *accident*. For many, road traffic injuries and death are simply the price we pay for mobility. While some progress has been made toward changing public perception about the predictability of injury and its preventable nature, more must be done. Public health professionals have been relatively successful in framing motor vehicle injuries in the context of other preventable causes of death and disease as we have seen in this chapter. The medical professions have been quick to recognize their role as advocates for motor vehicle safety with patients and policy makers, and the importance of emphasizing lifestyle changes that include safety behaviors. By framing motor vehicle injury as predictable and preventable, health practitioners will have a tool to educate the public and influence policy makers about a serious public health problem that can be reduced, just like many diseases. A culture of safety that provides for safe and accessible transportation can prevent injury and death, and improve the overall quality of life for populations. By improving traffic safety, we also improve public health.

REFERENCES

Bhalla, K., S. Shahraz, D. Bartels, and J. Abraham. 2009. Methods for developing country level estimates of the incidence of deaths and non-fatal injuries from road traffic crashes. *International Journal of Injury Control and Safety Promotion* 16 (4): 239–248. http://www.tandfonline.com/.

Bishai, D., A. Quresh, P. James, and A. Ghaffar. 2006. National road casualties and economic development. *Health Economics* 15 (1): 65–81.

Chen, C., H. Lin, and B. P. Y. Loo. 2012. Exploring the impacts of safety culture on immigrants' vulnerability in non-motorized crashes: A cross-sectional study. *Journal of Urban Health* 89 (1): 138–152.

Elvik, R. and A. B. Mysen. 1999. Incomplete accident reporting: A meta-analysis of studies made in thirteen countries. *Transportation Research Record* 1665 (1): 133–140.

Gururaj, G. 2008. Road traffic deaths, injuries and disabilities in India: Current scenario. *The National Medical Journal of India* 21: 14–20.

Institute of Medicine (IOM). 1999. *To Err Is Human: Building a Safer Health System.* Washington, DC: National Academy Press.

International Federation of Red Cross and Red Crescent Societies. 1998. World disasters report 1998. Geneva, Switzerland: IFRC.

International Transport Forum. 2011. IRTAD database coverage. http://internationaltransportforum.org/irtadpublic/coverage.html (accessed November 13, 2011).

Jacobs, G. D., A. Aeron-Thomas, and A. Astrop. 2000. Estimating global road fatalities. Transport Research Laboratory, Crowthorne, U.K.

Johansson, R. 2006. European Union issue 11 Transport: A new vision. http://www.vv.se/PageFiles/12660/eu11_roger_johansson_atl%5B1%5D.pdf?epslanguage=sv (accessed February 15, 2012).

Khayesi, M. and M. Peden. 2005. Road safety in Africa. *British Medical Journal* 331: 710–711.

Kopits, E. and M. Cropper. 2005. Traffic fatalities and economic growth. *Accident Analysis & Prevention* 37 (1): 169–178.

Loo, B. P. Y., W. S. Cheung, and S. Yao. 2011. The rural-urban divide in road safety: The case of China. *The Open Transportation Journal* 5: 9–20.

Loo, B. P. Y., W. T. Hung, H. K. Lo, and S. C. Wong. 2005. Road safety strategies: A comparative framework and case studies. *Transport Reviews* 25 (5): 613–639.

Loo, B. P. Y. and K. L. Tsui. 2007. Factors affecting the likelihood of reporting road crashes resulting in medical treatment to the police. *Injury Prevention* 13 (3): 186–189.

Mathers, C. D. and D. Loncar. 2005. Updated projections of global mortality and burden of disease, 2002–2030: Data sources, methods and results. Projected DALYs for 2005, 2015, and 2030 by country income group under the baseline scenario. Geneva, Switzerland: World Health Organization, October 2005, http://www.who.int/healthinfo/global_burden_disease/projections2002/en/ (accessed November 13, 2011).

Odero, W., M. Khayesi, and P. M. Heda. 2003. Road traffic injuries in Kenya: Magnitude, causes and status of intervention. *Injury Control and Safety Promotion* 10 (1–2): 53–61.

Peden, M., R. Scurfield, D. Sleet, D. Mohan, A. Adnan, E. Jarawan, and C. Mathers, eds. 2004. World report on road traffic injury prevention. Geneva, Switzerland: World Health Organization.

Sleet, D., T. Dinh-Zarr, and A. Dellinger. 2007. Traffic safety in the context of public health and medicine. In *Improving Traffic Safety Culture in the United States*, pp. 41–58. Washington, DC: AAA Foundation for Traffic Safety.

Turner, B., M. Tziotis, P. Cairney, and C. Jurewicz. 2009. Safe system infrastructure: National roundtable report. Research Report ARR 370. ARRB Group Ltd., Vermont South, Victoria, Australia.

Vasconcellos, E. A. 1999. Urban development and traffic accidents in Brazil. *Accident Analysis & Prevention* 31 (4): 319–328.

Ward, J. W. and C. Warren, ed. 2007. *Silent Victories: The History and Practice of Public Health in Twentieth-Century America*. New York: Oxford University Press.

Watkins, K. and D. Sridhar. 2009. Road traffic injuries: The hidden development crisis. In *Policy Briefing for the First Global Ministerial Conference on Road Safety*, Moscow, Russia, November 19–20, 2009.

Wramborg, P. 2005. A new approach to a safe and sustainable traffic planning and street design for urban areas. In *Road Safety on Four Continents Conference Proceedings*, Warsaw, Poland.

World Health Organization. 2008. The global burden of disease: 2004 update. Geneva, Switzerland: WHO.

World Health Organization. 2014. Road traffic deaths: Data by country, Global Health Observatory Data Repository. http://apps.who.int/gho/data/node.main.A997?lang=en (accessed October 29, 2014).

4 Risk and Socioeconomic Factors

4.1 RELATIONSHIPS AND RISK

In this chapter, our main concern is not to understand personal behavior and driver characteristics per se, but to understand *types* of people or groups in society and their association with road collisions. Road users involved in road collisions are affected not only by the environment and the behavior of other individuals but also by their own characteristics. The association between socioeconomic characteristics and road collision analysis is not new. Recent literature has shown the linkages between road collisions and many socioeconomic factors and related areas including health care, education, the family, cultural aspects, the physical environment, and geographic location. A conceptual framework is used to identify the mechanisms through which socioeconomic geography may interact in the determination of health inequalities relating to road traffic collisions. It is used as a structure for presenting the current evidence concerning socioeconomic differences in road traffic collision risk. Previous research has suggested that road traffic collision risk is higher for those people with a lower socioeconomic status. Whether the greater number of road collisions to people of lower socioeconomic status is a phenomenon attributed to the areas or a reflection of a wider pattern of road collisions affecting lower socioeconomic status groups is not clear. There is evidence of interaction between socioeconomic status, area, and risk. The mechanisms surrounding socioeconomic inequalities of road collisions in society require greater scrutiny. Further theoretical developments and empirical evidence are required in order to enable more effective targeting, both tactically and strategically.

In this chapter, we will first examine the nature of the relationships and risk associated with socioeconomic characteristics and road collisions. In the second section, we focus more specifically on certain aspects of road collision analysis, linking ethnicity, deprivation, children, and inequality to road collision risk. The third section of this chapter will look more specifically at measurement and analysis, how socioeconomic characteristics can be measured in relation to road collision risk, the use of geodemographics in road collision analysis, and future directions. The geography and scope of socioeconomic data for road collision analysis require explanation and understanding. While examining the nature of the data, we will look at the scale of data, focusing on the notion of "neighborhood," census tracts, and postcodes. Socioeconomic data come in many different formats and scales, and it is important to outline these and how it can assist in the understanding of road collisions. The final section of this chapter will look at policy and intervention, and how the data and analysis can be used to manage and reduce road collisions and educate people on the risk of being in a road collision.

4.2 SOCIOECONOMIC CHARACTERISTICS

4.2.1 Deprivation

The association between socioeconomic deprivation and road injury risk in England and Wales was identified 10 years ago in a study of individual social class–coded child death records (Roberts and Power 1996). The study found steep and widening social class gradients in injury mortality. The injury death rate for child pedestrians in the lowest social class was five times greater than that for children in the highest social class. The analysis was recently updated, and the results show that these inequalities in road injury risk persist, and indeed may have increased. What is evident from the research is the polarization of focus on children and deprivation. This is not an unworthy cause; however, it does mean that there is less understanding of the nature of road collision risk and adult deprivation and disadvantage. All people are exposed to the risk of injury on the road as part of their everyday life, but the burden of these injuries is not evenly spread across our society. Road traffic injuries disproportionately affect some groups more than others. Disadvantaged people and those living in deprived neighborhoods are much more affected than those living in more affluent areas; some age groups of vulnerable road users, such as children, young adults, and older people, bear a greater burden.

4.2.1.1 What Is Deprivation?

From studies within the United Kingdom and from the international literature, there is little agreement about what "social deprivation" means. Definitions of social deprivation can be based on the characteristics of geographical areas, such as wards or enumeration districts, and for the most recent census, super output areas. Definitions can also be based on the characteristics of individuals themselves, such as "low-income families" or the characteristics of the places in which they live, such as multi-occupancy housing. The UK government uses a composite indicator called the Index of Multiple Deprivation to describe disadvantage.

4.2.1.2 What Are the Influencing Factors?

The nature of people's living environment has an immediate bearing on their exposure to road traffic hazards. Generally speaking, people living in more deprived areas are more likely to be pedestrians. Dense living means that people as drivers and pedestrians are more likely to hit one another and cause damage. Deprivation is also closely linked with social and economic factors, and it is the combination of these factors that can conspire against someone and make people more vulnerable. These can include issues such as child supervision and ability of people, young and old, to manage hazards (this is also cultural and ethnically influenced, which will be discussed later in the chapter). Factors such as being a single parent, low levels of parental education, inadequate child care facilities, lack of money all come into play when assessing road collision risk.

Figure 4.1 highlights one example of linking no-car households and child pedestrians and cyclists. It provides a graphical illustration of an independent variable at the enumeration district level with the distribution of child injuries.

Risk and Socioeconomic Factors

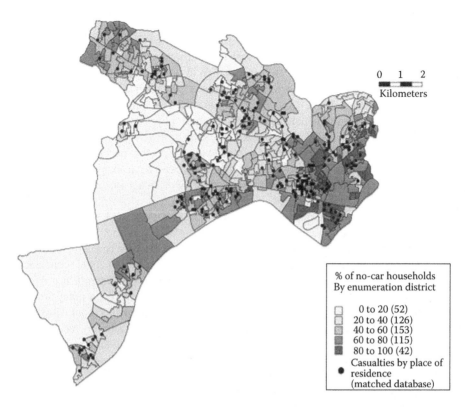

FIGURE 4.1 No-car households and child pedestrian/cyclist casualties. (Reprinted from *J. Transport Geogr.*, 8(3), Petch, R.O. and Henson, R.R., Child road safety in the urban environment, 197–211, Copyright 2000, with permission from Elsevier.)

Although that is not to say, deprived areas lack a sense of community. In fact, many areas often have a heightened sense of community and cohesion. There are also broader factors to consider, such as the wider economic and political processes, and the way society works. Road safety can play a significant role in the way in which a community can become stronger and safer. The next section identifies the role of deprivation in road collision causation, with particular attention to child pedestrians.

4.2.1.3 Child Pedestrians and Deprivation

There are many different factors that can contribute to the increased road collision risk for child pedestrians of lower social classes. The majority of the studies are small scale, with no national policies having been published. The most comprehensive report based in the United Kingdom on this subject was conducted by White et al. (2000). The main findings were that road traffic collision risk for child pedestrians is class related and that injuries of child pedestrians involved in collisions in lower socioeconomic areas are more severe than those that occur in higher socioeconomic areas. It was found that children of single mothers are twice as likely to be involved in a road collision as pedestrians than children living with two parents. Also, antisocial and overactive children are more likely to be involved in a road collision. Finally, child pedestrian collisions during journeys to and from school are more common in low socioeconomic status areas than in more affluent areas.

Many possible influencing factors can be identified as increasing the risk of child pedestrian collisions. These factors are determinants of social exclusion including variables such as unemployment levels, low incomes, poor housing, high crime environments, bad health, and family breakdown. Generally, children from low socioeconomic backgrounds have a greater exposure to hazards that may result in a higher risk of road collisions. It is clear however that few studies have sought to address the issues of family factors with reference to the relative affluence or deprivation of the area in which the household resides.

There is a notable relationship between the risk of child pedestrian collisions and the physical environment. The layout of residential environments influences the safety of child pedestrians, from dense urban areas near major roads to small villages and rural roads. A study in Australia by Robinson and Nolan (1997) concluded that over 79% of road collisions involving child pedestrians occurred in a driveway, carport, or garage of home addresses. Overall, there have been many studies into social deprivation, locational factors, and road collision involvement. However, MacIntyre et al. (1993) argued that there has been limited research into variations of socioeconomic and cultural features of areas that influence health and the likelihood of death, particularly in a road collision. There is conflicting evidence as to whether people of low socioeconomic status have poorer health due to areas in which they live in being health damaging, or whether ill health and mortality can be wholly explained by personal or socioeconomic factors.

Christie (1995a) surmises that children from lower socioeconomic backgrounds are more likely to be involved in pedestrian-related collisions, because their activities involve higher rates of risk than those of their counterparts from higher socioeconomic backgrounds. These activities are also more likely to be unsupervised and to take place in unprotected environments, namely, the streets where the children reside. This report highlighted the distinctive relationship between social class and risk of death of child pedestrians at a household level. Christie concluded that children from the lowest socioeconomic group are over four times more likely to be killed as a pedestrian than their counterparts in higher socioeconomic groups (Christie 1995a).

A more recent work in this area is highlighted in research from the Centre for Transport Studies at London's Imperial College (Grayling et al. 2002; Graham et al. 2005). The first of these looks at whether the level of socioeconomic well-being influences child pedestrian collision rates. The approach differs from previous studies in this area, whereby instead of taking the socioeconomic status of the victim, they take a small area-based approach of pedestrian collisions. The aim is to ascertain the complex relationship between deprivation and injuries. For example, it may be that deprivation is more commonly found in dense urban areas, but collisions also occur more frequently in high-population-density urban areas. In short, areas of similar density but different levels of deprivation suffer similar injury numbers, or it could be that areas of similar density have very different injury numbers depending on the levels of deprivation. They seek to disentangle the effects of *area* from the influences of personal/household characteristics. The study is UK-wide and uses STATS19 data and 1991 Census data by ward. Information is therefore aggregated at the ward level. It concluded that deprivation is only one of many factors that will influence the number of child pedestrian collisions. Other factors that will influence involvement are (1) absolute number of children in a given area, (2) volume of traffic flows, (3) physical nature of the environment, (4) characteristics of the local road infrastructure, and (5) *other* local specific factors (Grayling et al. 2002).

The complementary study by Noland and Quddus (2004) highlights a more general population picture of the United Kingdom in terms of socioeconomic status and road collisions. The research demonstrates that although there has been a significant reduction in road collisions in the past 30 years, this has been specifically because of improved vehicle design, safety belt usage, detailed safety audits after collisions, and comprehensive engineering measures. Some factors that have received very little attention have been the land use characteristics of an area, population densities, and urban development. In this study, the data are aggregated to the ward level, with a general reference that wholly urban wards experience a lower level of fatalities compared to wholly rural wards. It builds on two main findings from the literature: first from a research study by Sawalha and Sayed (2001) indicating that commercial land use (in Canada) is associated with a higher frequency of collisions. The second fundamental research findings by Ossenbruggen et al. (2001) that examine the location of shops and find that typical shopping sites are more hazardous than village-style shopping sites (generally because of lower vehicle speeds). This study concludes that urbanized areas are indeed more likely to have fewer collisions, while, in comparison, areas with higher unemployment densities are more likely to have a greater number of road collisions.

A study by Edwards et al. (2006) studied child pedestrian injuries in London from 1999 to 2004. A total of 5834 children who were injured were recorded with complete postcodes. Pedestrian injury rates within deprivation deciles ranged from 178 to 522 per 100,000 children. The ratio of the pedestrian injury rate among the most deprived 10th of London's children to that among the least deprived was therefore 522/178, which is 2.93. The pedestrian injury rate among the most deprived children was therefore also nearly three times as high as that among the least deprived (Table 4.1).

4.2.1.4 Scales of Factors Linking Deprivation, Disadvantage, and Road Collisions

There are differences between social characteristics at an individual basis and at the area level. Findings indicate that both deprived households and disadvantaged areas give rise to increased collision risk, although not necessarily to the same extent. With respect to road collisions in particular, research conducted at the household or individual level has indicated that the risk of death for child pedestrians is highly class related (Christie 1995b). Mortality statistics indicate that children in the lowest socioeconomic group are over four times more likely to be killed as pedestrians than their counterparts in the highest socioeconomic group. Motor vehicle collision fatalities involving child occupants, pedestrians, and cyclists constituted 51% of all

TABLE 4.1
Child Pedestrian Injury Rates within Deprivation Deciles in London, 1999–2004

Deprivation Deciles	Child Pedestrian Injuries with Postcodes	Child Population	Pedestrian Injury Rate per 100,000
1	243	136,485	178
2	314	129,670	242
3	348	130,125	267
4	432	128,831	335
5	514	134,212	383
6	583	137,061	425
7	722	145,681	496
8	776	157,501	493
9	955	167,258	571
10	947	181,420	522
	5834	1,448,244	

Source: Reprinted from Edwards, P. et al., Deprivation and road safety in London: A report to the London Road Safety Unit, LSHTM, London, U.K., 2006, p. 17. With permission.

child deaths from injury and poisoning in 1979–1983 and 44% of all such deaths in 1989–1992. For motor vehicle collisions, death rates in social classes I and II declined by 30% and 39% respectively, compared with declines of 18% and 1% in social classes IV and V, respectively.

Investigations into the importance of area-related factors were conducted by Abdalla (1997) and Abdalla et al. (1997a,b), who found that the injury rates among residents from areas classified as relatively deprived were significantly higher than those from relatively affluent areas. A database was created by merging road injury information and census data for the former Lothian Region, and the relationships between injury rates and social deprivation indicators for the road traffic victims' zones of residence were investigated. Similarly, Erskine (1996) argues that those who benefit least from the motor vehicle seem disproportionally likely, given their relative exposure to risk, to die in road traffic collisions. The incidence of traffic injury in deprived urban areas is greater than in more prosperous areas. Social class correlates highly with mortality for all ages by all causes of death and that child pedestrian death rates correlate closely with all causes of child deaths. Social gradients in injury mortality exceed those for any other cause of death in young people. The inequalities between social classes are even more extreme in relation to child pedestrian deaths than either all injury deaths or all causes of death.

There have been increasing studies that have looked at the family structure as playing a role in road collision risk. It is very difficult to accurately measure the amount of exposure experienced by certain individuals from different socioeconomic backgrounds. In many cases, it is not possible to separate the area and household effects.

Research has indicated that for individuals living in deprived households, family factors are linked to child pedestrian injury rates and overall injury for children. Lone parenthood in particular is a risk factor for children.

The children of lone mothers have the highest death rates of all social groups (Judge and Benzeval 1993), and lone parenthood is a risk factor for traffic injuries. The risk of pedestrian injury is over 50% higher. One of the main factors affecting exposure rates is the alternative modes of transport on offer to children. Consistent with this is the finding that lack of access to a car is associated with a doubling of the risk of injury as a pedestrian. Lack of access to a car is most likely among poorer households. Households of single elderly people or single parents have particularly low rates of car use (18% and 31% respectively), and single elderly people and single parents are predominately women (Erskine 1996). While children living in deprived households are exposed to greater collision risk, there is evidence that an area effect is also evident.

4.2.2 Ethnicity

Studies of road collision rates have found significant differences in collision risk rates based on ethnicity. Internationally, studies have found large disparities in road traffic injury rates by ethnic group (Schiff and Becker 1996; Stevens and Dellinger 2002; Braver 2003; Campos-Outcalt et al. 2003; Stirbu et al. 2006; Savitsky et al. 2007). Evidence in the United Kingdom is limited but suggests that injury rates are disproportionately high for some black and minority ethnic (BAME) road user groups (e.g., Lawson and Edwards 1991; Christie 1995a). While these international and British studies concur that ethnic minorities are at greater risk of road traffic injury, they provide conflicting evidence of who is at risk. In the international studies cited earlier, ethnic minorities described as "Hispanic," "American Indian," "non-Jewish," and of Turkish, Moroccan, Surinamese, or Antillean/Aruban origin have been found to have higher road traffic injury rates than the native population. Within the United Kingdom, both "Asian" and "non-White" groups have been found to be at increased risk of injury, depending on the timing and location of the study. This suggests that there is nothing fundamental about belonging to a particular minority ethnic group that causes traffic injury. Rather, perhaps there is something context specific about belonging to a particular ethnic minority within a particular environment that is associated with high road traffic injury rates. The reasons for ethnic differences in road traffic injury are unclear, but are likely to be at least partially explained by the strong association between ethnicity and socioeconomic status, particularly in London (Grayling et al. 2002; Edwards et al. 2006). It is also important to remember that any associations found between belonging to a particular ethnic group and road traffic injury are merely associations. Although we can assess how far differences are accounted for by socioeconomic factors (and, for instance, suggest that these do not account for all observed differences), we cannot control for all other differences between groups defined through ethnicity. Although there are probably no direct effects of ethnicity on road traffic injury risk, the interplay of ethnicity and environment does have a number of implications on risk exposure.

TABLE 4.2
Average Annual Pedestrian Injury Rates per 100,000 People in London, 1996–2006

Age Group	Sex	White	Black	Asian
0–4	M	45	95	68
	F	29	52	41
5–9	M	125	235	141
	F	72	135	69
10–14	M	254	313	136
	F	179	255	97
15–24	M	144	164	84
	F	122	148	69
25–34	M	84	124	61
	F	63	84	44
35–44	M	75	97	56
	F	46	62	38
45–54	M	68	106	61
	F	43	69	46
55–64	M	68	102	78
	F	49	82	49
65+	M	85	127	109
	F	68	101	58

Source: Reprinted from Steinbach, R. et al., Road safety of London's black and Asian minority ethnic groups: A report to the London Road Safety Unit, LSHTM, London, U.K., 2007, p. 20. With permission.

London is still one of the most ethnically diverse cities in the world, with a prominent research agenda on road accidents and the casual factors surrounding them. In a recent study by Steinbach et al. (2007), analysis was conducted on ethnicity (recorded as part of the national road accident database, STATS19) alongside rates of traffic collisions and socioeconomic status. The study differentiated between pedestrians, cyclists, and car occupants, and aggregated it by borough. Some of the most significant results of this study can be seen in Table 4.2. There was a total of 78,716 people injured as pedestrians in London between 1996 and 2006. Annual pedestrian injury rates within age–sex groups ranged from 29 to 313 per 100,000 people. Pedestrian injury rates appeared highest in *black* children and adults of all ages, males and females.

4.2.3 Exposure and Inequality

Exposure might be the hardest element of the road collision model to establish. Our exposure to vehicular traffic can change depending on the time of day, and although it is possible to link it with deprivation and ethnicity, its link is difficult to measure.

The link between social deprivation and the high collision rate of child pedestrians from lower socioeconomic group families can be explained in terms of increased exposure to hazardous environments (Christie 1995b). Such environments may relate to the areas immediately surrounding individual households and further afield. Hazards may include busy roads with a lack of safe crossing sites, the location of schools within the community, availability and access to safe play areas, and so on. Child-rearing practices of lower socioeconomic group families often, by necessity, involve less supervision with less time spent in shared activities. Christie (1995b) has also suggested that children from lower socioeconomic groups may be encouraged to take part in activities that involve greater physical risk, where competitive drives find an outlet in unsupervised activities in unprotected environments.

There is also evidence that exposure to road traffic injury risk varies between socioeconomic groups (Sonkin et al. 2006). Using data from the National Travel Survey, Sonkin et al. (2006) found that children from households without access to vehicles walk more than their counterparts in car-owning families. Per kilometer traveled, there are about 50 times more child cyclist deaths and nearly 30 times more child pedestrian deaths than there are deaths to child car occupants. These differences in risk by mode of travel are likely to contribute to the steep social class gradients in road traffic injury death rates. Although walking and cycling provide important benefits in terms of physical activity and have none of the adverse climate impacts of motorized travel, pedestrians and cyclists remain at greatest risk.

4.2.4 GEODEMOGRAPHICS

The use of geodemographics to analyze road safety has been largely unexplored. Its potential is supported by research linking road collision risk and socioeconomic variables such as unemployment, low income, area of residence, educational level, race, children, and marital status (Haepers and Pocock 1993; Christie 1995a; Kposowa and Adams 1998; Murray 1998; Abdalla 1999).

Road collision analysis has acknowledged the relationship between the social characteristics of the road collision drivers and injuries. However, there has been no use of geodemographics in terms of exploring road traffic injury risk. Social class as a discriminator for road collision risk has been addressed by a number of research papers (Hasselberg and Laflamme 2005; Laflamme et al. 2005). Research in Scotland has considered using deprivation indicators (variables included proportion of unemployed people, proportion of people with no car, proportion of people of pensionable age, and proportion of people in a lower social class) from the 1991 Scottish Census as an indicator for road collision involvement (Abdalla 1997). One of the key findings concluded that children who came from families in social classes IV or V (semiskilled or unskilled jobs) were overrepresented in the total number of child injuries (Abdalla 1997).

There have been recent reports from the Transport for London that demonstrate a relationship between deprivation and risk of road traffic injury in London, with pedestrians in particular at higher risk of injury in more deprived areas (Edwards et al. 2006). Sonkin et al. (2006) identified that exposure to road traffic injury risk varies between socioeconomic groups. This supports evidence from

Grayling et al. (2002), who found that children are three times as likely to be pedestrian traffic victims in the top 10 most deprived wards in England and Wales. Alongside the studies of deprivation and the disadvantaged socioeconomic classes, there have been a number of studies that convey large disparities between road traffic injury and ethnic group (Schiff and Becker 1996; Braver 2003; Campos-Outcalt et al. 2003; Stirbu et al. 2006; Savitsky et al. 2007). The influence and the effect of certain residential layouts and housing types have also been found to cause an overrepresentation in collisions involving children (Christie 1995a). Furthermore, research undertaken by Hasselberg and Laflamme (2005) presents results for Swedish young adults that show that drivers with a basic and secondary education experience a greater risk of collisions of all types than drivers with an experience of higher education. In addition, the study found that children of manual workers showed a 60% greater risk of being involved in any type of collision. These findings support the potential use of geodemographics as being a good indicator for understanding the "who" and the "where" of the people who may be at higher road injury risk.

There are two types of geodemographic profiling systems, commercially lead and census lead. The commercially lead systems are dominated by two leading commercial geodemographic providers, Experian Ltd. (Mosaic) and CACI Ltd. (A Classification of Residential Neighbourhoods, ACORN). For this analysis, Mosaic will be used to categorize the unit postcodes into neighborhood types, alongside an open-source UK Census geodemographic profiler. These geodemographic types are based on social and demographic proximity and built environment characteristics. Geodemographic classifiers cluster small areas on the basis of social similarity rather than locational proximity (Webber and Longley 2003). The core idea lies in the relationship between geodemographic attributes used to create the neighborhood types and how they can assist the profiling of road user specifically those who are involved in road accidents. Mosaic classifies 1.6 million British unit postcodes into 61 "lifestyle" types. These types describe sociocultural and socioeconomic behavior. There are more than 350 variables taken from sources such as the 2001 Census, Family Expenditure Survey, MORI's financial surveys, and Experian Lifestyle Surveys. These data are used in statistical cluster analysis to build the 61 neighborhood types that can be aggregated to 11 Mosaic groups.

Existing approaches to understanding road user risk in spatial terms have been centered on using census data, specifically deprivation indicators to determine a relationship between those people who experience high levels of deprivation and their overrepresentation in road collision statistics. The UK Census in 2005 commissioned its own geodemographic classification of output areas, which classifies small areas using 41 variables. These variables were clustered into 7 "groups," 21 "subgroups," and 52 "clusters" (more information on clustering and data used, see Vickers et al. 2005). This classification provides a useful open-source geodemographic tool that, when used in conjunction with Mosaic, can provide a deeper insight into the patterns of geodemographics and road user risk. Using geodemographics for road collision research enables the user not only to create a more succinct profile of the vulnerable (and less vulnerable) road user, but also to target reduction strategies more effectively because of the inclusion of information regarding the most commonly used media outlets and preferred retail chains used by each Mosaic type.

Since 1997, there has been a renewed interest in academia and government in the use of neighborhood classifications (Longley 2005). In policy terms, these developments have arisen from the opportunity to improve efficiency by targeting preventative communication programs toward those most at risk (Longley 2005). In recent years, these programs have centered on policing and health needs (Ashby and Longley 2005), and with these public service applications come the opportunity and methodological feasibility to apply geodemographics to road safety research. In response to the narrow research base is the issue that nearly all research in this domain is restricted to children and their socioeconomic risk as shown in the previous discussions. There have been a limited number of studies that have aimed to explore understanding the risks faced by adults within neighborhoods and what can be deemed their "risk exposure."

Research by Julien et al. (2002) in Paris stated that the majority of people who traveled on foot during the day were children, those not in paid work, and the elderly. She concluded that these pedestrians were at higher risk of being involved in a collision than other types of pedestrian. This study indicates that different levels of risk exposure do prevail between different groups in society, predominantly associated with mobility. Mobility, and constraints on mobility, has often been referred to with respect to the elderly and children. A person's mobility will in effect influence their exposure to traffic collision risk. Scheiner et al. (2003) summarize that certain lifestyle groups (based on employment and income) have specific forms of mobility. Mobility here refers to "short-term" mobility (travel) rather than long-term mobility (e.g., housing mobility), and in turn we can relate this mobility to differences in risk exposure.

4.3 MEASUREMENT AND ANALYSIS

4.3.1 DATA

Road traffic collisions take place at specific places and times. In other words, they create "snapshots" of time and space that can be used to evaluate the circumstances surrounding collisions. This representation of the real world is necessarily incomplete and often inaccurate to unspecified degrees (Longley et al. 2005). Traditionally, most studies of road collisions have relied on collision statistics to address a range of safety-related concerns such as the identification of road collision hot spots, the evaluation of safety programs, or the correction of irresponsible driver behavior. However, this chapter challenges this assumption, arguing that using road collision data alone is insufficient to identify the main causes of road collisions. In most cases, the cause of road collisions is not attributed to one single cause, but it is an outcome of a complex process of interaction involving the driver, the vehicle, and the road environment. Therefore, it is sometimes difficult to identify the main causes of a collision due to counts alone, which is why it is important to recognize that enrichment through use of socioeconomic data can create a more accurate picture of who is more likely to be involved in a collision and possibly why. Pasquier et al. (2002) indicate that collision frequencies segregated by location, time, and type are generally low. Given the low rate of occurrence and the statistical nature of the problem, they go on to argue that the task of drawing statistically significant inferences by merely examining collision counts may not be an easy one. However, this just goes to exemplify the importance of using enrichment data in the form of socioeconomic data analysis.

The most important dataset within the classification process remains the Census data with the most recent census in 2001 creating a surge of new classifications such as Experian's Mosaic (2004). Experian has the advantage when creating Mosaic of having access to a wide variety of data sources as it is the United Kingdom's largest originator and owner of consumer data. Just over 400 variables were used to create the current version of Mosaic based on the publication of the 2001 UK Census. Some 54% of the data used to build Mosaic are sourced from the 2001 UK Census. The other 46% comes from Experian's own Consumer Segmentation Database, including information about the Electoral Roll, Shareholders Register, House Price, Council Tax information, and ONS (Office for National Statistics) local area statistics. Figure 4.2 shows the types of data used to build Mosaic.

A discussion about scale is of particular importance when thinking about socioeconomics, risk, and road collisions. First and foremost, the issues of geography and scale are just as prominent as when analyzing road collisions on a network. When we talk about geography, we often use the phrase "neighborhood" as a scale to which we describe mechanisms within. A neighborhood has a geographical context within which urban social mechanisms are measured. These might range from tobacco use, mobility, crime, health, or indeed road collision injuries. Essentially, we can ask the question, is a person's risk of being in a road collision influenced not only by the composition of that area's population but also by the area's geographical context? The problem of scale and geography in road collision analysis is not a new one, but when looking at socioeconomics and analyzing the role of the road traffic victim, the element of geography becomes more complex. Moreover, for the main reason that unlike other themes studied within the structure, say of the *neighborhood* such as health, when we are studying road collisions and socioeconomics, we are analyzing not only the area in which the road traffic victim lives, but also the area of the road collision site itself. Although neighborhoods and their boundaries are sometimes obvious to local residents, it is more common to find considerable disagreement on the size and contents of a neighborhood. There have been many debates over the years that argue the changing nature of the meaning of neighborhood, and increasingly, it is being used in conjunction with *community*. However, there is a strong belief that the nature of *community* is far less geographic in nature than neighborhood. So, let us for a moment think about the notion of neighborhood and road collision analysis. There are many examples of how we constitute a neighborhood, and it is not our aim to discuss these debates here.

4.3.2 Database Construction

One of the major challenges of using socioeconomic data to underpin road collision risk exposure of different types of people in society is both data availability and combining those datasets. Literature shows a diversity of potential data that can be

Risk and Socioeconomic Factors 71

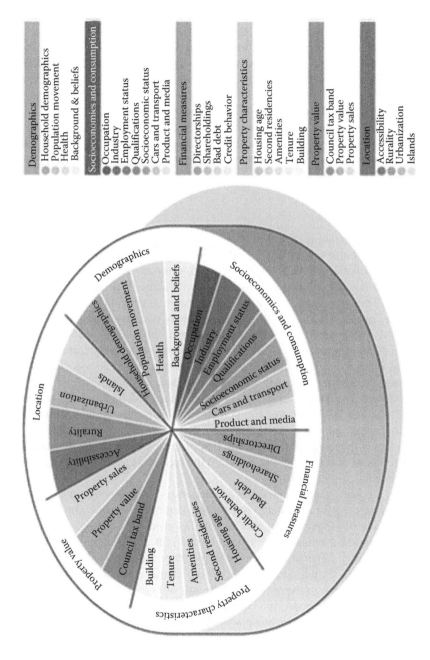

FIGURE 4.2 Mosaic UK data sources. (Courtesy of Experian, 2006; http://www.business-strategies.co.uk/, accessed 2006.)

used; however, accessibility always depends on the national or regional policy situation. Some countries collect detailed data with regard to information about road traffic victims and drivers (such as postcode information of their address, or ethnicity). There are two ways to approach this. First, we can rely on the road collision database that may or may not collect information regarding the socioeconomics of the road collision (including detailed information about the injuries and drivers). The second approach is mortality data. Most countries collect detailed information about mortality and socioeconomics (age, ethnicity, gender, address, etc.). Roberts and Power (1996) used data from the Office of Population Censuses and Surveys in England and Wales to understand mortality among children by parent's social class. Much of the data relating to socioeconomics and road collisions are aggregate data, for example, in the United Kingdom, socioeconomic data are collected at the smallest scale, the output area (small building blocks comprising of approximately 40 resident households).

4.3.3 METHODS

Methodologically speaking, should we be concerned with people or places? As the nature of this book suggests, we are predominantly concerned with the analysis of road collision locations; however, it is important to acknowledge the role of people, and how their locations influence their risk or likelihood of collision involvement. Therefore, looking at traffic injury rates alone would be misleading. For example, people who live in deprived areas may not have road collisions in those areas and vice versa. Research methodology in this area has become much more sophisticated in recent years and has moved on considerably from measuring rates of injuries in different age ranges and genders. What is important is to understand the road safety issues that local people face, whether this is where they live or elsewhere. Determining these groups at risk can be delineated in a number of ways. These can be determined by (1) their lifestyle (same religious establishment, school, park, carrying out similar high-risk activities), (2) age group, (3) characteristics of the nature of their deprivation (illiteracy), (4) cultural differences (different religious or cultural groups), and (5) the area in which they live (housing estate, street, etc.).

However, it is important to note that while delineating by groupings in society to better improve the ways road safety professionals can target specific groups, there can be drawbacks. By looking at smaller and smaller groups in areas or using injury rates, patterns become increasingly difficult to interpret. Research has also shown that people with very different lifestyles can experience different types of road collision risk. The next section outlines how we can characterize the different types of methods used to analyze the socioeconomics of road collisions. There are many different ways we could choose to classify these methods, for example, we could divide the locations by studying home address and road collision locations. More commonly, research has tended to focus on the home location of the road traffic victim. This book has made it clear that the relationship between the two locations is complex.

4.3.3.1 Qualitative Data Analysis

In a number of research studies, in order to supplement the road collision data, additional data are often collected. They are usually survey data that gather more detailed information on people's risk-taking behavior in the road environment and further socioeconomic information. For example, Dobson et al. (1999) used a survey to ask women about recent road collisions they had been involved in. Questions included information about occupation, education, number of working hours in a week, family status, health, money, and relationships. There are many issues surrounding this method, largely due to the low response rates of surveys.

4.3.3.2 Descriptive Statistics

The first stage of any analysis into the socioeconomic characteristics is usually descriptive statistics. This gives insight into the nature of the data and the basic patterns and processes. These will include percentages, standard deviations, Chi-squared, and simple summaries of the data.

4.3.3.3 Regression Analysis

Often, the most common method to understand the relationship between socioeconomic factors and road traffic collisions is regression analysis. This section intends to outline the regression analysis most suitable for road collisions and associated socioeconomic data. First, it is important to determine whether one is focusing on the location of the road traffic victim or the location of the collision itself. This is clearly an important distinction and one that can sometimes be overlooked. In the raw data, it is often clear that there are relationships between socioeconomics (more commonly deprivation) and road traffic collisions. However, unpicking the complexities can prove hard. There is no straightforward relationship between road collisions and socioeconomics.

The socioeconomic variables to describe people are based on a geographical unit, for example, in the United Kingdom, this would be an output area or, in Hong Kong, a district (although the two are very different in terms of size). Therefore, we make inferences about people only based on area data. The discrete nature of count data and the prevalence of zeros and small values means that the use of linear multiple regression can produce inconsistent and biased results. The formulations widely used to analyze models in which the dependent variable takes only nonnegative integer values corresponding to the number of events occurring in a given interval are based on the Poisson regression model (Cameron and Trivedi 1986, 1998). The Poisson model assumes that the conditional mean and conditional variance are the same. A related generalization of the Poisson regression is the negative binomial model, which does not require the assumption of equidispersion. It is also important to remember when using regression techniques for socioeconomic analysis of road collisions that there will be unobservable effects on collisions that may arise at a regional level. These could include differences in public policy decisions, public investment, climate, social habits, and other unknown effects. Often such differences are represented as dummy variables in the model. One of the key determinants of the likelihood of being in a road collision is relative exposure to traffic. Obtaining traffic volume data and linking this to socioeconomic characteristics of people are very complex and require less than perfect solutions. One of the main issues with regression analysis is that it often does not take into account geography. It is easy to use the census tracts as the building blocks for regression; however, by doing this, you are subject to the modifiable areal unit problem discussed earlier. Further analysis of road collisions and socioeconomics has been to use geostatistics.

4.3.3.4 Geostatistics

The use of geographical information system (GIS) and geospatial statistics to understand road collisions and socioeconomics is well established in the research. Often, the use of GIS will be used as an initial descriptive technique, to determine the spatial characteristics of the data. Spatial exploratory analysis in the form of general maps, choropleth maps, and so on are a common tool to understand socioeconomic patterns. Most of these methods will be explained at greater length later in the book. Gruenewald et al. (1996) used geostatistical modeling techniques to study the relationships between the spatial distribution of alcohol outlets and alcohol-related road collisions. These statistical techniques allowed the researchers to correct for spatial autocorrelations between rates of collisions due to their relative proximity in space, while providing a means to explore the relationships between the availability of alcohol in one area and rates of collisions in adjacent areas (i.e., spatial lags). Spatial autocorrelations between adjacent geographic units introduce bias into statistical analyses due to the violation of the assumption of unit independence. A more recent study by LaScala et al. (2000) focused on determining the relationship between the observed rates of pedestrian injuries and measures of environmental and demographic characteristics of San Francisco. It was expected that rates of pedestrian injury would be greatest in those areas of the city that provided greatest access to alcohol via restaurants, bars, and retail outlets. It was also expected that this relationship would be the strongest among pedestrians who had been drinking. Additional demographic and environmental measures of roadway complexity, traffic flow, and population density were included to control for known effects of these variables in geographic studies of traffic-related outcomes. The study used a spatial regression analysis, specifically an ordinary least squares. Studies using spatial regression techniques often find geographic relationships between demographic characteristics and road collisions. Generally speaking, even using spatial analysis of this nature to analyze the relationships between socioeconomics and road collisions, it can often be difficult to navigate through the results and unpick specific relationships. Most of the time, we can only make inferences about the relationship found to be significant.

4.3.3.5 Typology Analysis

There have been few attempts to create typologies from the data available into more meaningful groupings in society. For example, instead of just taking one variable such as unemployment, it is more useful to combine different variables and create a "typology." There are two examples of such analysis in the literature. The first is by Fontaine and Gourlet (1997), who focused on fatal pedestrian collisions in France. The research used a correspondence analysis with a factorial plan that outlined four distinct groups. The groups were outlined to be the following:

Risk and Socioeconomic Factors 75

1. Over 65 crossing the road in an urban area, more likely to be women
2. Rural road, nighttime walking, alcohol, walking in carriageway, men
3. Children running around in an urban area during the daytime
4. Secondary collision, on the pavement, loss of control

Coupled with this, the other example by Anderson (2010) uses geodemographics to create a spatial typology for road collisions in London. Further information can be found in the next section.

4.3.3.6 Case Study: Geodemographics in London, United Kingdom

This case study is based in London, United Kingdom, and its focus is threefold. First, a nonspatial analysis of postcode data of road traffic victims and geodemographics is presented. Second, the spatiality of the geodemographics and road traffic victims is demonstrated; and finally, here we consider the ways in which geodemographics might be used for road safety social marketing purposes.

The data used to determine the road traffic victim's and driver's locations were obtained from the road collision database STATS19, where information about the collision is recorded such as time of day, collision location, how many people were involved, and what class they are (in terms of driver, passenger, pedestrian, cyclist). This information is collected by the police and divided into three separate datasets that include attendant circumstances, casualty details, and vehicle details. Each postcode for the driver and casualty was subsequently linked to a postcode point dataset, which meant a point could be displayed on the map that represented a postcode.

Both the driver and casualty datasets were kept separate for the purpose of maintaining the structure of the original data collection procedure and to avoid any confusion between the two datasets. The data consisted of only postcode data Mosaic type for each postcode (which has been appended to the dataset), collision reference, and the easting and northing centroid point for the postcode in order to plot the residential location. The aim was to elaborate the understanding of the nature of people's propensities to be involved in a collision based on geodemographic indicators and to assess the potential for reducing collision risk. Accordingly, Mosaic codes were appended to individual records of all the driver and casualty postcode data within London for the years 1999–2003.

The first stage of the analysis entailed attaching each of approximately 100,000 postcodes for both drivers and injuries to a Mosaic type. The geodemographic classification was then used to analyze the incidence of drivers and injuries across both the 61 Mosaic neighborhood types and 11 groups. By standardizing around an index value of 100, the geodemographic codes can be compared across London, thereby comparing different neighborhoods or boroughs. A value over 100 indicates a higher-than-average propensity to be involved in a collision, while a score below 100 indicates a lower-than-average propensity to be involved in a collision. The postcodes and appended Mosaic type are then mapped using GIS to highlight the areas of high or low propensities. There are a few considerations that need to be addressed before the Mosaic patterns are outlined. There is no detailed outline of the typical vehicle ownership traits of each Mosaic type, and so inferences have been drawn from the

TABLE 4.3
Mosaic Types and Associated Population Percentages and Index Scores for Both Casualties and Drivers

Mosaic Type (by Highest Household Population%)	London Household%	London Index (Average Rate = 100)	Casualty Index	Driver Index
F36 Metro Multiculture	12.32	758	101	94
D27 Settled Minorities	11.38	680	113	109
E28 Counter Cultural Mix	9.12	751	91	84
E30 New Urban Colonists	6.64	519	77	75
C20 Asian Enterprise	6.46	508	105	109
C19 Original Suburbs	6.31	244	96	101
E29 City Adventurers	5.84	531	68	65
A01 Global Connections	5.31	734	42	41
H46 White Van Culture	4.70	143	135	133
A02 Cultural Leadership	4.65	438	63	66
C18 Sprawling Subtopia	3.70	112	111	118
D21 Respectable Rows	2.54	99	119	122
A03 Corporate Chieftains	2.22	161	86	103
J52 Childfree Serenity	2.00	163	81	80
A05 Provincial Privilege	1.90	104	82	88
E32 Dinky Developments	1.25	135	126	132
C15 Close to Retirement	1.20	38	146	151
B12 Middle Rung Families	1.16	38	151	161
D26 South Asian Industry	1.16	94	104	108

Source: Reprinted from Anderson, T.K., *Environ. Planning A*, 42(9), 2186, 2010, http://www.envplan.com. With permission from Pion Ltd., London, U.K., http://www.pion.co.uk.

Mosaic pen portraits regarding the types of cars (if any) that people are most likely to own and how they are likely to drive.

The results suggest evidence of people in specific Mosaic types to be overrepresented in being involved in a collision. This overrepresentation occurs when the index score is over 100, which is the *expected*. Therefore, if a Mosaic type has an index score of 200, one can determine that people who belong to that type are twice as likely to be involved in a collision as would *normally* be expected. Table 4.3 shows the results highlighting the Mosaic type, corresponding percentage of households in London within each type, the overall London index score for that particular type, and corresponding index scores for both the driver and casualty. This enables us to compare the driver and casualty outcome with the overall London index.

Arguably, geodemographics offers a worthwhile and suitable solution to a complex problem. Socioeconomic areal solutions to profile road traffic collisions have been slow in evolving. There has been skepticism with regard to the application of road collision data to geodemographics, especially in relation to the notions of time, mobility, and static versus dynamic risk. However, measuring a person's risk of being involved in a collision is difficult. There are many factors to consider other than geodemographics, such as mobility, or how the person interacts with the road (i.e., as a driver, pedestrian, or cyclist). This collision risk can change over time, and often does.

4.3.4 Methodological Issues

As with most studies of this nature, there are a number of methodological issues to take into account. An important possible source of bias is the so-called omitted-variable problem, which arises when a study does not account for all explanatory variables. In other words, analysis is usually based on the availability of variables that could explain, totally or partially, the variability in the outcome variables. Therefore, in most socioeconomic studies of road collisions, there are a number of variables that are often left out, largely due to the nature of measurement. It is impossible to include every socioeconomic indicator. These variables might be important explanatory variables whose effects can be seen at different geographical levels. As a measure of exposure to substance abuse in injured drivers, often it is easier to use data on prevalence of alcohol drinking but usually not the injured drivers' use of illicit drugs, because this is not routinely collected after a traffic collision. Another potential limitation of socioeconomic methods is the accuracy of the data. Often, the data sources are from established national institutes (e.g., the National Institute of Statistics and the National Institute of Health); one can argue that for some of them, possible errors could occur.

4.3.5 Can You Measure Risky Behavior?

The main area for contention when discussing methodological issues is the measurement of risky behavior. It is one thing to link socioeconomics with road collisions, but quantifiably measuring people's risk-taking behavior is very difficult. We know, generally speaking, younger people are more likely to take risky behavior (Hatfield and Fernandes 2009), and men more so than women, leading to more collisions relating to loss of control related to speed and nighttime driving.

4.4 POLICY AND INTERVENTION

The development of any road safety policy should be multidisciplinary in its approach. Usually, there is focus on generically reducing the number of injuries and deaths on the road. There have been many recent initiatives that have focused on reducing road injuries of vulnerable road users, such as children or elderly people. However, the implementation of road safety policy that specifically targets certain socioeconomic groupings in society has been slow to evolve. Largely, this has been due to the lack of evidence available to support which appropriate mechanisms for reduction would be successful. In this chapter, we have discussed the fact that there are significant variations in the incidence of injury and death that are related to a range of socioeconomic factors. Membership of a socially excluded group (e.g., one that does not have the same access to employment, education, good housing, transport, and amenities) increases the likelihood of being involved in some form of road traffic collision. Slowly, in terms of policy for road safety, with particular emphasis on socioeconomics, people are realizing that there is no "one size fits all" solution. Road safety policy must be tailor-made to local circumstances. The need for a multipronged approach to road safety continues to exist. Although highway engineering and enforcement measures can reduce injuries, the impact of both will be enhanced if the community is actively involved.

In 2002, the Department for Transport in the United Kingdom set up the Neighbourhood Road Safety Initiative, where 15 local authorities in England were allocated funds to develop schemes to reduce road injuries in their most deprived areas. The local authorities were encouraged to come up with new and innovative approaches to road safety to try to reach into the most deprived areas. This was the first "community-based" initiative in the United Kingdom, and it meant thinking about different ways of tackling certain issues and concerns, many of which required new sources of data or information. Alongside strategies for crime and health, the notion of "community" in the United Kingdom stands out as largely a "new geography" within which to tackle social issues. The question is, with regard to road safety, do we tackle the issue of deprivation and inequality first, or focus directly on road safety? Arguably, Whitehead (1995) outlines four ways in which deprivation can be tackled at a community level. These are (1) strengthening individuals, (2) strengthening communities, (3) improving access to services, and (4) encouraging broad economic and cultural change.

Strengthening individuals to cope with the road hazards around them, focusing on parent education, road crossings for children, or helping older people manage new traffic environments are some important measures. Strengthening communities is something that is already taking place at different levels, such as supervising child pedestrian journeys to and from school, developing safe play areas, and schemes that involve the whole family that are not just age related. Third, by improving people's access to services, such as health, employment, and education, would improve people's hazard management risk. Finally, tackling the broader issues of social and political change would mean changing society-wide attitudes to situations such as drink-driving, speeding, or taking drugs and driving. In addition, efforts should be made to make people aware and more accepting of other road users, such as pedestrians and cyclists. Western societies tend to be automobile oriented, with little acceptance of other road users sharing the road space. This is an issue that needs to change in order for all road users to use the space together in a harmonious manner.

How is geodemographics shaping road safety policy? A recent project has been undertaken by the Thames Valley Safer Roads Partnership, which uses a web-based interface and road collision data interlinked with geodemographic data. This tool allows practitioners to select data interactively and use it with the Experian's Mosaic software. It can be user customized and will offer "customer insight" (Road Safety Analysis 2010). This is the first attempt to successfully link geodemographics and road collision data for road safety practitioners. Figure 4.3 helps to show the promise

Risk and Socioeconomic Factors

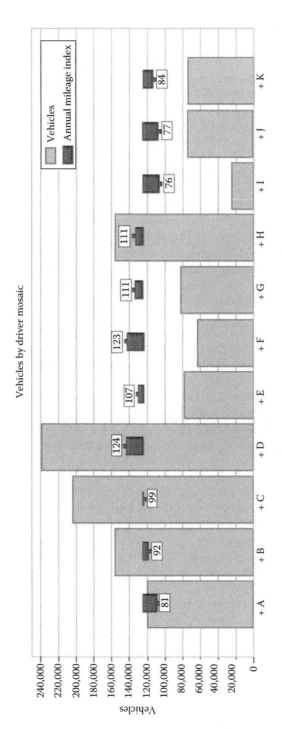

FIGURE 4.3 MAST online. (Data from http://www.roadsafetyanalysis.org/, generated October 13, 2010.)

of such tools in order to help analyze and communicate information to the public with regard to reducing road traffic injury.

According to Fleury et al. (2010), the interest of analyzing the sociospatial dimension of road safety lies in the objective of the action. One of the first questions that should be asked is about the uniformity of the area: are there any spatial characteristics that cause different risks justifying different actions? Fleury's research is one of the few that tackles the notion of targeting the population at risk in an area, no matter where they have road collisions.

As we take a moment to consider the nature of community in terms of road safety and socioeconomic policy, the community is one of the most, arguably, appropriate frameworks within which to place socioeconomic road safety policy. Actively engaging with the "community" can have important benefits for developing and strengthening social cohesion and implementing successful road safety projects. It might be useful to consider what we mean by community; in general, it means a group of people who live in an area and therefore experience a common road safety program. They may be represented by local residents' association or religious leaders. This notion of community, especially when dealing with issues such as road safety, is very important, and there have been proven benefits, particularly in crime and health.

This angle of community is facilitated by the framework of partnerships. Examples of organizations that can facilitate support in dealing with road safety and socioeconomics would include council departments, regeneration, youth services, housing, health professionals, schools and colleges, police, fire and rescue services, children's play centers, elderly organizations, community organizations, religious organizations, and private sector bodies such as retail outlets and larger retailers. The major overall issue in relating road safety policy and findings from socioeconomic studies is the understanding of how specific communities and areas can deal appropriately with targeting specific groups within society in the most effective way. There has been evidence that certain socioeconomic characteristics are linked with road collision propensity; however, there is little evidence of the types of intervention required to tackle these factors. However, often these factors are variables such as "unemployment" or "deprivation" that are difficult to tackle in society. Therefore, does it require society to make more efforts to underpin these major societal issues?

Let us consider obesity as a public health concern. The majority of the population are aware of the lifestyle choices and which choices pose a risk for the likelihood of falling into this category: we need to eat well, take exercise, and avoid unhealthy foods. Generally speaking, there has been much research into the lifestyle of the population and their likelihood of being obese. If we consider the likelihood of being in road collision akin to the likelihood of being obese, does society understand the risks as well? Research has posed the hypotheses that lifestyle does have an impact on involvement in road collisions. Many previous research studies have concentrated on a single aspect of a person's lifestyle such as deprivation, age, place of residence, and personality. However, there has been no research that has tried to encompass all these factors and determine a better understanding of people's road safety risk as a whole proportion of their lifestyle.

4.5 CONCLUSION

Nowadays, we are dealing with road safety in more and more diverse settings, dense urban areas with multicultural residents. These people interact and compete in the road environment daily, in many different roles. There are people who cycle, walk, use the car for business, use the car for pleasure, motorbike, visitors, and so on. This interaction of different road users is bound to cause conflict. The sheer density of urban areas, roads, and population mean that unpacking the patterns and processes that exist can be extremely complex.

People's perception of their own risk in the road environment is generally underestimated. You are more likely to die in a road collision than in an airplane collision. Yet, somehow this does not affect how we perceive our interaction with the road environment. This is partly because we interact on a daily basis, but the difference is that different people in society have different levels of risk. For example, young ethnic males driving in old sports cars are likely to be in a road collision. The use of socioeconomics to understand and gain insight into crime, health, and education has motivated this use in a road safety context. It is not the intention that this chapter answers all the questions of lifestyle and road collision involvement, but it offers a clearer picture of the types of people likely to be involved and, most importantly, where they are likely to be injured. In the past, the large proportion of research in road safety concentrated on, quite rightly, the location of the collision itself. However, by only looking at the collision location, there is a lot of information that is being left out of the equation. Although it is impossible to incorporate all the facets relating to collision involvement, there needs to be an understanding of the many spatial locations that are involved, particularly where the injuries and drivers reside.

The major difference between road collisions and, say, obesity is that there is a large proportion of chaos and uncertainty involved in the likelihood of being involved. Risks of being in a road collision are inherently subjective and incorporate many different players, including the road user, any other road user, the road environment, time, mood, and type of car. It would be impossible to accurately capture a person's risk of being involved in a road collision because of the difficulty of unpacking all these competing factors. There are so many facts banded around in the media proclaiming the high numbers of road collision deaths in relation to say something like terrorism. A study at the University of Otago determined that the body count from road collisions in developed economies is 390 times higher than the death toll in these countries from international terrorism. To put this into context, as many people died every 26 days on America's roads in 2001 as died in the terrorist attacks of 9/11.

REFERENCES

Abdalla, I. M. 1997. Statistical investigation and modelling of the relationships between road accidents and social characteristics. Napier University, Edinburgh, U.K.

Abdalla, I. M. 1999. Linking road accident statistics to census data. PhD dissertation, Napier University, Edinburgh, U.K.

Abdalla, I. M., D. Barker, and R. Raeside. 1997a. Road accident characteristics and socioeconomic deprivation. *Traffic Engineering & Control* 38 (12): 672–676.

Abdalla, I. M., R. Raeside, D. Barker, and D. R. D. McGuigan. 1997b. An investigation into the relationships between area social characteristics and road accident injuries. *Accident Analysis & Prevention* 29 (5): 583–593.

Anderson, T. K. 2010. Using geodemographics to measure and explain social and environment differences in road traffic accident risk. *Environment and Planning A* 42 (9): 2186–2200.

Ashby, D. I. and P. A. Longley. 2005. Geocomputation, geodemographics and resource allocation for local policing. *Transactions in GIS* 9 (1): 53–72.

Braver, E. R. 2003. Race, Hispanic origin, and socioeconomic status in relation to motor vehicle occupant death rates and risk factors among adults. *Accident Analysis & Prevention* 35 (3): 295–309.

Cameron, A. C. and P. K. Trivedi. 1986. Econometric models based on count data: Comparisons and applications of some estimators and tests. *Journal of Applied Econometrics* 1 (1): 29–53.

Cameron, A. C. and P. K. Trivedi. 1998. *Regression Analysis of Count Data*. London, U.K.: Cambridge University Press.

Campos-Outcalt, D., C. Bay, A. Dellapena, and M. K. Cota. 2003. Motor vehicle crash fatalities by race/ethnicity in Arizona, 1990–96. *Injury Prevention* 9 (3): 251–256.

Christie, N. 1995a. Road accident involvement of children from ethnic minorities. Department for Transport, London, U.K.

Christie, N. 1995b. Social, economic and environmental factors in child pedestrian accidents: A research review. TRL Project Report PR116. Transport Research Laboratory, Crowthorne, U.K.

Dobson, A., W. Brown, J. Ball, J. Powers, and M. McFadden. 1999. Women drivers' behaviour, socio-demographic characteristics and accidents. *Accident Analysis & Prevention* 31 (5): 525–535.

Edwards, P., J. Green, I. Roberts, C. Grundy, and K. Lachowycz. 2006. Deprivation and road safety in London: A report to the London Road Safety Unit. London, U.K.: LSHTM.

Erskine, A. 1996. The burden of risk: Who dies because of cars? *Social Policy & Administration* 30 (2): 143–157.

Experian. 2004. Experian UK website. http://www.experian.co.uk/ (accessed December 16, 2004).

Experian. 2006. Experian Business Strategies website. http://www.business-strategies.co.uk/ (accessed December 16, 2006).

Fleury, D., J. F. Peytavin, T. Alam, and T. Brenac. 2010. Excess accident risk among residents of deprived areas. *Accident Analysis & Prevention* 42 (6): 1653–1660.

Fontaine, H. and Y. Gourlet. 1997. Fatal pedestrian accidents in France: A typological analysis. *Accident Analysis & Prevention* 29 (3): 303–312.

Graham, D., S. Glaister, and R. Anderson. 2005. The effects of area deprivation on the incidence of child and adult pedestrian injuries in England. *Accident Analysis & Prevention* 37 (1): 125–135.

Grayling, T., K. Hallam, D. Graham, R. Anderson, and S. Glaister. 2002. *Streets Ahead: Safe and Liveable Streets for Children*. London, U.K.: Institute for Public Policy Research.

Gruenewald, P. J., P. R. Mitchell, and A. J. Treno. 1996. Drinking and driving: Drinking patterns and drinking problems. *Addiction* 91 (11): 1637–1649.

Haepers, A. S. and P. T. Pocock. 1993. Road traffic accident involving children from ethnic minorities. Leeds County Council, Leeds, UK.

Hasselberg, M. and L. Laflamme. 2005. The social patterning of injury repetitions among young car drivers in Sweden. *Accident Analysis & Prevention* 37 (1): 163–168.

Hatfield, J. and R. Fernandes. 2009. The role of risk-propensity in the risky driving of younger drivers. *Accident Analysis & Prevention* 41 (1): 25–35.

Judge, K. and M. Benzeval. 1993. Health inequalities: New concerns about the children of single mothers. *British Medical Journal* 306 (6879): 677–680.
Julien, A. and J. Carré. 2002. Risk exposure during pedestrian journeys. *Recherché Transports Sécurité* 76: 173–189.
Kposowa, A. J. and M. Adams. 1998. Motor vehicle crash fatalities: The effects of race and marital status. *Applied Behavioral Science Review* 6 (1): 69–91.
Laflamme, L., M. Vaez, M. Hasselberg, and A. Kullgren. 2005. Car safety and social differences in traffic injuries among young adult drivers: A study of two-car injury-generating crashes in Sweden. *Safety Science* 43 (1): 1–10.
LaScala, E. A., D. Gerber, and P. J. Gruenewald. 2000. Demographic and environmental correlates of pedestrian injury collisions: A spatial analysis. *Accident Analysis & Prevention* 32 (5): 651–658.
Lawson, S. D. and P. J. Edwards. 1991. The involvement of ethnic minorities in road accidents: Data from three studies of young pedestrian injuries. *Traffic Engineering & Control* 32 (1): 12–19.
Longley, P. A. 2005. Geographical information systems: A renaissance of geodemographics for public service delivery. *Progress in Human Geography* 29 (1): 57–63.
Longley, P. A., M. F. Goodchild, D. Maguire, and D. Rhind. 2005. *Geographic Information Systems and Science*. Chichester, U.K.: John Wiley & Sons.
Murray, A. 1998. The home and school background of young drivers involved in traffic accidents. *Accident Analysis & Prevention* 30 (2): 169–182.
MacIntyre, S., S. MacIver, and A. Sooman. 1993. Area, class and health: Should we be focusing on places or people? *Journal of Social Policy* 22 (2): 213.
Noland, R. B. and M. A. Quddus. 2004. A spatially disaggregate analysis of road injuries in England. *Accident Analysis & Prevention* 36 (6): 973–984.
Ossenbruggen, P. J., J. Pendharkar, and J. Ivan. 2001. Roadway safety in rural and small urbanized areas. *Accident Analysis & Prevention* 33 (4): 485–498.
Pasquier, M., C. Quek, W. L. Tung, D. Chen, and T. M. Yep. 2002. Fuzzylot II: A novel soft computing approach to the realisation of autonomous driving manoeuvres for intelligent vehicles. In *Seventh International Conference on Control, Automation, Robotics and Vision (ICARCV 2002)*, Singapore, Vol. 2, pp. 746–751.
Petch, R. O. and R. R. Henson. 2000. Child road safety in the urban environment. *Journal of Transport Geography* 8(3): 197–211.
Road Safety Analysis (MAST) website. http://www.roadsafetyanalysis.org/ (accessed March 31, 2010).
Roberts, I. and C. Power. 1996. Does the decline in child injury mortality vary by social class? A comparison of class specific mortality in 1981 and 1991. *British Medical Journal* 313 (7060): 784–786.
Robinson, P. and T. Nolan. 1997. Paediatric slow-speed non-traffic fatalities: Victoria, Australia, 1985–1995. *Accident Analysis & Prevention* 29 (6): 731–737.
Sawalha, Z. and T. Sayed. 2001. Evaluating safety of urban arterial roadways. *Journal of Transportation Engineering* 127 (2): 151–158.
Savitsky, B., L. Aharonson-Daniel, A. Giveon, The Israel Trauma Group, and K. Peleg. 2007. Variability in pediatric injury patterns by age and ethnic groups in Israel. *Ethnicity and Health* 12 (2): 129–139.
Scheiner, J. and B. Kasper. 2003. Lifestyles, choice of housing location and daily mobility: The lifestyle approach in the context of spatial mobility and planning. *International Social Science Journal* 55 (176): 319–332.
Schiff, M. and T. Becker. 1996. Trends in motor vehicle traffic fatalities among Hispanics, non-Hispanic whites and American Indians in New Mexico, 1958–1990. *Ethnicity & Health* 1 (3): 283–291.

Sonkin B., P. Edwards, I. Roberts, and J. Green. 2006. Walking, cycling and transport safety: An analysis of child road deaths. *Journal of the Royal Society of Medicine* 99: 402–405.

Steinbach, R., P. Edwards, J. Green, and C. Grundy. 2007. Road safety of London's black and Asian minority ethnic groups: A report to the London Road Safety Unit. London, U.K.: LSHTM.

Stevens, J. A. and A. M. Dellinger. 2002. Motor vehicle and fall related deaths among older Americans 1990–98: Sex, race, and ethnic disparities. *Injury Prevention* 8 (4): 272–275.

Stirbu, I., A. E. Kunst, V. Bos, and E. F. Van Beeck. 2006. Injury mortality among ethnic minority groups in the Netherlands. *Journal of Epidemiology and Community Health* 60 (3): 249–255.

Vickers, D., P. Rees, and M. Birkin. 2005. Creating the national classification of census output areas: data, methods and results. School of Geography, University of Leeds, Leeds, U.K.

Webber, R. and P. A. Longley. 2003. Geodemographic analysis of similarity and proximity: Their roles in the understanding of the geography of need. In *Advanced Spatial Analysis: The CASA Book of GIS*, eds. P. A. Longley and M. Batty, pp. 233–266. Redlands, CA: ESRI Press.

White, D., R. Raeside, and D. Barker. 2000. Road accidents and children living in disadvantaged areas: A literature review. Scottish Executive Central Research Unit, Edinburgh, U.K.

Whitehead, M. 1995. Tackling inequalities: A review of policy initiatives. Benzeval M, Judge K, Whitehead M. *Tackling Inequalities In Health*. An agenda for action. King's Fund, London, U.K., pp 22–52.

5 Road Collisions and Risk-Taking Behaviors

5.1 INTRODUCTION

What do we mean by a road safety problem? There are many answers to this question that all make sense. The risk that children run on their trip to school is a road safety problem. Drinking and driving is a road safety problem, driving in the dark is a road safety problem and young driver risk is a road safety problem. By making such lists of problems, it is possible to cover all areas of road safety. The snag is that the various problems on such a list tend to overlap. Children are at risk when travelling to school partly because young drivers have a high risk of collision involvement, partly because of drinking and driving, and partly because driving in the dark increases the risk of an accident. Drinking and driving is a major problem partly because it takes place in the dark and on roads where there are pedestrians and cyclists. These examples show how difficult it is to define road safety problems in an orderly and logical way. The difficulty is particularly relevant when we want to give an exhaustive definition of road safety problems.

Pedersen et al. (1982, 29)

The term "risk" is frequently used when discussing road traffic collisions. We talk about people being "at risk" from certain types of collisions, or different locations being "high risk." We research "risk factors" that contribute to road collisions. Risk is a phenomenon applied in many different contexts, financial, crime, health, and, in this instance, transport and road collision risk. According to the definition found most commonly in road safety literature, risk is the probability of an adverse future event multiplied by its magnitude (Adams 1995). It is true to say that past collision events are not entirely trustworthy "guides" to the future as people respond differently to risks as soon as they are confronted with them.

In the literature, risk has been defined as a "measure of the probability and severity of adverse effects" (Lowrance 1976, 94) and as "chancing of a negative outcome, which includes two defining components: the chance and the negativity" (Rescher 1983, 33). According to Delfino et al. (2005), the most widespread concept of risk that originates from the insurance business (the concept of "predicted loss") indicates risk as the undesired consequence of a particular activity in relation to the probability of the events occurring, or the "probability of events occurring and consequences of such events" (Stewart and Melchers 1997; Chapman and Ward 2003). According to Nilsen et al. (1998), the last two definitions use the quantitative analysis approach, which defines quantitative measures of risk function components.

Vlek and Stallen (1980) determine that the concept of risk can be open to various interpretations that focus on (1) the components, and (2) measuring the same components according to the environment concerned. The concepts of probability and outcome are the most common and are adapted to various disciplines. With regard to quantitative risk analysis in transport systems, some perceive risk as the product of the incident occurring multiplied by the effect of the incident (Eurotunnel 1994; Evans 1994; Tsai 1998). The function of risk within a transport system is applied for different purposes, including the assessment of the effects of measures to mitigate a certain event. For example, evacuation of densely occupied areas is a measure to mitigate a component of risk by reducing exposure (Russo and Vitetta 2004). When we think about the transport system as a whole, we incorporate not only roads but rail, shipping, and air transportation as well.

In the specific case of risk related to road collisions, a probabilistic approach is made. A unit of exposure corresponds to a (probabilistic) trial, and result of such a trial is the occurrence or nonoccurrence of a collision (Hauer 1980), or a circumstance that must be present to have a road collision (Tobey et al. 1983). At times, the methods applied to risk assessment are not specifically typical of engineering but more of medical disciplines (Roberts et al. 1995; Agran et al. 1996), which in some cases focus particularly on collisions involving weaker road users (Posner et al. 2002; Vaganay et al. 2003).

Delfino et al. (2005) formulate a table of the classification of papers in relation to risk type, the summary of which can be found in Table 5.1. Fuller (2005) distinguishes between three basic uses of the term risk: objective risk, subjective risk, and the feeling of risk. In the first usage, objective risk may be defined as the objective probability of being involved in a collision. This is usually determined in a post hoc way from the analysis of collision data. This concept of risk has been referred to elsewhere as "statistical risk" (Grayson et al. 2003). Subjective risk estimate refers to the driver's own estimate of the (objective) probability of collision. Such estimates of risk represent the output of a cognitive process, while the feeling of risk represents an emotional response to a threat, a distinction previously clarified, for example, by Haight (1986) and Summala (1986). Under certain conditions, subjective estimate of risk and feelings of risk may be closely associated, such as when a driver has lost control of a vehicle on an icy road and is about to collide with another road user. However, this association may apply only after subjective estimates of risk have exceeded some critical value.

The analysis of risk and its role in road collisions has a strong element of human behavior, that is, why drivers take the risk they do. Adams (1999) outlines the strong argument for "risk compensation" within our society, whereby people perceive themselves as safer or better equipped against danger and are therefore more likely to take risk. This theory has been applied to the use of seat belts and speed. However, there are a number of risks that are not apparent to the naked eye while interacting with the road environment. For example, these might include being a certain age. Teenagers are a high-risk group within our society. In terms of risk and road collisions, risk comes in many different forms (Adams 1999). Although the propensity to take risks and be of increased risk is widely assumed to vary with circumstances and individuals, there is no way of testing this assumption by direct measurement. If a road has many collisions, it might fairly be called dangerous, but using past collision rates to estimate future risks can potentially be misleading. There are many dangerous roads that may have good collision records because they are seen to be dangerous,

TABLE 5.1
Classification of Papers Proposed for Risk Analysis of Road Collisions

	Qualitative	Quantitative Descriptive	Quantitative Behavioral
Individual		Bernard et al. (2001), Brenac et al. (1996), Fleury and Brenac (2001), Brenac and Megherbi (1996), Tsai (1998), Hauer (1980), Tobey et al. (1983), Cowley and Salomon (1976), Evans (1991a), Grayson (1979), Grayson and Howard (1982), Roberts (1993, 1995), Macpherson et al. (1998), Roberts et al. (1995), Vaganay et al. (2003), Posner et al. (2002), Keall (1995)	Russo and Vitetta (2003)
Societal	Lowrance (1976), Rescher (1983), CCPS (1995), Vlek and Stallen (1980), Stewart and Melchers (1997), Chapman and Ward (2003)	Nilsen et al. (1998), Evans (1994), Agran et al. (1996)	Russo and Vitetta (2004)

Source: Reprinted from Delfino, G. et al., Risk analysis in road safety: A model and an experimental application for pedestrians, in: *Proceedings of European Transport Conference*, Strasbourg, France, Association for European Transport, Glasgow, U.K., 2005. With permission from Association for European Transport.

and therefore children are not allowed to cross them or elderly people are afraid to drive on them. The good collision record is purchased at the cost of community severance due to road safety intervention at dangerous collision locations in order to make them safer. By the time there is a good safety record, a number of collisions would have already occurred for this intervention to take place. The good collision record gets used as a basis for risk management.

It is clear therefore that people are exposed to different risks in the road environment for many different reasons. For example, on a crowded local shopping street on a Saturday morning as cyclists, pedestrians, cars, lorries, and buses all compete for the same road space. Not all the dangers confronting the participants are able to be seen. Choices about risk also occur at the community, regional, and national levels. As a society, we decide how much loss we are willing to accept in exchange for how much freedom and mobility. A large number of factors influence what drivers do, ranging from behavioral genetics to visual perception of the economy (Lonero 1998).

Risk compensation is an ethological term whereby people tend to adjust their behavior in response to perceived changes in risk. The road collision literature offers many examples of risk compensation. A basic example being all motorists will reduce their speed when they come to a sharp bend in the road. Adams (1999) explains that road collision statistics provide a misleading measure of safety and danger. In the United Kingdom, half as many children are killed on the road as there were in 1922; however, traffic has increased 25-fold. The explanation, as Adam explains, is not due to roads becoming safer, but that roads have become so dangerous that children have been progressively withdrawn from them. Risk compensation is essentially a premise based on a metaphor. It assumes we all have "risk thermostats." The premise proposes the following (Adams 1999):

1. Everyone has a propensity to take risks.
2. The propensity varies from one individual to another.
3. The propensity is influenced by the potential rewards of risk taking.
4. Perceptions of risk are influenced by experience of collision losses, one's own and others.
5. Individual risk-taking decisions represent a balancing act in which perceptions of risk are weighed against propensity to take risks.
6. Collision losses are by definition a consequence of taking risks.
7. The more risks an individual takes, the greater on average will be both the losses he incurs and the rewards he reaps.

The theory of risk compensation suggests that safety measures that reduce risk to levels below the setting of the "risk thermostat" will be countered by behavior that reasserts the levels with which people were originally content. If the propensity (or willingness) to take risks is the principal determinant of the collision rate, this rate can be reduced only by measures that reduce the propensity. There is increasing evidence that the effect of risk compensation has been to shift part of the burden of the risk from people in vehicles to what we term "vulnerable road users" on the outside of vehicles (Hillman et al. 1990).

Some kinds of people are more at risk of being involved in a collision than others (Standish 2003). For example, a strong implication of whether someone is more likely to be involved in a collision is their age. In particular, children aged 12–16 are at high risk from being involved in a road collision (Think Road Safety 2005). The reasons for this increased risk are subject to debate, as it is difficult to underpin the exact causes for collisions. However, 1 in 10 teenagers across the United Kingdom involved in a collision say they were not paying attention (Think Road Safety 2005). Road use is highly prone to risk consciousness because other people are perceived as a threat in what have been dubbed our "risk societies" by Ulrich Beck (1992). The development of risk consciousness is an outcome of profound social change implying that society has problems that cannot be resolved, only managed (Furedi 1997). People tend to think that the risks of driving come from other road users. However, transport safety does not exclude our own roles as road users. The key issue surrounding this notion of risk is that when choosing a mode of transport, individuals look toward their own "perceived risk level" instead of the objective risk level when making their decisions.

The traffic environment is one that is constantly changing. It has been suggested that the greatest factor contributing to collision severity is an underestimation of the level of risk a traffic environment presents. All road safety research places a static risk level or understanding on individuals or areas in what is a dynamic traffic environment. The road traffic environment is constantly being referred to as dynamic, and this is because it is constantly changing, varying from second to second. In other words, someone's chances of being involved in a road collision regardless of who they are and where they are from can change within seconds. At an urban city-wide scale, this static measurement is useful in determining a wide-range understanding of the risk patterns in a spatial environment.

Often, when measuring and trying to manage risk, road safety analysts categorize road collisions and those involved in terms of severity of the collision. This method however according to Adams (1995) does not provide the best allocation of risk measurement. This is partly due to the small numbers of actual fatal collisions that occur, since they are both infrequent and scattered across space and time. Adams (1995) summarizes the argument that there is a higher proportion of minor collisions in London compared to the rest of the UK urban road network and attributes this to the fact that London is so congested and traffic speeds are so slow that there are large numbers of minor collisions, but that high-speed collisions resulting in more serious injury are more rare. Adams (1999) also notes the uniqueness of London's road user risk, as it presents the highest urban UK proportion of cyclist and pedestrian-related collisions. This presents a strong rationale for a broad societal risk analysis and evaluation.

Collision risk (or "collision proneness") has been a long established discipline within the road safety domain. Extensive work by academics, such as Wilde (1982), Hauer (1980), Janssen and Tenkink (1988), Adams (1995), and McKenna (1983), has illustrated the propensities for risk within the road environment. A collision has often been likened to a stochastic event; however, human participation leads to an elevated element of prediction and anticipation. The pioneering statistical contributions of Greenwood and Woods (1919), Greenwood and Yule (1920), Newold (1926, 1927), and Greenwood (1951) take this approach. The basic thesis was "when discarding chance is possible in the statistical data of accidents, then the human factor is the single cause of accidents" (Blasco et al. 2003, 482). In a similar vein, "some people have many more crashes than can be expected by chance, so these people are crash-prone" (Blasco et al. 2003, 482).

In the field of traffic research, continued emphasis has been placed on identifying factors that contribute to increased driving risk, with the goal of reducing the frequency and impact of traffic violations, collisions, and fatalities. Two major attempts that utilized the concept of risk to explain behavioral adaptation to changes in traffic systems were proposed in the 1970s: one by Näätänen and Summala (1974, 1976) and the other by Wilde (1976). The former proposed a threshold model on the assumption that in the dynamic driving situation, drivers actually control safety margins rather than some specific risk measure, and only when the risk or fear threshold is exceeded or expected to be exceeded, does it influence behavior. They postulated a "subjective risk" or "fear monitor" (Summala and Näätänen 1988) that alarms and influences driver decisions when safety-margin thresholds are violated. One aspect of this approach is that with repeated confrontations, drivers adapt to situations that at first elicited a "risk response" and drive most of the time with overlearned habitual patterns based on safety margins, with no concern for risk: hence the label "zero-risk theory" (Näätänen and Summala 1976; Summala 1986, 1996).

In his 1942 book *Why We Have Automobile Accidents?*, De Silva noted that "the degree of hazard to which a driver is subjected, or, expressed differently, the extent of his exposure, is determined by how much, and where, and when he drives" (De Silva 1942, 11). Thus, it is necessary to split exposure by type of road, visibility, weather conditions, time of day, and day of week and so on. In reducing collisions, we do not always need risk measures (collision rates), but it is practical to search for hot spots in the collision mass—accumulations of collisions that indicate where effort to save lives may be concentrated. The search for hot spots continues to be a major safety activity among road and traffic engineers. Although not always based on proper understanding of causes and at risk of directing resources on a random basis (e.g., Hauer 1986), cumulative efforts by road and traffic engineers together with improved design standards have gradually made this approach less and less useful.

Along with this "geographical" hot spot method, however, it is essential to continue searching for hot spots in the collision mass by disaggregating it into smaller units by type, road and traffic conditions, time of week, and, where possible, the characteristics of road users involved. Well-defined hot spots or peaks in the collision mass, which may be widely distributed geographically, are of practical and theoretical significance to collision prevention even without corresponding exposure measures. An example of a major hot spot that is widely distributed geographically is young male drivers' collisions at night on weekends. These hot spots necessarily reflect exposure: both of populations at risk and their behavior. For example, motivational (lifestyle) factors on weekend nights especially tend to get young male drivers into problems in contrast to older drivers who appear to avoid impaired conditions (e.g., night driving). The degree of hazard to which a driver is subjected is thus determined not only by how much, where, and when he or she drives, but also by how he or she drives—which is related to the level of control over potential dangers.

Figure 5.1 shows a risk typology in the shape of a Venn diagram, highlighting three different types of risk. Risks in the "perceived directly" circle are managed using judgment. We do not undertake a formal probabilistic risk assessment before crossing the road; some combinations of instinct, intuition, and experience see us to the other side of the road. In the second "risk perceived though science" circle, this dominates the risk management literature. The central science here is statistics, and this is where we find most of the published work on road safety. The circle labeled "virtual risk" contains contested hypotheses, ignorance, uncertainty, and unknown unknowns. It is an issue that cannot be settled by science and numbers. In a sense, some feel free to argue their beliefs, prejudices, and superstitions. Virtual risks may not be real, but people's beliefs about them have consequences. Road safety is a

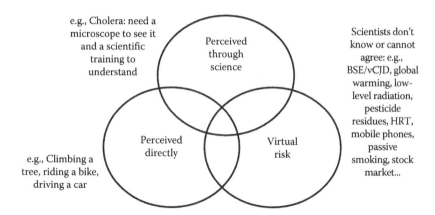

FIGURE 5.1 Different kinds of risk. (Reprinted from Adams, J., Risk, freedom and responsibility, in: *The Risk of Freedom: Individual Liberty and the Modern World*, Institute of United States Studies, London, U.K., 1999. With permission from School of Advanced Study, University of London.)

comprehensively and intensively studied subject. It is awash with numbers, statistics; however, most of the debate about road collisions can be consigned to virtual risk. After decades of road safety interventions, we are still uncertain about what works and what is best; however, one thing we do know is our own risk. At this point in road safety, we should not be uncertain about what works in terms of seat belts, air bags, engineering solutions, and Adams (1999) identifies the reason for this is because there are two different sets of risk managers. Adams outlines two managers, the first being "institutional risk managers" (legislators and regulators who enforce rules that govern transport safety); also in this group would include the engineers (highway and vehicle) concerned with making our roads safer and our vehicles safer. The second set of managers are what Adams identifies as the worldwide road users managed directly perceived risks through individual judgment. Academics use what we call a "risk thermostat" that shows a model of the risk management process (Figure 5.2).

Empirical research has identified a very large number of risk factors that are statistically associated with road collision occurrence, that is, factors whose presence increases the probability of collisions. In principle, one might try to *explain* road collisions by listing these factors, perhaps adding information on their relative importance. This would at best be only the beginning of a theory of collision causation. A list of risk factors can be informative and useful, but it begs more basic questions, like the following: Why is factor x a risk factor for collisions? Why does risk factor y appear to be more important in explaining collisions than risk factor x? What we need is, in other words, an account of mechanisms that explain why a certain factor becomes a risk factor.

Elvik (2006) outlines what he refers to as "laws" that govern the risks involved in collision causation. The number of risk factors that influences collisions is vast. Nobody can enumerate all these risk factors, yet their effects on collisions may display striking regularities. The ability of a road user to recognize risk factors and

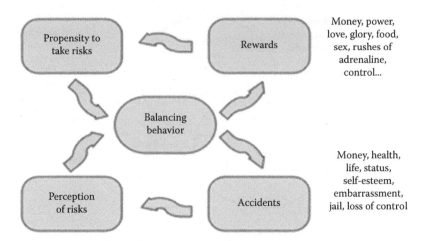

FIGURE 5.2 Risk thermostat. (Reprinted from Adams, J., Risk, freedom and responsibility, in: *The Risk of Freedom: Individual Liberty and the Modern World*, Institute of United States Studies, London, U.K., 1999. With permission from School of Advanced Study, University of London.)

prevent them from leading to collisions is likely to be strongly influenced by the experiences made when using the transport system.

- *The universal law of learning*: The ability to detect and control traffic hazards improves continuously as a result of exposure to these hazards.
- *The law of rare events*: The more rarely a certain traffic hazard is encountered, the greater is its effect on collision rate.
- *The law of complexity*: The more potentially relevant items of information a road user must attend to per unit of time, the higher the probability of collisions.
- *The law of cognitive capacity*: The more cognitive capacity approaches its limits, the greater the increase in the rate of collisions.

5.2 WHAT IS *RISK-TAKING* BEHAVIOR?

Risky driving behavior may include self-assertive driving, speeding, and rule violations. Speeding as a risky driving behavior has been studied by many researchers (Jonah 1997; Lam 2003; Aarts and Van Schagen 2006). Excessive driving speed for the road conditions is considered one of the most important contributors to road collisions, regardless of driver age and level of skill (Elliott et al. 2004). Even when aware of the potential consequences for speeding, drivers in Australia still indicate involvement in speeding behavior (Brown and Cotton 2003). Clarke et al. (2002) also suggested that speed was the most common factor involved in driving offence among young drivers. West and Hall (1997) found that speed was a significant contributor to specific kinds of collisions (i.e., active shunts, right-of-way violations, active reversing, and loss of control collisions) along with both (poor) attitudes toward driving and social deviance. McKenna and Horswill (2006) suggested that involvement in speeding behavior may also be due to a low probability of negative outcome. For example, individuals may consider involvement in a collision as less likely than being caught by the police.

In relation to driving behavior, risk perception refers to "the subjective experience of risk in potential traffic hazards" (Deery and Fildes 1999, 226). Therefore, risk perception is considered a precursor of actual driving behavior. Many researchers have indicated that risk perception is negatively related to risk behavior in general (Cohn et al. 1995). That is, a higher level of perceived risk for a particular behavior is associated with a lower chance that an individual would take part in that behavior. There is some controversy about the direction of effect between risk perception and driving behavior. Horvath and Zuckerman (1993) indicated that a sense of competence may increase with involvement in risk behavior that does not produce negative consequences, such as injury or penalty. In that sense, risk perception may be a consequence, not a cause of behavior.

Rundmo and Iversen (2004) suggested that most research has emphasized a cognitive or belief-based component of risk perception, which focuses on the way young drivers perceive and process information (see Horvath and Zuckerman 1993; Deery and Fildes 1999; Brown and Cotton 2003). However, when measuring perceived risk, Rundmo and Iversen (2004) considered it was important to distinguish between cognitive-based and affective-based subjective assessments. Affective components of risk perception such as worry and concern have also been found to be a predictor of risky driving behavior. However, McKenna and Horswill (2006) found that worry and concern appeared to have less influence than other variables (e.g., legal constraints, mood, passengers, journey time, economics, and thrill) and accounted for only 2% of the variance in both speeding and driving violations.

The role of personality in risk research remains unclear despite a plethora of related research (Iversen and Rundmo 2002). Although a great deal of research has considered the problem of traffic psychology (Signori and Bowman 1974; McGuire 1976; Golding 1983; Hansen 1988; Hilakivi et al. 1989; Arthur et al. 1991; Evans 1991b; Lester 1991; Elander et al. 1993; Peck 1993), the contribution of psychology to the traffic policies has been repeatedly neglected in some European countries such as Portugal (Santos et al. 1995). According to Manstead (1993), in the analysis of rule infringement, socio-cognitive variables such as attention (Theeuwes 1993), perception (Owsley et al. 1991; Manstead 1993), and judgment processes (Cavallo and Laurent 1988) should be considered. Additionally, a study with a sample of 1000 drivers revealed that 11 variables (3 perceptive and 8 psychomotor) were valid predictors of car collisions since they explained 85% of total variance of road collisions (Alves and Silva 1993). Besides, Özkan et al. (2006) revealed that safety skills (e.g., "conforming to the speed limits") and perceptual-motor skills (e.g., "fluent driving") are important predictors of the number of road collisions across different countries.

The WHO outlines the following risk factors for road traffic collisions:

1. Factors influencing exposure to risk. They include (a) economic factors such as level of economic development and social deprivation; (b) demographic factors such as age and sex; (c) land use planning practices that influence

length of trip and mode of travel; (d) a mixture of high-speed motorized traffic with vulnerable road users; and (e) insufficient attention to integration of road function with decisions about speed limits, road layout, and design.
2. Risk factors influencing collision involvement. They include (a) inappropriate and excessive speed; (b) presence of alcohol, medicinal, or recreational drugs; (c) fatigue; (d) being a young male; (e) having youths driving in the same car; (f) being a vulnerable road user in urban and residential areas; (g) traveling in darkness; (h) vehicle factors—such as braking, handling, and maintenance; (i) defects in road design, layout, and maintenance, which can also lead to unsafe behavior by road users; (j) inadequate visibility because of environmental factors (making it hard to detect vehicles and other road users); and (k) poor eyesight of road users.
3. Risk factors influencing collision severity. They are (a) human tolerance factors; (b) inappropriate or excessive speed; (c) seat belts and child restraints not used; (d) collision-helmets not worn by users of two-wheeled vehicles; (e) roadside objects not collision protective; (f) insufficient vehicle collision protection for occupants and for those hit by vehicles; and (g) presence of alcohol and other drugs.
4. Risk factors influencing post-collision outcome of injuries. They include (a) delay in detecting collision and in transport of those injured to a health facility; (b) presence of fire resulting from collision; (c) leakage of hazardous materials; (d) presence of alcohol and other drugs; (e) difficulty in rescuing and extracting people from vehicles; (f) difficulty in evacuating people from buses and coaches involved in collision; (g) lack of appropriate pre-hospital care; and (h) lack of appropriate care in hospital emergency rooms.

5.3 MEASURING RISKY BEHAVIOR

The question we should ask is this: can risk be measured? Lord Kelvin has been quoted as saying "Anything that exists, exists in some quantity and can therefore be measured" (Adams 1995, 10). Some people believe that risk is culturally constructed as opposed to risk being "objective risk" (Kelvinists). Britain's Department of Transport lies in the latter camp. It measures the safety of danger of a road by its injury record, the consequence of real collisions. It draws a clear line between actual danger and perceived danger. The department is willing to spend money to relieve only the actual danger. If a road does not have a fatality rate significantly above the "normal," it will not receive funds for measures to reduce the danger. The objective of collision analysis is collision prevention.

Measuring risky behavior is a complex matter. There is no uniform or decided measurement that is used. This section will attempt to highlight the complexities surrounding measuring risky behavior and the different methods. In his influential book *Risk* (Adams 1995), Professor John Adams focuses on the issue of measuring risk in relation to road collisions. When it comes to measuring risk, there are many factors that need to be taken into account. One of the major criticisms road safety academics face is ultimately the challenge of measuring risk. Any analysis of the

TABLE 5.2
Types of Behavioral Variables Related to Collision Risk

Variable	Skill/Ability	Style/Trait
Measures specific to driving	Advanced driver training	Speed choice
	Hazard perception	Traffic violations
		Gap acceptance
Extrinsic measures	Detection of embedded figures	Type A behavior
	Attention switching	Antisocial attitudes
	Visual acuity	

Source: Reprinted from Elander, J. et al., *Psychol. Bull.*, 113(2), 279, 1993. With permission.

causes of road collisions always leads to the conclusion that they are a stochastic or probabilistic phenomenon. As Adams (1995) refers, they result from a conjunction of circumstances to which the victim had, a priori, in most cases assigned a negligible probability. The number of fatal road collisions each year for the United Kingdom (and most Western countries) is low in comparison to other causes of death. Especially considering the fact that deaths from road collisions were spread over hundreds of thousands of kilometers of roads and millions of motor vehicles. Risk in this sense measures the connection between the potential of a fatal road collision occurring and an actual fatal road collision occurring (Table 5.2).

In road safety literature, we talk about risk a lot. There is collision risk, health risk, individual risk, and others. However, it is hard to break down these different risks and measure them in isolation. In this section, we refer to risky behavior, so one would interpret that as individual risk. However, a person's individual risk on the road (or in the road environment) may have nothing to do whether they are injured or killed in a road collision. A person driving a vehicle down a street may have tendencies of risky behavior, speeding or driving erratically. This person could in term hit person x cycling down the same street. Person x may have less risk-taking tendencies but will still be the victim of risky behavior.

We have discussed measuring risky behavior, but what about mapping risky behavior? It is important to note that risk maps based on collision rates do not show the extent to which the behavior of a specific road user might result in the risk being higher or lower than the average. They also do not show the extent to which the road user can make a mistake and recover from it without serious injury. What they do illustrate is the risk of an individual road user, or to the community as a whole, being involved in a road collision, providing that they are behaving within acceptable boundaries of road use, for example, not intoxicated, not using a mobile phone, and obeying speed limits.

Risk maps show essentially individual and collective risk. Individual risk is the rate at which people are being killed or injured. For example, over 3 years, 30 people are killed and seriously injured on a 30 km stretch of motorway carrying 100,000 vehicles a day, and 30 people are also killed and seriously injured on a 30 km stretch of single carriageway carrying 10,000 vehicles per day. It means that the risk to the individual is 10 times greater on the single carriageway. Collision rates per kilometer traveled on a road can show the likelihood of a particular type of road user being involved in a collision. Their main purpose is to inform the road user how and where their behavior needs to be modified to minimize risk and, in doing so, enable them to recognize the sources of risk on different types of road.

Collective risk maps show the density, or total number, of collisions on a road over a given length. Risk rates shown in these maps are the result of the interaction between all elements of the road system, that is, road users, vehicles, and roads. Community risk can be mapped in three different ways:

1. *Collision density*: showing collision rates per kilometer of road, illustrating where highest and lowest numbers of collisions occur within a network.
2. *Collision rate in relation to similar roads*: comparing the collision rate of similar roads with similar traffic flows, illustrating which road sections have a higher rate. Separate road groups are considered, for example, motorways, main roads with traffic flows below 10,000 vehicles per day, main roads with daily traffic flow between 10,000 and 20,000 vehicles per day, and main roads with daily traffic flow 20,000 vehicles per day.
3. *Potential for collision reduction*: providing information on the number of collisions that might be saved if collision rates of road sections, with risk above the average roads of a similar flow, were reduced to the average. This information can be used for considering investment decisions, providing authorities and policy makers with a valuable tool for estimating the total number of collisions that could potentially be avoided if safety on a road were improved. Used with cost information, this map can indicate locations where the largest return on investment can be expected.

5.4 AGE AND GENDER DIFFERENCES

The relationship between the age and gender of road users and their collision risk has been studied extensively. One of the major reasons for this is that detailed information on age and gender is readily available from collision records, whereas many human factors are not recorded in official collision statistics. Elvik et al. (2009) presents findings from a combination of studies referring to the age and gender of car drivers (Figure 5.3). Research has shown that both the young (under 18) and the elderly (over 65) are at greater "risk" of being in a road traffic collision as a pedestrian. As we have cited in previous chapters, the proportion of young/elderly and male/female involved in road collisions varies considerably. Evidence from the National Health Service (NHS) in the United Kingdom has shown that males between the age of 15 and 19 are of particular concern. In the United Kingdom in 2010, 29% of the total 5605 killed or seriously injured on the roads were under 16. Perhaps more worryingly, 58% of the moped riders killed or seriously injured were under 17 years old.

Deaths from road traffic collisions are much more prevalent among the under 25-year olds than other causes of death that often reported by the media such as hangings, shootings, stabbings, alcohol, or drug abuse. Between the ages of

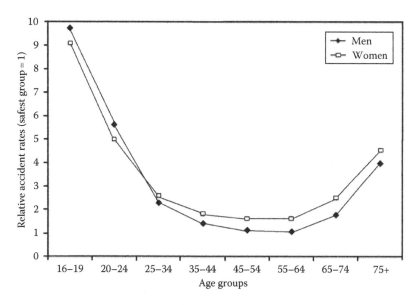

FIGURE 5.3 Relative rates of involvement in injury collisions by driver age and gender. (Reprinted from Elvik, R. et al., *The Handbook of Road Safety Measures*, 2nd edn., Emerald Group Publishing, Bingley, U.K., 2009, p. 63. With permission from Emerald Group Publishing Limited.)

15 and 24, a young person is twice more likely to die from a road traffic collision than be fatally assaulted by firearms, a sharp/blunt object, or intentional self-harm via hanging combined. Those in the 15–24 age category are also four times more likely to die from a road traffic collision than from drug, alcohol, or other substance poisoning.

Different groups of people have different exposures to risk. As populations change over time, so their overall exposure will change. Fluctuations in the relative sizes of different population groups will have a strong effect on the road traffic toll. For instance, in high-income countries, young drivers and riders are currently overrepresented in casualty figures. Demographic changes in these countries over the next 20–30 years, however, will result in road users over 65 years of age becoming the largest group of road users. The physical vulnerability of older people places them at high risk for fatal and serious injuries. Despite the rising number of older people holding driving licenses in high-income countries, their declining driving ability as well as possible financial constraints will mean that many of them will have to give up driving. This may differ from many low-income countries where older people may never have driven in the first place. In low-income countries in general, the expected demographic evolution suggests that younger road users will continue to be the predominant group involved in road traffic collisions. Worldwide, a large proportion of older people will be dependent on public transport or will walk. This illustrates the importance of providing safe and short pedestrian routes, and safe and convenient public transport.

5.5 CULTURE AND ETHNICITY

What Adams (1995) goes on to think about is the changing risk and therefore exposure, and how it differs with regard to cultural influences. Adams (1995) discusses the idea of a "risk thermostat" by which everyone perceives their own risk, and how this varies between groups of people, cultures, and obviously over time. These cultural filters select and construe evidence to support established biases. There are some threats on which all cultures can agree, for example, all drivers slow down when they come to a sharp bend in the road, and they are in general agreement about the nature of the risk. However, some subcultures would slow down more than others. Adams (1995) argues that this filtering process works both directly (through our five senses) and indirectly (through reactions to stories, news reports, statistics, and research). These are in effect pre-filters, by which no government or research institution can gather even a fraction of this evidence. Adams (1995) summarizes the road safety policy of the Western world (including the United Kingdom) as having just two aims: making motoring safer for motorists and getting everyone else safely out of the way.

Over many decades, research, policy, legislation, education, and highway engineering have all focused strongly on the safety of people in vehicles, to the neglect of welfare of safety of vulnerable road users (on foot and bicycle). These safety measures have created vehicles that are safer to have collisions in and road environments more forgiving of reckless driving. The measures that have been adopted in the interests of the safety of pedestrians usually take the form of movement-restricting barriers, which make people on foot travel further through the use of footbridges and tunnels. For road safety education, we tell children nothing of their rights as road users but told to fear the road environment and be at the mercy of vehicles (Adams 1995). Everything in terms of road safety favors the vehicle.

Adams (1995) interestingly comments that if road safety has the twin aims of making motoring safer for motorists and getting everyone else safely out of the way, there is little need for detailed research into the activities of those who are to be displaced and consequently little is found. There is little research into focusing on the "risky behavior" of pedestrians or cyclists. Politically, they are seen as vulnerable road users. However, we know from research that both pedestrians and cyclists can be as much to blame in a road collision due to risky behavior as motorists.

What is also important and not studied in greater depth is the interpretation of different cultures to the risks on the road they face. Can people understand the government's statistics of "risk to car occupants per vehicle-mile traveled"? Different people will respond to road safety measures in different ways, and this ultimately has not been explored. This research is only one cog in the machine of trying to investigate this phenomenon further. By understanding that different people have different levels of exposure to risk, we can start to think about how this might be managed. Therefore, an overrepresented level of risk exposure can be expected, but what we are interested in here is the level of exposure. No one knows how risk varies according to subgroups in society, because detailed data on activity patterns are not collected for extensive and representative samples of people. This study presents itself as a valuable exploratory investigation into the "what is" analysis of variation in risk. Knowing they are likely to be overrepresented is one thing, but understanding the spatial pattern is also important.

5.6 DRINK-DRIVING

The effects of alcohol on risk of road collisions mean that any driver and motorcyclist with a blood alcohol content higher than zero is at a higher risk of being in a road collision. For the general population, as the blood alcohol level continues to rise, so does the risk of being involved in a road collision. Inexperienced young adults driving with a blood alcohol content of 0.05 g/dL have 2.5 times the risk of a collision compared with more experienced drivers (Mathijssen 1998). If a blood alcohol content limit is fixed at 0.10 g/dL, this will result in three times the risk of a collision than that at 0.05 g/dL, which is the most common limit in high-income countries (Peden et al. 2004). If the legal limit stands at 0.08 g/dL, there will still be twice the risk than at 0.05 g/dL (Peden et al. 2004). The risk of a road collision when a driver is alcohol impaired varies with age. Teenagers are significantly more likely to be involved in a fatal collision than older drivers. At almost every blood alcohol level, the risk of collision fatality decreases with increasing driver age and experience. Teenage drivers who are alcohol impaired are at increased risk of having a road collision if they have passengers in the vehicle, as compared with those driving alone. A low expectation of getting caught with a blood alcohol content above the legal limit has been shown to lead to an increased risk of a collision.

The risk factor status of a unit of risk is often measured at two levels "exposed" or "not exposed." However, it is possible to have both ordinal and categorical outcomes of the risk status. For example, when analyzing "trip purpose," you can have, that is, leisure, work, shopping, and so on, all of them are measured equally. Risk factors can also be a continuous variable such as age. However, when you group the age, it can often lead to a loss of information. From a conceptual and technical point of view, many risk factors have the common feature that they can be measured in several different ways. For example, the trip- and driver-related characteristic of driving under the influence of alcohol may be measured precisely by a blood alcohol test, or even more simply asking the driver whether he/she has consumed alcohol prior to the trip. Frequently, the possibilities of assessing risk factors are limited due to the nature of the data recorded. With regard to alcohol, the data often available are only for those associated with collisions that have occurred (and police have been involved). It is very difficult to assess the general risk associated with alcohol unless it occurs during seasonal or directed campaigns (where police stop every driver for a breath test). Randomized breath tests are a common procedure in many Western countries and Hong Kong with police focused on seasonal periods and often Friday and Saturday nights near large urban centers.

When the breathalyzer was introduced in 1967 in the United Kingdom, there was (unsurprisingly) a drop in the number road collision fatalities. However, it was noted that the effect was only temporary (Adams 1985). It has been tricky over the years to determine the effect of the breathalyzer and drink-drive laws on road collision statistics. Many of the analytical research papers focus on small-scale studies involving questionnaires or previous collision data where alcohol was a component of the collision. Epidemiological data from a landmark study conducted by Borkenstein et al. (1964) have been used in the assessment of collision risk due to alcohol for almost four decades. Since those data were collected, however, driving and drinking environments have changed, and the changes possibly have altered the risk of a collision. In a sense here, we are talking about alcohol as a risk component to a road collision. In terms of post-collision management, one can look back at the road collision data. From this, it is important to understand who drinks and drives, and where they drink. By understanding the social and locational dimensions, it becomes easier to pinpoint countermeasures to deal with drink-driving. We know from the literature that certain people are more likely to drink and drive; however, what are the spatial patterns of this? In countries such as New Zealand and Australia, there are vast media campaigns surrounding drinking and driving directed specifically at young male adults. While this is successful direct targeting, data from Western countries such as the United Kingdom and Australia show that there are an increasing number of older adults (50–60) who are drinking and driving. Although this represents a smaller proportion of the road collision fatalities, it is a growing trend.

5.7 DRUG-DRIVING

While the literature on road collisions and alcohol is extensive, there are less data and information on the prevalence and patterns of the use of drugs and road collisions. One of the main issues surrounding drugs and road collisions is the data. There is no requirement by the police (unless they see fit) to test for drugs at a road collision. If the driver has been hospitalized, there may be an option for drug testing; however, it is not mandatory. In a recent paper by Elvik (2013), he conducts a meta-study of the risk of road collisions associated with the use of drugs. In his study, summary estimates of the odds ratio of collision involvement are presented for amphetamines, analgesics, antiasthmatics, antidepressives, antihistamines, benzodiazepines, cannabis, cocaine, opiates, penicillin, and zopiclone (a sleeping pill). For most of the drugs, a small or moderate increase in road collision risk is associated with the use of the drugs. Information about whether the drugs were actually used while driving and about the doses used was often imprecise. Most studies that have evaluated the presence of a dose–response relationship between the dose of drugs taken and the effects on collision risk have confirmed the existence of a dose–response relationship. Use of drugs while driving tends to have a larger effect on the risk of fatal and serious injury collisions than on the risk of less serious collisions. He noted that the quality of the studies that have assessed risk varied greatly. There was a tendency for the estimated effects of drug use on collision risk to be smaller in well-controlled studies than in poorly controlled studies. Evidence of publication bias was found for some drugs. The associations found cannot be interpreted as causal relationships, principally because most studies do not control very well for potentially confounding factors.

5.8 CONCLUSION

In this chapter, we are concerned with measuring the human element in road collisions. Most of the book focuses on statistics and geography; however, in terms of measuring risk-taking behavior, the process becomes much more multifaceted. It is claimed that over 90% of road collisions are due to driver error (Elander et al. 1993). Everyone responds to risk in different ways (something we will go on to discuss in the next two sections); however, measuring this risk-taking behavior requires drawing from different disciplines, including psychology and sociology, to understand factors associated with people behavior in a road environment.

REFERENCES

Aarts, L. and I. Van Schagen. 2006. Driving speed and the risk of road crashes: A review. *Accident Analysis & Prevention* 38 (2): 215–224.

Adams, J. 1985. *Risk and Freedom: The Record of Road Safety Regulation.* London, U.K.: Transport Publishing Projects.

Adams, J. 1995. *Risk.* London, U.K.: Routledge.

Adams, J. 1999. Risk, freedom and responsibility. In *The risk of freedom: Individual liberty and the modern world*, ed. University of London, pp. 33–58. London: Institute of United States Studies.

Agran, P. F., D. G. Winn, C. L. Anderson, C. Tran, and C. P. Del Valle. 1996. The role of the physical and traffic environment in child pedestrian injuries. *Pediatrics* 98 (6): 1096–1103.

Alves, J. A. and J. Silva. 1993. Avaliação psicológica e prognóstico de desempenho. In *Factores humanos ao tráfego rodoviário*, ed. J. A. Santos, pp. 227–237. Lisboa, Portugal: Escher.

Arthur, W., G. V. Barret, and R. A. Alexander. 1991. Prediction of vehicular accident involvement: A meta-analysis. *Human Performance* 4 (2): 89–105.

Beck, U. 1992. *Risk Society: Towards a New Modernity.* London, U.K.: Sage Publications Limited.

Bernard, A., Y. Bussière, and J. P. Thouez. 2001. Analyse comparative des accidents de la route survenus durant l'hiver et l'été au Québec, 1989–1996. *Recherche Transports Sécurité* 73: 31–41.

Blasco, R. D., J. M. Prieto, and J. M. Cornejo. 2003. Accident probability after accident occurrence. *Safety Science* 41 (6): 481–501.

Borkenstein, R. F., R. Crowther, R. Shumate, W. Ziel, and R. Zylman. 1964. The Role of the Drinking Driver in Traffic Accidents. Bloomington, IN: Department of Police Administration, Indiana University.

Brenac, T., J. Delcamp, S. Pelat, and G. Teisseire. 1996. Scénarios types d'accidents de la circulation dans le département des Bouches du Rhône. Institut National de Recherche sur les Transports et leur Sécurité, Arcueil, France, *Rapport MA* 9611–9612.

Brenac, T. and B. Megherbi. 1996. Diagnostic de sécurité routière sur une ville: intérêt de l'analyse fine de procédures d'accidents tirées aléatoirement. *Recherche Transports Sécurité* 52: 59–71.

Brown, S. L. and A. Cotton. 2003. Risk-mitigating beliefs, risk estimates, and self-reported speeding in a sample of Australian drivers. *Journal of Safety Research* 34 (2): 183–188.

Cavallo, V. and M. Laurent. 1988. Visual information and skill level in time-to-collision estimation. *Perception* 17 (5): 623–632.

Center for Chemical Process Safety (CCPS). 1995. *Guidelines for Chemical Transportation Risk Analysis.* New York: American Institute of Chemical Engineers.

Chapman, C. and S. Ward. 2003. *Project Risk Management: Processes, Techniques and Insights.* Chichester, U.K.: John Wiley & Sons.

Clarke, D. D., P. Ward, and W. Truman. 2002. In-depth accident causation study of young drivers. TRL Report 542. Transport Research Laboratory, Crowthorne, U.K.

Cohn, L. D., S. Macfarlane, C. Yanez, and W. K. Imai. 1995. Risk-perception: Differences between adolescents and adults. *Health Psychology* 14 (3): 217–222.

Cowley, J. E. and K. T. Solomon. 1976. An overview of the pedestrian accident situation in Australia. In *Proceedings of International Conference on Pedestrian Safety*, Haifa, Israel.

Delfino, G., C. Rindone, F. Russo, and A. Vitetta. 2005. Risk analysis in road safety: A model and an experimental application for pedestrians. In *Proceedings of European Transport Conference*, Strasbourg, France. Glasgow, U.K.: Association for European Transport.

Deery, H. A. and B. N. Fildes. 1999. Young novice driver subtypes: Relationship to high-risk behavior, traffic accident record, and simulator driving performance. *Human Factors: The Journal of the Human Factors and Ergonomics Society* 41 (4): 628–643.

De Silva, H. R. 1942. *Why We Have Automobile Accidents*. New York: John Wiley & Sons.

Elander, J., R. West, and D. French. 1993. Behavioral correlates of individual differences in road-traffic crash risk: An examination of methods and findings. *Psychological Bulletin* 113 (2): 279–294.

Elliott, M. A., C. J. Armitage, and C. J. Baughan. 2004. Applications of the theory of planned behaviour to drivers' speeding behaviour. In *Proceedings of Behavioural Research in Road Safety 2004: Fourteenth Seminar*, pp. 157–169. London, U.K.: Department of Transport.

Elvik, R. 2006. Laws of accident causation. *Accident Analysis & Prevention* 38 (4): 742–747.

Elvik, R. 2013. Risk of road accident associated with the use of drugs: A systematic review and meta-analysis of evidence from epidemiological studies. *Accident Analysis & Prevention* 60: 254–267.

Elvik, R., A. Høye, V. Truls, and M. Sørensen, eds. 2009. *The Handbook of Road Safety Measures*, 2nd edn. Bingley, U.K.: Emerald Group Publishing Limited.

Eurotunnel. 1994. *The Channel Tunnel: A Safety Case*. Folkestone, U.K.

Evans, A. W. 1994. Evaluating public transport and road safety measures. *Accident Analysis & Prevention* 26 (4): 411–428.

Evans, L. 1991a. Factors influencing pedestrian and motorcyclist fatality risk. *Journal of Traffic Medicine* 19 (2): 69–73.

Evans, L. 1991b. *Traffic Safety and the Driver*. New York: Van Nostrand Reinhold.

Fleury, D. and T. Brenac. 2001. Accident prototypical scenarios, a tool for road safety research and diagnostic studies. *Accident Analysis & Prevention* 33 (2): 267–276.

Fuller, R. 2005. Towards a general theory of driver behaviour. *Accident Analysis & Prevention* 37 (3): 461–472.

Furedi, F. 1997. Conflict of values and the consciousness of risk. Paper presented at *the 1997 Annual Meeting of the Society for Risk Analysis – Europe*, Stockholm.

Golding, A. P. 1983. *Differential Accident Involvement: A Literature Survey*. Johannesburg, Africa: Council for Scientific and Industrial Research.

Grayson, G. B. 1979. Methodological issues in the study of pedestrian behaviour. University of Nottingham, Nottingham, U.K.

Grayson, G. B. and C. I. Howarth. 1982. *Evaluating Pedestrian Safety Programmes*. Chichester, U.K.: John Wiley & Sons.

Grayson, G. B., G. Maycock, J. A. Groeger, S. M. Hammond, and D.T. Field. 2003. Risk, hazard perception, and perceived control. Transport Research Laboratory, Crowthorne, U.K.

Greenwood, M. 1951. Accident proneness. *Biometrika* 37 (1): 24–29.

Greenwood, M. and H. M. Woods. 1919. The incidence of industrial accidents upon individuals with special reference to multiple accidents. Industrial Fatigue Research Board, London, U.K.

Greenwood, M. and G. U. Yule. 1920. An inquiry into the nature of frequency distributions representative of multiple happenings with particular reference to the occurrence of multiple attacks of disease or of repeated accidents. *Journal of the Royal Statistical Society* 83 (2): 255–279.

Hansen, C. P. 1988. Personality characteristics of the accident involved employee. *Journal of Business and Psychology* 2 (4): 346–365.

Hauer, E. 1980. Traffic conflicts and exposure. In *Proceedings of International Symposium on Risk-Exposure Measurement*, Aarhus, Denmark.

Hauer, E. 1986. On the estimation of the expected number of accidents. *Accident Analysis & Prevention* 18 (1): 1–12.

Haight, F. A. 1986. Risk, especially risk of traffic accident. *Accident Analysis & Prevention* 18 (5): 359–366.

Hilakivi, I., J. Veilahti, P. Asplund, J. Sinivuo, L. Laitinen, and K. Koskenvuo. 1989. A sixteen-factor personality test for predicting automobile driving accidents of young drivers. *Accident Analysis & Prevention* 21 (5): 413–418.

Hillman, M., J. Adams, and J. Whitelegg. 1990. *One False Move...: A Study of Children's Independent Mobility*. London, U.K.: Policy Studies Institute.

Horvath, P. and M. Zuckerman. 1993. Sensation seeking, risk appraisal, and risky behavior. *Personality and Individual Differences* 14 (1): 41–52.

Iversen, H. and T. Rundmo. 2002. Personality, risky driving and accident involvement among Norwegian drivers. *Personality and Individual Differences* 33 (8): 1251–1263.

Janssen, W. and E. Tenkink. 1988. Risk homeostasis theory and its critics: Time for an agreement. *Ergonomics* 31 (4): 429–433.

Jonah, B. A. 1997. Sensation seeking and risky driving: A review and synthesis of the literature. *Accident Analysis & Prevention* 29 (5): 651–665.

Keall, M. D. 1995. Pedestrian exposure to risk of road accident in New Zealand. *Accident Analysis & Prevention* 27 (5): 729–740.

Lam, L. T. 2003. Factors associated with young drivers' car crash injury: Comparisons among learner, provisional, and full licensees. *Accident Analysis & Prevention* 35 (6): 913–920.

Lester, J. 1991. Individual differences in accident liability: A review of the literature. Transport and Road Research Laboratory (TRRL), Crowthorne, U.K.

Lonero, L. P. 1998. Risk mentality: Why drivers take the risks they do (and how can we help?). Paper presented at the *8th World Traffic Safety Symposium*, New York.

Lowrance, W. W. 1976. *Acceptable Risk*. Los Altos, CA: William Kaufmann.

Macpherson, A., I. Roberts, and I. B. Pless. 1998. Children's exposure to traffic and pedestrian injuries. *American Journal of Public Health* 88 (12): 1840–1843.

Manstead, A. S. 1993. Social psychological factors in drive behavior. In *Factores humanos ao tráfego rodoviário*, ed. J. A. Santos, pp. 69–90. Lisboa, Portugal: Escher.

Mathijssen, M. P. M. 1998. Rijden onder invloed in Nederland, 1996–1997: Ontwikkeling van het alcoholgebruik van automobilisten in weekendnachten [Drink-driving in the Netherlands, 1996–1997]. Report number R-98-37. Leidschendam, the Netherlands: SWOV Institute for Road Safety Research.

McGuire, F. L. 1976. Personality factors in highway accidents. *Human Factors: The Journal of the Human Factors and Ergonomics Society* 18 (5): 433–441.

McKenna, F. P. 1983. Accident proneness: A conceptual analysis. *Accident Analysis & Prevention* 15 (1): 65–71.

McKenna, F. P. and M. S. Horswill. 2006. Risk taking from the participant's perspective: The case of driving and accident risk. *Health Psychology* 25 (2): 163–170.

Näätänen, R. and H. Summala. 1974. A model for the role of motivational factors in drivers' decision-making. *Accident Analysis & Prevention* 6 (3): 243–261.

Näätänen, R. and H. Summala. 1976. *Road-User Behaviour and Traffic Accidents*. Amsterdam, the Netherlands: North-Holland Publishing Company.

Newold, E. M. 1926. A contribution to the study of the human factors in the causation of accidents. Industrial Fatigue Research Board, London, U.K.

Newold, E. M. 1927. Practical applications of the statistics of repeated events' particularly to industrial accidents. *Journal of the Royal Statistical Society* 90 (3): 487–547.

Nilsen, T., O. T. Gudmestad, J. I. Dalane, W. K. Rettedal, and T. Aven. 1998. Utilisation of principles from structural reliability in quantitative risk analysis: Example from an offshore transport problem. *Reliability Engineering & System Safety* 61 (1): 127–137.

Owsley, C., K. Ball, M. E. Sloane, D. L. Roenker, and J. R. Bruni. 1991. Visual/cognitive correlates of vehicle accidents in older drivers. *Psychology and Aging* 6 (3): 403–415.

Özkan, T., T. Lajunen, J. E. Chliaoutakis, D. Parker, and H. Summala. 2006. Cross-cultural differences in driving skills: A comparison of six countries. *Accident Analysis & Prevention* 38 (5): 1011–1018.

Peck, R. C. 1993. The identification of multiple accident correlates in high risk drivers with specific emphasis on the role of age, experience and prior traffic violation frequency. *Alcohol, Drugs & Driving* 9 (3–4): 145–166.

Peden, M., R. Scurfield, D. Sleet, D. Mohan, A. Adnan, E. Jarawan, and C. Mathers, eds. 2004. World report on road traffic injury prevention. Geneva, Switzerland: World Health Organization.

Pedersen, T. O., E. Elvik, and K. T. Bérard-Andersen. 1982. *Oversikt over virkninger, kostnader og offentlige ansvarsforhold for 73 trafikksikkerhetstiltak*. Oslo, Norway: Transportøkonomisk institutt.

Posner, J. C., E. Liao, F. K. Winston, A. Cnaan, K. N. Shaw, and D. R. Durbin. 2002. Exposure to traffic among urban children injured as pedestrians. *Injury Prevention & Analysis* 8 (3): 231–235.

Rescher, N. 1983. *Risk: A Philosophical Introduction to the Theory of Risk Evaluation and Management*. Washington, DC: University Press of America.

Roberts, I. 1993. Why have child pedestrian death rates fallen? *British Medical Journal* 306 (6894): 1737–1739.

Roberts, I. 1995. Methodologic issues in injury case-control studies. *Injury Prevention* 1: 45–48.

Roberts, I., R. Marshall, and T. Lee-Joe. 1995. The urban traffic environment and the risk of child pedestrian injury: A case-crossover approach. *Epidemiology* 6 (2): 169–171.

Rundmo, T. and H. Iversen. 2004. Risk perception and driving behaviour among adolescents in two Norwegian counties before and after a traffic safety campaign. *Safety Science* 42 (1): 1–21.

Russo, F. and A. Vitetta. 2003. Disaggregate road accident analysis for safety policy and measures: Theoretical aspects and application. In *Proceedings of European Transport Conference*, Strasbourg, France. Glasgow, U.K.: Association for European Transport.

Russo, F. and A. Vitetta. 2004. Models for evacuation analysis of an urban transportation system in emergency conditions. In *Proceedings of European Transport Conference*, Strasbourg, France. Glasgow, U.K.: Association for European Transport.

Santos, J. A., M. F. Correia, J. F. Gomes, P. Z. Caldeira, and M. P. Cunha. 1995. Investigação em psicologia do tráfego e transportes em Portugal (1984–1994). *Análise psicológica* 3 (13): 243–248.

Signori, E. I. and R. G. Bowman. 1974. On the study of personality factors in research on driving behavior. *Perceptual and Motor Skills* 38 (3c): 1067–1076.

Standish, W. 2003. Categorising risk in road safety. *Journal of Safety Science* 23: 987–999.

Stewart, M. G. and R. E. Melchers. 1997. *Probabilistic Risk Assessment of Engineering Systems*. London, U.K.: Chapman & Hall.

Summala, H. 1986. Risk control is not risk adjustment: The zero-risk theory of driver behavior and its implications. University of Helsinki Traffic Research Unit, Helsinki, Finland.

Summala, H. 1996. Accident risk and driver behaviour. *Safety Science* 22 (1): 103–117.
Summala, H. and R. Näätänen. 1988. The zero-risk theory and overtaking decisions. In *Road Use Behaviour: Theory and Practice*, eds. J. A. Rothengatter and R. A. de Bruin, pp. 82–92. Assen, the Netherlands: Van Ggorcum & Comp BV.
Theeuwes, J. 1993. Visual attention and driver behavior. In *Factores humanos ao tráfego rodoviário*, ed. J. A. Santos, 103–23. Lisboa, Portugal: Escher.
Think Road Safety. 2005. THINK! Road Safety website, UK Department for Transport. http://www.thinkroadsafety.gov.uk/ (accessed December 25, 2005).
Tsai, M. C. 1998. Delay and risk at automatic level crossings in Britain. *Traffic Engineering & Control* 39 (9): 492–498.
Tobey, H. N., E. M. Shunaman, and R. L. Knoblauch. 1983. Pedestrian trip making characteristics and exposure measures. Federal Highway Administration, Office of Safety and Traffic Operations Research and Development, Washington, DC.
Vaganay, M., H. Harvey, and A. R. Woodside. 2003. Child pedestrian traffic exposure and road behaviour. In *Proceedings of European Transport Conference*, Strasbourg, France. Glasgow, U.K.: Association for European Transport.
Vlek, C. and P. J. Stallen. 1980. Rational and personal aspects of risk. *Acta Psychologica* 45 (1): 273–300.
West, R. and J. Hall. 1997. The role of personality and attitudes in traffic accident risk. *Applied Psychology* 46 (3): 253–264.
Wilde, G. J. S. 1976. Social interaction patterns in driver behavior: An introductory review. *Human Factors: The Journal of the Human Factors and Ergonomics Society* 18 (5): 477–492.
Wilde, G. J. S. 1982. Critical issues in risk homeostasis theory. *Risk Analysis* 2 (4): 249–258.

6 Road Collisions and Urban Development

6.1 URBAN LANDSCAPE AND ROAD SAFETY

Interactions between land use and urban development and transport decisions play a huge role in a person's propensity to be involved in a collision, depending on where they live and interact with the built environment. Many studies have explored the combined effect of roadway geometries and environmental factors on road safety. The spatial environment can be understood in many ways. For example, land use was considered by Petch et al. (2000), Ivan et al. (2000), and Ossenbruggen et al. (2001). Land use, infrastructure, and transport networks play a significant role in determining road user risk. Its changing role and dynamic within a city has been discussed in detail by Batty and Longley (1994). The growth of a city outward will almost never be exactly concentric and even, cities usually organized into neighborhoods enough to support educational and retail functions (Batty and Longley 1994). These patterns in turn will affect the propensity and location of road collisions. The significant environmental and spatial factors that relate to the changing city attributes with distance from a city center include changing land use and changing road network (usage and density) (Anas et al. 1998). Every motorized city's land use and infrastructure is unique with respect to urban transport history and growth, which produces an agglomeration whose road network, land use, and city center have experience continued growth and change.

There has been a strong interest in the relationship between road collisions and the characteristics of roads and local environmental conditions. There have been many studies that have examined the relations between vehicle collisions and roadway geometrics (Agent et al. 1975; Zegeer et al. 1990; Ivan et al. 1999; Martin 2002). What is being considered here, however, is the nature of the nearby environment of the roads. This can be analyzed in a number of ways. Hamerslag et al. (1982) studied the location of bus stops, parking bays, and so on, whereas Henning-Hager (1986) examined the relationship between residential development and road safety, and Abdalla et al. (1997) researched the relationship between road collisions and the effects of areal characteristics (which has been discussed in Chapter 4).

Urban environments, although rich with many unique resources and opportunities, are often "hot spots" for pedestrian collisions. The nature of urban design contributes to highly condensed and heavily motorized areas, as they are usually the business/commercial centers of the surrounding area, as well as hubs for entertainment and residence. Daily trips in an urban environment will take place over an ever-expanding perimeter covering several neighborhoods and even small towns (Millot 2004a). Interactions define the urban area; with this interaction and traffic function of urban roads, there can be conflicting road uses. The research literature shows there has been limited investigation into the effects of urban growth and road safety; however, this will be explored later in the chapter. Planning decisions regarding land use and the road network have a significant impact on the volume and nature of road collisions. Without strategic land use planning practices, residential, commercial, and industrial land use will evolve ad hoc as will the road network to meet its needs. The effects of this lack of planning can and would result in inappropriate interaction between road users and land use. For example, there would be increased volumes and speeding traffic through residential areas, increasing the risk of road collisions, due to the increased potential for interaction with pedestrians. Urban areas since the 1970s have been experiencing a movement or migration of residents from the inner districts to the suburbs (suburbanization). Socioeconomic changes have brought about an increase in out-of-town shopping centers and shopping "belts"; for example, Brent Cross and Bluewater in the United Kingdom have led to a decrease in local shops and the notion of the "high street." This in turn has increased traffic volume on roads surrounding these shopping centers, with the predominant form of transport to get there being the private car. This increases the risk of collisions due to the high volume of traffic and minimum public transport. Rumar (1999) has summarized that the following strategies can reduce exposure to road collision risk in a planning and land use environment focusing on the road network:

- Reducing the volume of traffic flow of motor vehicle traffic by means of better land use
- Providing efficient networks, where the shortest or quickest routes coincide with the safest routes
- Encouraging people to switch from higher risk to lower risk modes of transport
- Placing restrictions on motor vehicle users within the road infrastructure

6.2 CHANGING URBAN POPULATION AND ROAD COLLISIONS

In Western motorized countries, since the 1960s, we have witnessed rapid urbanization. However, coupled with this is counter-urbanization or urban–rural migration. The emergence of "suburbia" has been fuelled by the growth of the private car. This low-density "out-of-town" housing has led to an increase in road collisions. It goes without saying that lower traffic density, higher speed limits, and low-density housing means a greater risk for road collisions. One must not assume that counter-urbanization is the same as suburbanization. The two are closely similar; however, counter-urbanization implies a movement to a more rural location rather than suburban. Both phenomena are subject to higher road collision risk. The development of these decentralized cities or towns as well as other significant trends in society and the economy has led to increased use of the car. Coupled with this, the length and number of commuting trips has also increased. Many countries and cities have led to formulate transport policies that focus on public transport, and less use of the car. There are few research papers that address the changing nature of the urban population and road collisions. To say that globally we are becoming a more urbanized world is clear; however, there are clear linkages between the "speed" of urbanization and road collisions. The significant changing urban population in countries such as China highlight that not only urbanization alone increases road collisions but the "speed" also. In terms of numbers, currently approximately 3.3 billion people live in cities, which is just over half of the global population (United Nations Population Fund 2007). By 2030, over 60% will live in cities, compared to only 2% and 30% in 1800 and 1950, respectively. The most rapid changing urban population is occurring in Africa and Asia. Traffic increases, leading to more congestion and more road collisions. According to the WHO, 1.2 million people die and as many as 50 million are injured in urban traffic collisions in developing countries each year. Victims are mostly poor pedestrians and bicyclists. Those who survive are often left disabled. For example, in Bangladesh, it is reported that nearly 50% of hospital beds are occupied by road collision victims.

6.3 URBAN SPRAWL

Many people move to the suburbs to escape the "ills of the city." They move out of the city to get closer to the country air, to have more space, or to get away from the noise and congestion of the city. While suburban life has benefits, a growing body of evidence suggests there are significant public health costs of spread-out urban development, often called "urban sprawl." Injury and death caused by traffic collisions is one of the effects of sprawling, for car-dependent communities. Spread-out suburban communities make car travel the fastest, most convenient way to get around. The often long distances separating suburban homes from workplaces/recreational facilities and schools mean that people spend a significant amount of time each day on busy roads. The more hours people spend driving or riding in cars, the more likely they will be injured or killed in a car collision.

Urban sprawl is poorly planned development characterized by low-density, car-dependent communities, typically built on the outskirts of an urban area. People living in sprawling communities are often too spread out to make public transportation convenient or effective. That means people depend on their cars to get around—everything from getting to work or school, running errands, or going shopping. With greater dependence on cars comes an increased risk of death or injury on the roads. One of the outcomes of suburban expansion is the necessity of a much greater private vehicle use than is needed by urban dwellers, which leads to an increase in death and injury on the roads. Road collisions represent a unique social phenomenon. It is something that most people are aware of, but of which few have a realistic perception. Road fatalities and injuries are an inevitable consequence of the increased annual distance people must drive, the multiplicity of separate trips to engage in different activities, and the dynamics of suburban commuting such as multiple lanes, high speeds, tractor trailers, conflicting and competing needs of different drivers, multiple access points and exits on both sides of the road, distracting advertisements and signs, parked cars, and, of course, pedestrians. This is a consequence of the mobility of today's society.

In the U.S., there are over 40,000 deaths per year with 3.4 million injuries (Centers for Disease Control and Prevention 1999; NHTSA 2000). Although road collisions per mile driven have decreased over the years due to the physical improvements previously mentioned, the number of miles driven per person has increased inexorably, due to both increasing affluence coupled with expanding suburbanization/sprawl. There is considerable difference in distance driven, depending on location, such that in dispersed cities like Atlanta, the value has been estimated at 35.1 miles/day on average (TTI 2001), compared with concentrated cities, for example, Philadelphia, 16.7; Chicago, 19.7; and San Francisco, 21.1. Much of the research relating to the influence of various factors on road collision rates has come from extensive research in the U.S., including the effects of urban sprawl as a factor. There is general information suggesting a difference between driving on urban and rural roads, which we have discussed in the previous section. For example, 60% of fatalities occur on rural roads, which account for only 39% of the vehicle-miles traveled (NHTSA 2002). Different types of roads have different risk factors because of the way they are used (Lourens 1999). Suburban sprawl roads are characterized by many conflicting driving operations such as turning, stopping, and heavy straight-through volumes (Ossenbruggen et al. 2001). Overall statistical information is lacking for collision rates on suburban and urban roads separately.

In two studies (Ewing et al. 2002, 2003), an analysis was made of traffic collisions using a more complex indicator of the sprawl for various communities across the U.S. Sprawl is considered to be characterized by four main factors that generally occur together: low residential density; rigid zoning separation of residential from commercial and industrial uses; the absence of high activity town centers; and a stretched-out network of roads with limited access into and out of residential areas (low street accessibility), which makes extensive vehicle use essential to access services. A sprawl index was devised for a total of approximately 83 urban and suburban regions, which accounts for about two-thirds of the U.S. population. Using census data, various measures of population density were combined with measures of block size and street accessibility to generate a composite number, called an index. The higher the sprawl index, the more compact the locality. To examine the relationship between urban sprawl and road collisions, researchers in the United States took a look at 450 counties, about two-thirds of the total population. Researchers found that the 10 most compact, dense communities (New York, Philadelphia, Boston, and San Francisco) had fewer deaths from traffic collisions than the 10 least dense communities (Cleveland, Atlanta, and Minneapolis). In fact, the more spread-out cities had a death rate from car collisions almost five times that of more dense cities. Overall, the relationship indicates that a 1% increase in the sprawl index, which signifies increasing density, is associated with a 1.5% decrease in fatality rate.

When compared with fatality rates for the 83 districts, the sprawl index was found to vary inversely. Using the same concepts, about 450 counties were examined in detail, covering about two-thirds of the U.S. population. The 10 most compact communities examined, which included some of the densest counties in large cities such as New York, Philadelphia, Boston, and San Francisco, had an average sprawl index of 218 units and a fatality rate of 5.6 per 100,000. Conversely, for the 10 least dense areas in dispersed cities such as Cleveland, Atlanta, and Minneapolis, the corresponding numbers were 69 units and 26 per 100,000, respectively (almost five times the fatality rate). Overall, the relationship indicates that a 1% increase in the sprawl index function—that is, increasing density—is associated with a 1.5% decrease in the fatality rate. Another way of looking at the risk is as a function of distance driven. Road collision statistics are often quoted per unit of distance driven, daily or annually. This does not, of itself, prove that risk is proportional to distance driven, but the widely used statistic suggests that it is plausible. It is also likely that people who live in low-density suburbs must, of necessity, drive a greater distance annually.

6.4 EFFECTIVE LAND USE PLANNING

The organization of land use affects the number of journeys people need to take, means of transport, length of trip, and route taken. In short, different land use creates a different set of traffic patterns. Hummel (2001) outlines the main aspects of land use that influence road safety:

- The spatial distribution of origins and destinations of real journeys
- Urban population and density in pattern of urban growth
- The configuration of the road network
- The size of residential areas
- Alternatives to private motorized transport

Hummel (2001) suggests that land use planning practices and "smart growth" land use policies, coupled with the development of high-density, compact buildings with easily accessible services and amenities, can lessen the risk exposure to road users. The creation of clustered mixed use community services, for example, can cut the distances between commonly used destinations, cutting the need to travel, and reducing dependence on private motor vehicles.

In the Netherlands, the integration between traffic safety and land use planning is taking place in three stages:

1. Urban concept
2. Site design
3. Public space

The urban design stage can be considered as the foundation of the definite urban design. The master plan is designed at this stage. The next stage is the site design phase, and this is where the global interpretation of the plan takes place. At this stage the provision of facilities, the network design, the extent of mixed areas, and parking requirements are included. The third stage is the public space the final urban plan is completed.

According to Hummel (2001), the Netherlands plan uses two hypotheses focusing on the relationship between the degree of policy makers' interest in traffic safety issues on one hand and on the other hand the real impact of policy interest on traffic safety. First, the study dictates that the policy makers' interest is strongest at the public space phase. It is at this stage that detailed developments about the infrastructure are made. For example, questions about whether to build roundabouts, pedestrian crossings, speed ramps, cycle paths, and so on are discussed. At this stage, the contribution of transport and traffic issues is very transparent. The stage of site design deals with the degree of density within a certain area, the location of the site itself as well as the dimensions of the facilities, and the routing of public transport. At this stage the distribution of traffic demand within the site and also by trip lengths and modal choices are influenced. The stage of urban concept is concerned with site selection, the proximity of the town center, the regional capacity, and the function of public transportation. At this stage traffic demand is strongly influenced.

Goldman and Gorham (2006) and Cervero and Gorham (1995) conclude on the basis of a comparison between Stockholm and San Francisco that land use at a regional stage has a stronger impact on mobility than land use at the local stage. For example, it is worthwhile for urban planners to think about the location of facilities from the perspective of traffic safety. When facilities are situated in closer proximity to each other, we need less mobility; however, when facilities are less densely situated, we will need to transport ourselves across larger distances in order to fulfill the same pattern of activities.

The relationship between land use and mobility has always been a topic of hot debate. According to Gorham (1998, 4) "urban planners and designers have argued for most of [the last] century the need for better co-ordination of land use—the decisions of what goes where in an urban region—with transport policy making and investment (Mumford 1938)." One reaction to this is that land use policy has no influence on mobility. Others state that the land use policy is of great impact on mobility, and it makes sense to develop urban designs that will lead to sustainable mobility.

We can assume traffic behavior results from three categories of factors:

1. *Land use factors*: This relates to both the distribution of locations for activities over space and relationship between these activities. These locations might include dwellings, offices, workplaces, shops, services, recreational facilities, and education facilities.
2. *The desires, wants, needs, and possibilities of people*: The desires, wants, and needs of people have several complex aspects. At first it is widely recognized that income is a factor in traffic behavior. People with higher incomes have more possibilities to fulfill travel desires. For example, they often own more cars and fly more. Also culture is especially relevant, given that a country such as the U.S. is a very "pro-automobile" culture compared to several European and Asian countries.
3. *The transport resistance*: It is generally recognized that the faster, the cheaper, the safer, and the more comfortable travel, the more kilometers people travel. On average, people seem to have a more or less constant travel time budget. Therefore, if infrastructure and transport service allow people to travel faster, they will travel more often.

The three categories of factors influence each other in all directions. These interactions imply that land use does not only a direct but also an indirect impact on travel behavior.

There are three categories of effects that are important:

1. Accessibility effects
2. Safety effects
3. Effects on the environment, livability, and risks

Accessibility effects are influenced by transport resistance, the locations of activities, and the wants, needs, and possibilities of people. The greater the transport resistance, the lesser the accessibility. The closer the locations of activities such as work and recreational activities, the higher the level of accessibility. First, safety, environmental, livability, and risk depend on the number of kilometers driven per mode. Second, the location of these kilometers as well as the locations of people that face these externalities are of importance. For example, road collision rates (numbers of people killed per billion vehicle-kilometers) are much lower on motorways than on lower-order roads in rural areas.

- Land use may affect overall mobility levels as well as the modal shift and, therefore, through effects of traffic, have safety and environmental impacts.
- The locations of infrastructure and the division of vehicle-kilometers are relevant.
- If dwellings and other buildings as well as recreational facilities are built at greater distances of roads, effects such as concentrations of pollutants and noise nuisance will be less.
- Related to transport distances and the distribution of infrastructure types, factors such as speeds, accelerations, and breaking may be influenced by land use factors.
- The advantages of a big fast car may be less if cars are used less frequently at congestion-free motorways.

Urban planners are increasingly urged to question how their urban forms and development will influence road safety. Questions such as "should we separate means of transport (cycle paths, pedestrian paths)?" or "should we share the street?" arise. It is clear from the research that while the influence on urban planning on road safety is a major issue, there has been little written on the subject especially concerning spatial analysis. The relationship between urban space and road safety is a complex one. Some studies have shown that road safety cannot be viewed simply as a consequence of "causes" due in part to the morphology and organization of the city, but must also be subject to regulations within the complex urban system (Millot 2004a). It therefore involves taking into account the dynamic nature of urban spaces and in particular improvements that can be made to the road network. Therefore, we can conclude that the linkages between urban space and road safety cannot be studied in such a limited vacuum.

First, let us consider the influence of urban forms on road safety. One of the foremost research studies into urban form, land use, and road safety is by Millot (2004a). Millot (2004a) considers the complexity of urban forms on road safety and uses an urban analysis method that consists of breaking up urban forms into their properties and analyzing these properties and how they relate to one another, with particular emphasis on road safety. The road network has often been studied in its influence on road safety, but other characteristics of urban spaces (land use, housing density, etc.) may also have an influence. Millot's method consists of three stages:

1. Identifying the properties of urban forms that influence road safety
2. Analyzing this influence
3. Studying the road safety of urban forms through their properties

Millot (2004a) uses four typical urban forms found in France (which can be accredited to other similar Western countries), which are

1. Traditional area
2. High-density area
3. Single family housing area
4. French "new town"

The traditional area comprises of individual attached houses along the street. There is no traffic segregation and no hierarchical organization of roads. All the roads have the same layout and transit traffic can cross the area. The high-density area comprises of multistory housing set back from the street. There is a large amount of public space (green space). This means that different modes of transport are separated in particular pedestrian paths running through public spaces. The road network organization has a hierarchy and there are three types of street: arterial streets, collector streets, and local streets. The single family housing area comprises of separate individual houses set back from the street. There is traffic segregation and hierarchical road organization. The local streets are often dead ends or encircling parking areas. Finally what Millot (2004a) describes as a "French new town" comprises of individual houses and multistory housing. The means of traffic are completely separated; there are not only pedestrian paths but also pedestrian bridges to cross arterial roads. Road organization prevents transit traffic from crossing the area. There is also hierarchical organization of roads with dead-end streets.

Millot (2003) conducted the study by examining all the characteristics of the areas: location in the town, morphology (housing types, density, etc.), structural aspects (road hierarchy, density of streets), and functional aspects (characteristics of population, means of transport, shops). Millot (2003) then used police road collision data to look at the causes of the road collisions and the characteristics of the areas. In total, 12 areas were studied. They compared results and found general characteristics of urban forms that influence road safety. They refer to these as "properties":

- Road network organization
- Distribution of road users in public space
- Public space organization
- Visual characteristics of the road environment
- Parking space organization
- Arterial road layout

Road Collisions and Urban Development

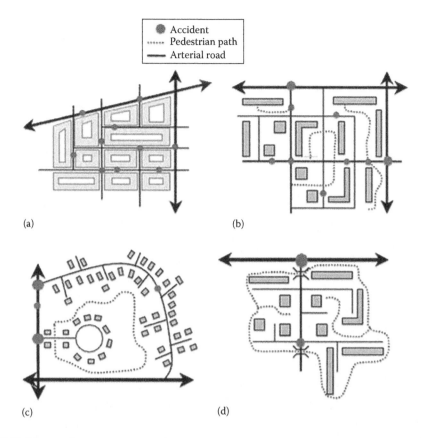

FIGURE 6.1 Collision locations in the four urban forms: (a) traditional area, (b) high-density area, (c) single-family housing area, and (d) French "new-town." (Reprinted from Millot, M., The influence of urban planning on road safety, in: *Proceedings of European Transport Conference*, Association for European Transport, London, U.K., 2004b. With permission.)

The study found that each of the properties exerted a different kind of influence. For example, road network organization impacts on the location of collisions and their distribution among different road types. The visual characteristics of the road have an influence on driver perception and behavior. Public space organization has an influence on the type of road user involved. Millot (2003) then proceeded to present the main features of the road safety problems for each of the typical urban forms (Figure 6.1). This is presented in a snapshot in the following text:

Traditional areas: Road safety problems are dispersed among the neighborhoods' interior streets. The urban planning concept is based on sharing space (no segregation). Transit traffic can cross the neighborhoods and can create conflict with local users (pedestrians, children). These problems can be handled using speed management tools (e.g., traffic calming areas).

High-density areas: In these areas, collisions occur not only on arterial roads but also on local roads. The urban planning concept is based on medium segregation and a medium hierarchy of roads. For example, local streets can support transit. On arterial roads, collisions are concentrated at a few junctions. Many tools already exist for reducing collisions at junctions; however, on local and collector streets, the road safety problems are dispersed. They often involve car drivers and vulnerable road users such as pedestrians or cyclists. It is more difficult to manage these areas because the layout is favorable to cars and their higher speeds. Roads tend to be wider and houses are set back from the street, and there are often large green spaces, which promote speed.

Single family housing: Collisions tend to be concentrated on a few junctions between arterial roads and local streets. These areas are built based on the separation of traffic modes. For example, local dead ends prevent transit traffic crossing the neighborhoods. Road safety problems tend to be similar to "hot spot" problems. Many tools already exist to manage safety in this type of area.

In the French-style "new towns," road collisions are also concentrated at the major junctions. The urban planning concepts are complete traffic segregation and strict hierarchical organization. This concentration of road collisions makes management easier.

Millot (2004a) identified a novel and interesting way to analyze road collisions based on urban form and morphology, a method that had been neglected in the past. We can see that the different urban forms do not influence road safety and collisions in the same way and understand the advantages and disadvantages of separating traffic in different environments.

6.4.1 Pedestrian Land Use Planning

Pedestrian safety measures are the most comprehensive and most closely associated with urban planning. There are two types of policies: (1) area-wide speed reduction or traffic calming schemes, and (2) provision of an integrated walking network.

These are two complementary measures, which can be implemented together without conflicting. Not only do they apply to different parts of the urban form, but they also address different objectives. Area-wide schemes (the most widespread of which is the 30 km/h zone) are aimed at reducing vehicle speeds and therefore allowing for a safer integration between pedestrians and motor traffic. Integrated walking networks (usually centered in or around an inner-city pedestrian zone) serve to remove and/or reduce conflicts between pedestrians and vehicles and to provide or improve crossing points.

The same basic planning principles that apply for pedestrians apply for cyclists. Due to the fact that cycling is suitable for travel over greater distances than walking, it is necessary to distinguish a flow and an access function. The same is required with motorized traffic; a network for the flow function is required. However, this network cannot follow the network for through-motor traffic easily, since the mesh of the routes of the cycling network is smaller. Provisions for cycling should not be seen as additional features of the traffic structure for motor traffic. Rather, they require a network of their own. Some cities are more successful with regard to cycle planning than others. An excellent example of cycle planning is in Copenhagen, Denmark. According to government data, one in three Copenhageners cycles to work. Constructing the cycle network has taken over a century and for the first time Copenhagen has published a "cycle policy," one of the few Western cities to do so. Cycle planning consists of not only better network faculties for cyclists but also the provision of parking. Many cities including London, Paris, and Shenzhen (mainland China) have rolled out a bicycle sharing system. Known in London as "Boris Bikes" (after the Mayor of London, Boris Johnson), the Barclays Cycle Hire system operates 5000 bicycles across approximately 17 square miles. In Paris, the "Velib," or English for "freedom," system for bicycle sharing has approximately 18,000 bicycles and is now complemented by an electric car sharing system. While both these schemes have had general success, they have been issues of financing and repair of bicycles. There is also an interesting spatial issue that has yet to be explored, which is of locational demand at bicycle pickup and drop-off points.

When facilities for cyclists are being designed, five criteria are important if their needs are to be met:

1. *Safety*: For large parts of the population in Europe (the perception of) road safety problem is a key reason for not cycling. Improvement of the safety of cyclists on the road is therefore a precondition for promotion of cycling.
2. *Coherence*: Continuity, consistency of quality, recognizability, and completeness. It is obvious that cycling will be restricted if the cycle network is not complete or coherent. These are mainly features at network level.
3. *Directness*: Mean travel time, detours, and delays.
4. *Comfort*: Smoothness of road surface, curves, gradients, number of stops between starting point and destination, complexity of rider's task.
5. *Attractiveness*: Visual quality of the road, survivability, variety of environment, and social safety.

6.4.2 Land Use Planning Risks

One of the most contentious issues surrounding land use planning and road collisions is traffic flow. This is a consequence of the land use planning itself and has a direct and indirect impact of the volume and nature of road collisions. Little has been researched on road collisions and traffic flow, for a number of reasons, largely data. Accurate data are limited only to large cities, and then it can be difficult to aggregate. Traffic flow is largely linked to speed, which often means that when traffic flow is slow, the likelihood of an injury collision is low. However, when traffic flow is fluid, the speeds are higher and likelihood of collision risk is higher.

Generally speaking, studies of relationships between collisions and traffic flow can be divided into two types: (a) aggregate studies, in which units of analysis represent counts of collisions or collision rates for specific time periods (typically months or years) and for specific spaces (specific roads or networks), and traffic flow is represented by parameters of the statistical distributions of traffic flow for similar time and space; and (b) disaggregate analysis, in which the units of analysis are the collisions themselves, and traffic flow is represented by parameters of the traffic flow at the time and place of each collision. Disaggregate studies are relatively new and are made possible by the proliferation of data being collected in support of intelligent transportation systems developments. Transportation management centers routinely archive traffic flow data from sensor devices such as inductive loop detectors, and these data can, in principle, be matched to the times and places of collisions.

Traffic congestion and road collisions are two important externalities created by road users. Increased travel time caused by traffic congestion imposes costs to road users, both in terms of economic loss and also the reduced quality of life and mobility. The costs of road traffic collisions to individuals, property, and society in general have also been significant. Traffic congestion and collisions both impose a burden to society, and as such it is important to reduce their impacts. An ideal solution would be to reduce them simultaneously but this may not be possible; however, it is speculated that there may be an inverse relationship between traffic congestion and road collisions (Shefer and Rietveld 1997). Shefer and Rietveld (1997) hypothesize that in a less congested road network, the average speed of traffic would be normally high, which is likely to result in more serious injuries or fatalities. On the other hand, in a congested road network, traffic would be slower and may cause less fatalities and serious injuries. This increased traffic congestion may lead to more collisions due to increased traffic volume; however, those collisions may be less severe. This suggests that the total external cost of collisions may be less in a congested situation relative to an uncongested situation. This poses a potential dilemma for transport policy makers since it would appear that traffic congestion can improve road safety; however, traffic congestion reduces mobility, which subsequently decreases economic productivity.

Land use is a principal determinant when making trips and is the main influencing factor for road-based environments and its related variables including traffic flows, speed limits, and pedestrian activities (Lupton et al. 1999). According to urban safety management theory, land use policy is one of the strategies used to prevent and reduce road collisions (The Institution of Highways and Transportation 1997). Different types of land use generate different types of trips and encourage different types of driver behavior, which in turn can lead to the potential of a road collision. If we take a wider perspective, many other aspects of road collision analysis are also associated with different land uses and associated activities.

Noland and Quddus (2005) reiterates the issue that although congestion (and reduced traffic flow) leads to more collisions, these collisions are less severe. However, they point out that this poses a policy dilemma for decision makers. As Noland and Quddus (2005, 738) summarizes, "[e]xternal costs associated with congestion may be off-set by external benefits associated with fewer traffic fatalities due to congestion." Noland's study of congestion and road safety in London found inconclusive results, speculating that the speeds are generally low in both inner and outer London, and those areas are congested already and have infrastructure already in place, which mitigates the safety effects of high-speed traffic (Noland and Quddus 2005).

Special mention must be given to China with regard to land use, road collisions, and increasing urban population. Rapid motorization in China especially private cars increases about 15%–20% each year. This increase creates undesirable environmental and social problems. In China, there are 160 cities with a population over one million; it is safe to say land use and rapid urbanization followed by motorization has a large role to play in the increasing number of road collisions. According to official studies, there are about 450,000 car collisions on Chinese roads each year, which cause about 470,000 injuries and 100,000 deaths. The total cost of these collisions was 2.4 billion dollars. The study concluded that 92% of these collisions were due to bad driving skills. These figures are disputed by a WHO study, which reported that the actual number of fatalities on China's roads is more than twice the official figure, or about 250,000 killed each year. This study estimates that 45,000 people are injured and 680 killed on China's roads each day. Road traffic collisions are believed to be the leading cause of death for people 15–45 years old. The direct and indirect costs of these collisions are estimated at between 12 and 21 billion dollars, or about 1.5% of China's GNP. This collision rate means that roughly 20% of the world's fatal car collisions take place in China (Zhang et al. 2008).

One of the major issues for China is the speed of urbanization and land use planning, residential and associated amenities are car-centric, building on the U.S. style of car-based infrastructure. With the increasing number of individual vehicles, road space and traffic control measures cannot keep up; the result is traffic congestion, safety, and parking problems. Although the rapid conversion of land from rural to urban areas is mostly due to financial incentives, it provides more transport infrastructure opportunities, which could indirectly increase the demand for automobiles. In turn, higher demand for automobiles could also result in suburban development, leading to long-term urban sprawl in low-density regions accessible only by individual vehicles, as public transport cannot afford to provide service when densities are low. Providing gasoline for private cars and mopeds and diesel for trucks also leads to rising air pollution as well as increasing greenhouse gas emissions.

Since motorization began to increase some two decades ago, road safety has been recognized as a major issue in China. During the period from 1975 to 1998, traffic fatalities increased 243% (Kopits and Cropper 2005). Currently, road collisions in China account for about 300 deaths per day (CRTAS 2006). The increase in the number of automobiles is one of the main factors contributing to the increase in traffic fatalities. Even with relatively low specific motorization rates, China already had 104,372 road-collision-related deaths in 2003. Put in perspective, the U.S., with about six times the number of motor vehicles (three if two-wheelers/mopeds are counted), has only 45% as many fatalities. Both the high number of pedestrians/cycles/two-wheelers as "targets" and poor safety per se account for this huge difference in collisions per vehicle, which would be even starker if the greater distances/vehicle characterizing the U.S. over China were used in the comparison.

The WHO estimates that 1.2 million people are killed in traffic collisions globally every year (Peden et al. 2004), implying that almost 3300 people are killed every day, of which approximately 10% occurred in China. The significant number of traffic fatalities in China is most likely to be caused by the sudden increase in motor vehicles in dense urban areas with infrastructure principally built for nonmotorized transport users. Moreover, the increase in the inflow of people from rural to urban areas could also result in the high fatality rate, as people are still adapting to the relatively new and rapidly developing city traffic flows.

Rapid motorization often leads to inequities in transport mobility and accessibility. Motor vehicles have clearly benefited people who can afford to own private cars, but traveling conditions for pedestrians and cyclists have deteriorated over the past few decades, mainly due to the significant reductions in sidewalks and bike lanes.

Benefits, therefore, to a proportion of society have come at a high cost to the majority, as cars have invaded the space of others. Problems caused by such changes have not been solved, nor have government authorities started looking into these issues yet. Numerous reports across major Chinese cities—including Beijing, Shanghai, Shenzhen, Nanjing, Fuzhou, Guangzhou, Hangzhou, and Shenyang—describe the takeover of pavements and bike lanes by motor lanes, and subsequent public dissatisfaction with these developments. As these reports on road-space distribution suggest, the integration of social issues in current transport planning is generally lacking. Currently, there is no legal regulation that forbids motor vehicles to travel on traffic lanes not specifically designed for their usage. Motor vehicles generally have the right-of-way, which cyclists and pedestrians find threatening. Bike-free streets in Beijing first appeared in the late 1990s, and Shanghai urban transport officials were known to project the role of bus transit as a replacement for walking. Rapid motorization in China presents individuals and authorities with a daunting challenge. Chinese car penetration is well below 100/1000 people, except in Beijing and a few other coastal cities. In a sense, it threatens to play the role of the third scenario in the original work used for the present scenarios, which was called "car collapse." What car collapse—gridlock in major Chinese cities—threatens to do is to deprive China and its people of the true benefits of individual motorization and the utility of owning cars. If that were to push China's automobile industry itself off the road to healthy growth and success, the costs would be very high. Many of the social problems associated with motorization could be mitigated if there were more time. The speed at which cars became popular in Beijing, however, suggests there is not much time.

Wedagama et al. (2006) investigated the relationship between pedestrian injuries and certain land use types, for example, retail, offices, leisure, and junction density, on weekdays and weekends. Furthermore, the analysis intended to derive a relationship between different land use types or trip attractors and temporal variation of pedestrian and cyclist injuries. However, the analysis by Wedagama et al. (2006) did not disaggregate the pedestrians by age. Graham et al. (2005) investigated the influence of area deprivation on child and adult pedestrian injuries, considering England as a case study. This study concludes that the residential areas are likely to be safer for children than mixed use areas in inner cities.

The issue of land use and road collisions has often been focused on pedestrians and cyclists. In the U.S., pedestrians are the second largest population group to die in motor-vehicle-related collisions (LaScala et al. 2000). In the United Kingdom, The House of Commons Transport Committee (1996) examined ways to reduce risk to pedestrians and cyclists. In addition, the Department for Transport (DfT) is promoting a modal shift to walking and cycling for shorter journeys and aims to make these modes safer. An increasing degree of urbanization is associated with an increase in nonmotorized transport injuries through demographic factors (e.g., population density) and road and traffic environment factors such as road length, junction density, and land use. For instance, the risk of a pedestrian collision is up to five times greater for children living in urban areas than for those living in rural settlements (Petch and Henson 2000).

A few previous studies have been published on the links between land use and road traffic collisions. For example, Levine et al. (1995) investigated the relationship between zonal land use and road traffic collisions. They found that residential population density, manufacturing, retail trade, and services industry were positively related to the number of road traffic collisions. A study of child road safety in Salford, England (Petch and Henson 2000) examined child injuries by Enumeration District zones, covering an area of about 100 km^2 over a 40-month period. Factors related to land use, principally the number of trip attractors or trip generators, percentage of terraced housing, and amount of open space, were found to be significant explanatory variables. Ivan et al. (2000) identified the number of driveways to the public highway on each road segment as a significant predictor of single-vehicle and multi-vehicle collisions. Researching pedestrian injuries, LaScala et al. (2000) suggested giving priority to efforts to prevent pedestrian alcohol impairment and reducing neighborhood alcohol availability. This was based on the relationship among the spatial location of alcohol retail outlets, demographic factors, and the road environment with pedestrian injuries. In relation to land use, the number of bars, off-licenses, and restaurants per kilometer location of activities and their spatial density have a significant role in determining the relationship between land use and transport. Land use patterns tend to be due to local decisions to respond to market demand for housing, employment, and services. In theory, land use should be planned to minimize road traffic conflict, particularly between motorized and nonmotorized transport. This can be achieved by reducing the need for motorized transport, for example, by locating shops and schools within walking distance of homes. In other words, locating trip generators and trip attractors close to each other could reduce the need to travel by car, thus encouraging walking and cycling.

Kim and Yamashita (2002) compared the vehicle-to-vehicle collisions with vehicle-to-pedestrian and vehicle-to-bicycle collisions per acre land use category for Honolulu over a 10-year period. They found that vehicle-to-vehicle collisions were highest in commercial and industrial areas (6.62/10 acre year), visitor lodging (5.15/10 acre year), and manufacturing and industry (3.67/10 acre year). The order was somewhat different for vehicle-to-pedestrian collisions with visitor lodging (0.43/10 acre year), commercial and industrial (0.30/10 acre year), and public services (0.20/10 acre year) being the three highest categories. The vehicle-to-bicycle collisions followed the similar pattern, although the collision rates were typically less than a half (visitor lodging 0.22/10 acre year).

6.5 PLANNING FOR SAFETY AWARENESS

Road safety considerations are central to planning, design, and operation of the road network. Altering the design and layout of the road and the road network to accommodate human characteristics and human error, road safety engineering can be successful in injury prevention and reduction. Ross and Britain (1991) outlines a framework for the systematic management of road safety, which is outlined as follows:

- Classifying the road network according to its primary function
- Setting appropriate speed limits according to this function
- Improving road layout and design to encourage better usage

The built environment, including road infrastructure, pedestrian infrastructure, and streetscape, has a strong influence on pedestrian safety. It can provide buffers between pedestrians and motorists, such as refuge islands. It can also encourage motorists to keep a safe speed through the inclusion of traffic calming measures, such as speed humps, traffic circles, and road narrowing. The built environment, such as crosswalk signs, may provide pedestrians with more visibility as well.

There have been several studies of the associations between the built environment and aspects of pedestrian and driver safety. In many of these studies, collisions are aggregated to some spatial area, such as intersections, census blocks, or buffers of specific pedestrian generators, and related to the environmental conditions in that spatial unit. Recently, more and more studies have been able to acquire and analyze more microscale built environment data and evaluate their impacts on pedestrian collisions. This microscale built environment data allow for a more disaggregate analyses of the collision data and ensures more variation within the variables. Retting et al. (1995) provides a review of the literature of examining the effect of microscale built environment measures on collisions and concludes that changes to the built environment that can impact the occurrence of pedestrian collisions. These measures include speed control, separation of pedestrians from the roadway, and increased visibility of pedestrians. Their analysis suggests that physical changes to the built environment can significantly decrease the rate of pedestrian–vehicle collisions. Clifton and Kreamer-Fults (2007) examined both risk exposure and injuries sustained in child pedestrian–vehicle collisions in the vicinity of public schools and found significant associations with several built environment and design characteristics. Specifically, they found that the presence of a driveway or turning bay decreases both the likelihood of collision occurrence and the severity of injuries sustained in a collision. Similarly, research has shown that more pedestrian collisions occur in areas with educational facilities and higher percentages of commercial land use (Loukaitou-Sideris et al. 2007). Clifton et al. (2005) also investigated the interaction between gender and land use on pedestrian collisions and found that land use characteristics had no impact on overall collision rates; however, a higher percentage of collisions involving female pedestrians occurred in areas with high pedestrian activity. Graham and Glaister (2003) developed a negative binomial spatial model to examine the role of scale, density, and land use mix on the incidence of pedestrian fatalities. Their results show that the characteristics of the local environment have a significant effect on pedestrian collisions that resulted in fatalities. In particular, they found that fatalities were more likely in residential areas than in commercial areas. They also found a negative correlation between population density and the number of collisions in an area, suggesting that traffic control devices, vehicle speeds, and more pedestrian accommodation in these areas increase pedestrian safety.

6.5.1 UNIVERSITY CAMPUSES

Urban university campuses face unique challenges when dealing with pedestrian safety issues. Densely packed street networks, combined with the assemblages of student pedestrians that navigate them, create corridors for pedestrian collisions. There is limited research that looks specifically at university campuses as a specific area for road collisions. University campuses are either in the center of big cities or outside of major urban areas. It goes without saying that inner-city campuses have a higher proportion of road collisions as they are integrated into the urban area. It becomes very difficult to separate road collisions to university students from any other urban dweller. One of the major limitations on studying road collisions on university campuses is data. It is very difficult to differentiate university students in the data from other members of society, and as with a lot of campuses around the world, they can be integrated into major cities.

6.5.2 Driveways

The risks of driveway collisions are high specifically for children. While the number of collisions is not high, they are consistent. One of the major factors in recent years has been the increase in size of car, which has led to a more limited line of sight in the driveway setting. Often in these collisions, the person driving the car is a family member or friend, which increases the level of preventable tragedy occurring. The two main types of collisions that occur in driveways are backovers (where children are killed or injured from the car being backed out of the driveway) and frontovers (where a small child will be hit from the front of the car). There are two types of literature on child injuries and fatalities in driveways, first medical and second academic/analytical. While in this book we are less concerned with the medical trauma of such incidents, it is important to note the fact that the majority of these incidents occur to toddlers, and the trauma and devastation is more pronounced. There has been limited research produced in this area of collision studies; however, most of the research conducted has taken place in the United States, New Zealand, and Australia (Figure 6.2).

In the U.S., it is estimated that at least 50 children per week are being backed over; of these, approximately 48 are injured and 2 are fatal. The predominant age for victims is between 12 and 23 months, and over 60% of incidents involve a larger size vehicle. It is also estimated that in approximately 70% of all these incidents, a parent or close relative is behind the wheel (KidsAndCars 2012a).

The possibility that there might be a special possibility that smaller children might be at particular risk as "pedestrians," struck by motor vehicles moving on private property, appears to have first appeared in the research literature over 20 years ago. In 1979, a Canadian group set out to examine the characteristics of 452 child pedestrian fatalities (Buhlman et al. 1979). The study sample was divided into three groups of victims, age 1–4, 5–7, and 8–14 years. Children aged 1–4 years were found to be involved disproportionately in daytime collisions occurring on or near a private driveway. Frequently, the child had been run over by a motor vehicle set in motion from a parked position. In distinction, pedestrians aged 5–7 years were involved disproportionately in collisions occurring in the hours immediately preceding or following school, and crossing road between parked cars. It was soon recognized that conventional police reporting of traffic collisions was not picking up a substantial proportion of these injuries, with a Baltimore study showing that underreporting of child pedestrian collisions was in the order of 20%. Later studies confirmed the extent of underreporting of these incidents.

124 Spatial Analysis Methods of Road Traffic Collisions

FIGURE 6.2 U.S. child fatalities in driveways. (Courtesy of KidsAndCars.org [KAC], Chart, statistics, graphics, 2012b, http://www.kidsandcars.org/back-overs.html, accessed April 17, 2012.)

In 1991, a Californian study (Agran et al. 1996) determined it was the youngest category of children who were most at risk from driveway collisions. They divided the children into two key age groups, the first being 0–2 years old (toddlers) and the second group being 3–4 years old (preschoolers). A high proportion of the toddler collisions occurred in residential driveways. The preschooler children were often injured while crossing mid-road on residential streets near their home. By the mid-1990s, it was clear there was an international problem occurring.

Data on child injuries and fatalities in residential driveways are not as accurate as normal traffic collisions. A 1993 study in Virginia used alternative sources to generate their data (country and rescue and fire departments). The research team found that between the age of 0 and 5 years, nontraffic collisions accounted for almost 50% of all child-pedestrian–motor-vehicle collisions. Collisions in commercial parking lots produced the most injuries, but home driveways produced the most fatalities. Most of the data comes from health authorities or hospitals. As with many issues with road collision data, there is no centralized collection of child driveway collisions, as the police are not always called to attend the scene. The definition of location in the data might also be misconstrued. Often the location is put as "recreation" or "house," which can be misleading.

For the United Kingdom, data on child driveway collisions are difficult to obtain. These types of collisions are not recorded in the STATS19 recording system. However, some data are available from the HASS/LASS database system (HASS, the Home Accident Surveillance System; and LASS, the Leisure Accident Surveillance System). A study in the United Kingdom in 2002 took a sample for this data (home and leisure collisions, not including road or workplace collisions) from 16 to 18 collision and emergency departments in the United Kingdom. The data were used for 1–14-year-olds and in 2002 there were 202 cases (giving a UK estimate of approximately 4100 hospital admissions each year). Half the injuries were to children under the age of 4 years, and interestingly (compared to the Australian findings) the genders were split equally. Not all of the cases are driveway vehicle incidents, during this period it was recorded that there were eight incidents of child fatalities due to driveway vehicle incidents. All of the children were aged between 1 and 2 years.

A 3-year study completed in 2002 by the Queensland Ambulance Service and CARRS-Q (Centre for Accident Research and Road Safety–Queensland) into pedestrian collisions in Queensland revealed an increase in the incidence of parents reversing a vehicle over a child in the family driveway. The study determined that low-speed runover is the third most frequent cause of death or injury in Queensland children aged between 1 and 4 years. They found young boys to be at greater risk than girls, and although only 2.3% of Australia's population is of indigenous status, 10% of all Australian driveway runover fatalities involve indigenous toddlers. In 80% of all incidents in Australia, the driver of the vehicle was male. Four-wheel-drive vehicles are overrepresented, and possible linkages are demonstrated between the increasing popularity of four-wheel drives and an increase in driveway collisions. Four-wheel drives have blind spots, whereby, a toddler less than 3 m from the vehicle is not visible to the driver. In Queensland cases, 41% of vehicles were four-wheel drives, despite four-wheel drives making up only 6% of passenger vehicles in the state. Utes, vans, and trucks are also involved more often than sedans, due to poor visibility from them. In 2000, the Henderson Report, commissioned by the Motor Collisions Authority of New South Wales, sparked a call to review safety for large vehicles.

In 2009, a study in New Zealand came to some interesting conclusions; first, looking at hazards in and around the driveway. New Zealand driveways are relatively long, creating several danger zones with lack of visibility. Second, there is a significant number of shared driveways servicing multiple homes, and fencing driveways is not a mandatory requirement. Like many other countries, the children see the driveway as a place to play. Another factor implicated in this type of collision is the increasing prevalence of larger vehicles (people movers) and four-wheel-drive-type vehicles. As these vehicles increase in size, the reversing visibility reduces, resulting in more "blind zones." The third factor the New Zealand study identifies is the human factor. This could encompass general awareness, parenting practices, driving behaviors, or socioeconomic factors. Media analysis suggests that "awareness" occurred at multiple levels. Most drivers were unaware of the actual size of the blind spot of their vehicle. Some drivers were not aware that children were in the driveway at all, while others were unaware that a child had moved into the driveway after they had checked for the whereabouts of any children on the property. In the time it takes to physically exit a vehicle, check behind, and reenter the vehicle, a child can come from nowhere and place him/herself in the path of the vehicle while remaining invisible to the driver, due to their physical size and the blind spots in vehicles.

Collisions are more likely to occur in the summer months and during the late afternoon hours (4–7 p.m.), rather than in the morning (Beasley 2009). Summer afternoons, of course, are times when families are more likely to be active outdoors. Daylight saving and warm weather result in collisions peaking in December (Beasley 2009). Technical and environmental factors also contribute to driveway runovers. For example, variations in vehicles' visibility index are a key factor, especially when drivers are not aware of the extent of the blind spot of their particular model of vehicle. This exacerbates a more general lack of awareness of the risk involved in not knowing exactly where children are while reversing down driveways (Beasley 2009). The lack of fencing within properties has also been noted as a key contributor to runovers, since on many New Zealand sections there is no practical distinction between play areas and driveways.

Cowley et al. (2005) noted that New Zealand has one of the highest rates of driveway runovers in the urbanized world and noted that children from lower socioeconomic groups are five times more at risk. Cowley also concluded that collecting data on driveway runovers was littered with problems predominantly associated with categorizing and coding the data. Often runovers are categorized as pedestrian injuries, which is misleading when analyzing the data. Two years later, Chambers (2007) reiterated a lot of Cowley's findings especially focusing on the three common factors: environment, human, and vehicular. He noted than vehicle design and poor reversing skills were key factors in collision causation. The Chambers report in 2007 in New Zealand made some very specific recommendations, which included calling for territorial authorities to have improved regard for the risk of child driveway collisions.

Because the victim is a small child, driveway runover survivors tend to suffer major trauma and often serious long-term effects from the collisions (Cowley et al. 2005; Chambers 2007). Hsiao et al. (2009) found that there were nine driveway fatalities involving children less than 15 years of age in the Auckland region between November 2001 and December 2005. Chambers (2007) notes that there are on average four fatalities per year involving runovers on private driveways, with an average of two children being hospitalized every week; over the last 10 years, rates of runover incidence have remained steady. According to Chambers (2007), the lack of public/community awareness is a major cause of collisions. Improving public awareness is therefore a primary consideration of this research.

In Victoria, Australia, slow-speed pedestrian nontraffic incidents to children have for some years been identified as an important cause of death to children, accounting for 14% of collision deaths from all causes in Victorian children under 5 years of age between 1985 and 1995, and 12% of pedestrian deaths of all ages. Age-related differences in patterns of fatal injury to child pedestrians were identified in 1988 in North America. All pedestrian vehicle collision fatalities to children less than 5 years of age in Washington State were evaluated for a 5-year period using death certificates, coroners' reports, and police records. For these very young children, unlike the pattern for older children injured in "dart-out" collisions, the child was often backed over in the home driveway by the family van or light truck driven by a parent.

Arguably one of the most comprehensive studies done in this area has been in Auckland, New Zealand. Despite a growing amount of literature and research, there are still many unanswered questions as to why boys are overrepresented, and understanding the casual factors in more depth. The literature presents numerous options for the prevention of driveway runovers, and these tend to fall into three main categories—modifying behavior, modifying the environment, and modifying vehicles. The most common environmental factor associated with runover collisions is the failure to separate driveways from children's play areas (Cowley et al. 2005; Chambers 2007; Hsiao et al. 2009). Housing design in the Auckland and Waikato areas is similar. Houses are located at the front of sections and garaging is located at the rear, necessitating long driveways. In both regions, driveway fencing is rare. This is evident by surveying suburban areas, in both Auckland and Waikato, on Google Maps. Beasley (2009) found long driveways to be high-risk areas precisely because they are so inviting for children to use as a play area, and there is usually no physical barrier separating the house entrance and the driveway. Chambers (2007) notes that access to the driveway is often obtainable from both the front and the back of properties, and cites the findings of Hsiao et al. (2009) wherein a salient feature was the existence of driveways shared between dwellings and running the length of the property. Again, the driveway tended to be merged as part of the children's play area.

The Auckland study found that the absence of physical separation of the driveway from the children's play area was associated with a threefold increase in the risk of driveway-related child pedestrian injury (odds ratio = 3.50; 95% confidence interval 1.38, 8.92%). Children living in homes with shared driveways were also at significantly increased risk (odds ratio = 3.24; 95% confidence interval 1.22, 8.63%). The population-attributable risk associated with the absence of physical separation of the driveway from the children's play area was 50.0% (95% confidence interval 24.7, 75.3%). The authors concluded that the fencing of residential driveways as a strategy for the prevention of driveway-related child pedestrian injuries was an issue that deserved further attention.

The specific issue of child deaths and injuries in driveways is an identifiable subset of the wider problem. Compared to child pedestrian injuries generally, it is an easier problem to address because the location is strictly constrained, the pattern of injury causation—reversing vehicles running over the children—is quite narrowly defined, and the ages of the children mostly affected are in the toddler category. It is, however, in many ways a harder problem to address because the mass epidemiological data is poorly defined. Thus, the *risk* associated with specific behaviors, vehicles, and environments is not clearly identified. This makes it difficult to develop a program that is founded on a sound theoretical model for the injury control problem and factors in which interventions can be based. Further, driveway injury reporting systems are insufficiently developed to allow the collection of adequate data and the use of meaningful analysis to either predict or evaluate changes in the present situation. As reiterated in this chapter in several sections, a highly important factor in collisions where a vehicle has reversed into a child is the size and configuration of that vehicle. The kinds of vehicles identified in local studies as having characteristics predisposing to these collisions are four-wheel-drive passenger cars and light commercial vans. Both have very poor visibility to the rear, particularly for objects relatively close to the vehicle. This problem has been widely recognized, although not widely publicized. Subaru is perhaps the only vehicle manufacturer that has put that recognition into practice by having an internal company standard that minimizes the blind spot by defining the uppermost point of the center of the rear window. The shallow "V" along the bottom of the rear window of the Subaru Liberty range of station wagons is a demonstration of this.

6.5.3 Schools

First, let us make the distinction here; this section discusses specifically road collisions occurring around school locations. The large majority of people killed or injured in this area will be schoolchildren; however, there are other members of society who are also at risk. Transportation between home and school has been under much debate in recent years due to the rising increase of road safety risks. Far fewer children are walker to school as a result of the changing dynamic of the road environment (and also schools often being further away). The risk exposure for children who walk and cycle to school is therefore higher. The actual number of children who walk and cycle to school has been declining. Pucher and Dijkstra (2003) outline that walking rates are highest amongst children of lower socioeconomic status. Children walking to school in lower income areas are more likely to be involved in a collision resulting in severe injury than children in higher income areas. Children in lower income areas tend to travel twice as far to get to school as their more affluent counterparts and are more likely to be required to cross a major road during that commute (Macpherson et al. 1998).

Clifton et al. (2007) examines pedestrian–vehicle collisions in the vicinity of public schools, the severity of injuries sustained, and their relationship to the physical and

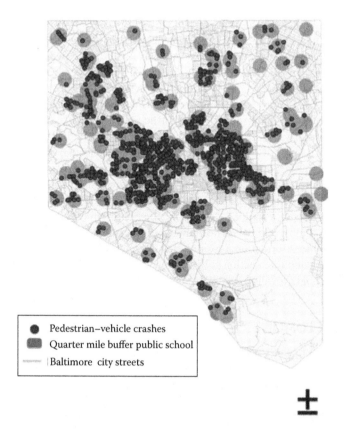

FIGURE 6.3 Pedestrian–vehicular collisions located within quarter mile buffer of Baltimore City public schools, 2000–2002. (Reprinted from *Acc. Anal. Prev.*, 39(4), Clifton, K.J. and Kreamer-Fults, K., An examination of the environmental attributes associated with pedestrian–vehicular crashes near public schools, 708–715, Copyright 2007, with permission from Elsevier.)

social attributes near the schools. The study was based in Baltimore City, Maryland. The results show that the presence of a driveway or turning bay on the school entrance decreases both collision occurrence and injury severity. Conversely, the presence of recreational facilities on the school site is positively associated with collision occurrence and injury severity of collisions. Findings related to neighborhood characteristics were mixed, but the significant variables—transit access, commercial access, and population density—are generally associated with increased pedestrian demand and should be interpreted with care. Figure 6.3 shows part of the research design of creating buffers around the schools for further analysis.

Although there is a large body of research that we have already addressed in earlier chapters concerning child pedestrians and road collisions, there is limited research involving spatial analysis of road collisions in the vicinities of schools and how the urban form and development around the schools might contribute to higher road collision risk. LaScala et al. (2004) uses an ecological approach to focus on the location of schools and the relationship with child pedestrian collisions. LaScala et al. (2004) highlights in the U.S. that investigations that consider the geography of communities have been primarily descriptive, identifying pedestrian injury "hot spots" (Braddock et al. 1991, 1994; Agran et al. 1996; Lightstone et al. 2001).

There have been studies that have focused on risk factors in urban environments that increase the likelihood of injury to child pedestrians. Stevenson et al. (1995) found that the risk of injury increased as traffic volume and average vehicular speed increased. Similarly, Agran et al. (1996) found that residential streets with a higher proportion of multifamily residences, more curb side parking, and greater vehicular speed were associated with higher rates of pedestrian injury. Bass et al. (1995) found a strong relationship between pedestrian injury and the number of unsupervised children playing or running errands in their immediate neighborhood. In a geographic analysis of motor vehicle collisions with child pedestrians, Lightstone et al. (2001) compared intersection and mid-block incident locations. This paper used Geographic Information System (GIS) to identify high-risk environments with both mid-block and intersection collisions occurred more often in census tracts with greater household crowding.

6.6 CONCLUSION

There has been recently a large amount of research into urban planning and design and road safety, especially in Europe. One of the most important goals has been to integrate aspects of traffic and transportation as early as possible into the development of regional and local land use and urban planning. Often, however, in reality this duality can be challenging in practical terms. An important conclusion of many of the studies has been that urban designs of new housing developments are still given priority over the integration between transport and land use. There is also evidence of lack of cooperation between urban planners and traffic planners, and when the traffic planners are consulted, it is often after the most important decisions are made.

REFERENCES

Abdalla, I. M., R. Raeside, D. Barker, and D. R. D. McGuigan. 1997. An investigation into the relationships between area social characteristics and road accident injuries. *Accident Analysis & Prevention* 29 (5): 583–593.

Agent, K. R. and R. C. Deen. 1975. Relationships between roadway geometrics and accidents. *Transportation Research Record* 541: 1–11.

Agran, P. F., D. G. Winn, C. L. Anderson, C. Tran, and C. P. Del Valle. 1996. The role of the physical and traffic environment in child pedestrian injuries. *Pediatrics* 98 (6): 1096–1103.

Anas, A., R. Arnott, and K. A. Small. 1998. Urban spatial structure. *Journal of Economic Literature* 36 (3): 1426–1464.

Bass, D., R. Albertyn, and J. Melis. 1995. Child pedestrian injuries in the Cape metropolitan area—Final results of a hospital-based study. *South African Medical Journal* 85 (2): 96–99.

Batty, M. and P. A. Longley. 1994. *Fractal Cities: A Geometry of Form and Function.* San Diego, CA/London: Academic Press.

Beasley, S. W. 2009. Driveway accidents in New Zealand. *Journal of the New Zealand Medical Association* 123 (1298): 5–7.

Braddock, M., G. Lapidus, E. Cromley, R. Cromley, G. Burke, and L. Banco. 1994. Using a geographic information system to understand child pedestrian injury. *American Journal of Public Health* 84 (7): 1158–1161.

Braddock, M., G. Lapidus, D. Gregorio, M. Kapp, and L. Banco. 1991. Population, income, and ecological correlates of child pedestrian injury. *Pediatrics* 88 (6): 1242–1247.

Buhlman, M. A., R. A. Warren, and H. M. Simpson. 1979. Profiles of fatal collisions involving child pedestrians. Paper presented at *the 23rd Conference of the American Association for Automotive Medicine (AAAM)*, Louisville, U.S.

Centers for Disease Control and Prevention. 1999. Motor–vehicle safety: A 20th century public health achievement. *Morbid Mortal Weekly Report* 48 (18): 369–374.

Cervero, R. and R. Gorham. 1995. Commuting in transit versus automobile neighborhoods. *Journal of the American Planning Association* 61 (2): 210–225.

Chambers, J. 2007. Understanding and acting to prevent driveway injuries to children. Position paper, Safekids New Zealand. http://www.safekids.org.nz/Downloads/ (accessed August 16, 2009).

Clifton, K. J., C. Burnier, and K. Kreamer-Fults. 2005. Women's involvement in pedestrian–vehicle crashes: Influence of personal and environmental factors. In *Transportation Research Board Conference Proceedings*, p. 35. Washington, DC: Transportation Research Board.

Clifton, K. J. and K. Kreamer-Fults. 2007. An examination of the environmental attributes associated with pedestrian–vehicular crashes near public schools. *Accident Analysis & Prevention* 39 (4): 708–715.

Cowley, N, M. Nicholls, and H. Parkinson. 2005. Preventing child pedestrian injuries and deaths arising from vehicle–child accidents in domestic driveways. University of Waikato, Hamilton, New Zealand.

CRTAS. 2006. China road traffic accidents statistics. Traffic Administration Bureau of China State Security Ministry, Beijing, China.

Ewing, R., R. Pendall, and D. Chen. 2002. *Measuring Sprawl and Its Impact*. Washington, DC: Smart Growth America.

Ewing, R., R. A. Schieber, and C. V. Zegeer. 2003. Urban sprawl as a risk factor in motor vehicle occupant and pedestrian fatalities. *American Journal of Public Health* 93 (9): 1541–1545.

Goldman, T. and R. Gorham. 2006. Sustainable urban transport: Four innovative directions. *Technology in Society* 28 (1): 261–273.

Gorham, R. 1998. Land-use planning and sustainable urban travel: Overcoming barriers to effective co-ordination. Background paper for the *OECD-ECMT Workshop on Land-Use for Sustainable Urban Transport: Implementing Change*, Linz, Austria, September 23–24, 1998.

Graham, D. and S. Glaister. 2003. Spatial variation in road pedestrian injuries: The role of urban scale, density and land-use mix. *Urban Studies* 40 (8): 1591–1607.

Graham, D., S. Glaister, and R. Anderson. 2005. The effects of area deprivation on the incidence of child and adult pedestrian injuries in England. *Accident Analysis & Prevention* 37 (1): 125–135.

Hamerslag, R., J. P. Roos, and M. Kwakernaak. 1982. Analysis of accidents in traffic situations by means of multiproportional weighted Poisson model. *Transportation Research Record* 847: 29–36.

Henning-Hager, U. 1986. Urban development and road safety. *Accident Analysis & Prevention* 18 (2): 135–145.

House of Commons. 1996. Risk reduction for vulnerable road users. Transport Committee Third Report, Paper HC 373, printed June 19, 1995, HMSO, London, U.K.

Hsiao, K. H., C. Newbury, N. Bartlett, R. Dansey, P. Morreau, and J. Hamill. 2009. Paediatric driveway run-over injuries: Time to redesign? *Journal of the New Zealand Medical Association* 122 (1298): 17–24.

Hummel, T. 2001. Land use planning in safer transportation network planning. SWOV Institute for Road Safety Research, Leidschendam, the Netherlands.

Ivan, J. N., R. K. Pasupathy, and P. J. Ossenbruggen. 1999. Differences in causality factors for single and multi-vehicle crashes on two-lane roads. *Accident Analysis & Prevention* 31 (6): 695–704.

Ivan, J. N., C. Wang, and N. R. Bernardo. 2000. Explaining two-lane highway crash rates using land use and hourly exposure. *Accident Analysis & Prevention* 32 (6): 787–795.

KidsAndCars. 2012a. KidsAndCars.org (KAC) website. http://www.kidsandcars.org (accessed April 17, 2012).

KidsAndCars.org (KAC). 2012b. Chart, statistics, graphics. http://www.kidsandcars.org/back-overs.html (accessed April 17, 2012).

Kim, K. and E. Yamashita. 2002. Motor vehicle crashes and land use: Empirical analysis from Hawaii. *Transportation Research Record* 1784: 73–79.

Kopits, E. and M. Cropper. 2005. Traffic fatalities and economic growth. *Accident Analysis & Prevention* 37 (1): 169–178.

LaScala, E. A., D. Gerber, and P. J. Gruenewald. 2000. Demographic and environmental correlates of pedestrian injury collisions: A spatial analysis. *Accident Analysis & Prevention* 32 (5): 651–658.

LaScala, E. A., P. J. Gruenewald, and F. W. Johnson. 2004. An ecological study of the locations of schools and child pedestrian injury collisions. *Accident Analysis & Prevention* 36 (4): 569–676.

Levine, N., K. E. Kim, and L. H. Nitz. 1995. Spatial analysis of Honolulu motor vehicle crashes: II. Zonal generators. *Accident Analysis & Prevention* 27 (5): 675–685.

Lightstone, A. S., P. K. Dhillon, C. Peek-Asa, and J. F. Kraus. 2001. A geographic analysis of motor vehicle collisions with child pedestrians in Long Beach, California: Comparing intersection and midblock incident locations. *Injury Prevention* 7 (2): 155–160.

Loukaitou-Sideris, A., R. Liggett, and H. G. Sung. 2007. Death on the crosswalk: A study of pedestrian–automobile collisions in Los Angeles. *Journal of Planning Education and Research* 26 (3): 338–351.

Lourens, P. F., J. A. Vissers, and M. Jessurum. 1999. Annual mileage, driving violations, and accident involvement in relation to drivers' sex, age, and level of education. *Accident Analysis & Prevention* 31: 593–597.

Lupton, K., M. Wing, and C. Wright, C. 1999. Conceptual data structures and the statistical modelling of road accidents. In *Mathematics in Transport Planning and Control: Proceedings of the Third IMA International Conference on Mathematics in Transport and Planning Control*, pp. 267–276. Oxford, U.K.: Elsevier Science.

Macpherson, A., I. Roberts, and I. B. Pless. 1998. Children's exposure to traffic and pedestrian injuries. *American Journal of Public Health* 88 (12): 1840–1843.

Martin, J. L. 2002. Relationship between crash rate and hourly traffic flow on interurban motorways. *Accident Analysis & Prevention* 34 (5): 619–629.

Millot, M. 2003. Développement urbain et insécurité routière: l'influence complexe des formes urbaines. PhD dissertation, Ecole des Ponts ParisTech, Champs-sur-Marne, France.

Millot, M. 2004a. Urban growth, travel practices and evolution of road safety. *Journal of Transport Geography* 12 (3): 207–218.

Millot, M. 2004b. The influence of urban planning on road safety. In *Proceedings of European Transport Conference*. London, U.K.: Association for European Transport.

Mumford, L. 1938. *The Culture of Cities*. New York, U.S.: Harcourt, Brace & Company.

NHTSA. 2000. Traffic safety facts 1999: A compilation of motor vehicle crash data from the fatality analysis reporting system and the general estimates system. DOT-HS-809-100. National Highway Traffic Safety Administration, U.S. Department of Transportation, Washington, DC.

NHTSA. 2002. Traffic safety facts 2002: Rural/urban comparison. DOT-HS-809-739. National Highway Traffic Safety Administration, U.S. Department of Transportation, Washington, DC.

Noland, R. B. and M. A. Quddus. 2005. Congestion and safety: A spatial analysis of London. *Transportation Research Part A: Policy and Practice* 39 (7): 737–754.

Ossenbruggen, P. J., J. Pendharkar, and J. N. Ivan. 2001. Roadway safety in rural and small urbanized areas. *Accident Analysis & Prevention* 33 (4): 485–498.

Peden, M., R. Scurfield, D. Sleet, D. Mohan, A. Adnan, E. Jarawan, and C. Mathers, eds. 2004. World report on road traffic injury prevention. World Health Organization, Geneva, Switzerland.

Petch, R. O. and R. R. Henson. 2000. Child road safety in the urban environment. *Journal of Transport Geography* 8 (3): 197–211.

Pucher, J. and L. Dijkstra. 2003. Promoting safe walking and cycling to improve public health: Lessons from the Netherlands and Germany. *American Journal of Public Health* 93 (9): 1509–1516.

Retting, R. A., A. F. Williams, D. F. Preusser, and H. B. Weinstein. 1995. Classifying urban crashes for countermeasure development. *Accident Analysis & Prevention* 27 (3): 283–294.

Roberts, I., R. Marshall, and T. Lee-Joe. 1995. The urban traffic environment and the risk of child pedestrian injury: A case-crossover approach. *Epidemiology* 6 (2): 169–171.

Ross, A. and G. Britain. 1991. Towards safer roads in developing countries: A guide for planners and engineers. Transport and Road Research Laboratory, Crowthorne, U.K.

Rumar, K. 1999. Transport safety visions, targets and strategies: Beyond 2000. European Transport Safety Council ETSC, Brussels, Belgium.

Shefer, D. and P. Rietveld. 1997. Congestion and safety on highways: Towards an analytical model. *Urban Studies* 34 (4): 679–692.

Stevenson, M. R., K. D. Jamrozik, and J. Spittle. 1995. A case-control study of traffic risk factors and child pedestrian injury. *International Journal of Epidemiology* 24 (5): 957–964.

The Institution of Highways and Transportation. 1997. *Transport in the Urban Environment*. London, U.K.: IHT.

TTI. 2001. Urban mobility report. Texas Transportation Institute, Arlington, TX.

United Nations Population Fund. 2007. State of world population 2007: Unleashing the potential of urban growth. UNFPA, New York.

Wedagama, D. M. P., R. N. Bird, and A. V. Metcalfe. 2006. The influence of urban land-use on non-motorised transport injuries. *Accident Analysis & Prevention* 38 (6): 1049–1057.

Zegeer, C., R. Steward, D. Reinfut, F. Council, T. Neuman, E. Hamilton, T. Miller, and W. Hunter. 1990. *Cost Effective Geometric Improvements for Safety Upgrading of Horizontal Curves*. Chapel Hill, NC: University of North Carolina.

Zhang, W., O. Tsimhoni, M. Sivak, and M. J. Flannagan. 2008. Road safety in China: Challenges and opportunities. Transportation Research Institute, University of Michigan, Ann Arbor, MI.

7 Nature of Spatial Data, Accuracy, and Validation

7.1 INTRODUCTION

Road collision data are subject to the same issues of any geographical data, namely uncertainties, inaccuracies, quality, scale, validation, and error. Road collision data can be fraught with challenges in terms of understanding the very nature of how it was collected, spatial resolution, accounting for inaccuracies, and ensuring data quality. This chapter reports on the nature of spatial road collision data and looks at issues of measurement, boundaries, transformation, time, coverage, scale, relevance, positional accuracy, and classifications. Inherently, road collision data are collected for administrative purposes rather than for scientific research; hence there is a need to validate and edit the data before conducting scientific analysis. Spatial concepts that are associated with road collisions are similar to those associated with many other geographical databases. However, the analysis of road collision data relies critically upon data quality and consistency in order to monitor and reduce road collisions. This chapter will outline some issues and examples on spatial data collection and use within the larger collision database.

7.2 CONCEPTUALIZING COLLISIONS AS NETWORK PHENOMENA

How do collision data differ from most other point spatial data like retail locations, tornado touchdown points, epicenters, or homes of people infected with malaria or other diseases? Traffic collisions do not happen in a typical Euclidian or two-dimensional (2D) space, that is, a "comprehensive" area with boundaries (polygons), by chance. Almost all traffic collisions occur on road networks, which are typically represented by arcs (lines) and nodes (points). Each arc is formed by a line joining a start and an end node on a one-dimensional (1D) space. These nodes and arcs, in turn, can be differentiated by *geometric features* like width, surface materials, curvature, and slope and *nongeometric features* like traffic volume, travel directions, and speed.

For spatial analysis, this fundamental understanding means that traffic collisions are network phenomena. Traffic collisions as spatial events are often mapped as points. However, these points do not occur "freely" or "randomly" in a 2D planar space, but are confined to the road network. Within a road network, these points can be distributed "randomly" but their spatial pattern in a 2D map will not be random. Figure 7.1 illustrates the difference. Figure 7.1a shows a random pattern of 50 traffic collisions using the point pattern method, that is, generating 100 random

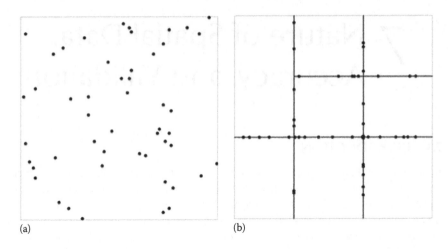

FIGURE 7.1 Random points in 2D and 1D space: (a) an example of 50 random points in a 2D space and (b) an example of 50 random points in a 1D space.

numbers within the maximum spatial extent on the x- and y-axes ($MaxX$ and $MaxY$). Figure 7.1b shows a random pattern of 50 traffic collisions using the point-in-network method, that is, generating 50 random numbers within the maximum length of the road network ($MaxL$).

As traffic collisions are network-constrained phenomena, they are best analyzed spatially as points-in-networks. It is rather meaningless for us to look at a collision pattern without reference to the road network pattern, as demonstrated in Figure 7.1a. Conversely, having a road network map of a city generally gives one a good idea of the collision patterns. Figure 7.2a shows the collision patterns in Hong Kong in 2008–2010. Figure 7.2b shows the road network in the city during this period. It can be seen clearly that the two spatial patterns bear remarkable resemblance at the city level. Describing and analyzing the spatial collision patterns in a 2D space without reference to the road network can hardly be considered appropriate from a theoretical point of view and useful from a policy perspective. At this point, the main argument is not so much on the statistical validity of analyzing spatial patterns of collisions without considering risk and exposure (more fully discussed in Chapters 10 through 12) but the lack of "value-added" in conducting direct 2D spatial analyses, given that information on the spatial pattern of the road network can be easily obtained and is extremely reliable in most developed economies. If one can identify clusters of collisions by identifying areas of dense roads, why bother with the spatial analysis of collisions at all? As suggested by Bailey (1994, 15), spatial analysis refers the "general ability to manipulate spatial data into different forms and extract additional meaning as a result." Therefore, the spatial analysis of collisions should focus on analyzing collision patterns within the network, so as to yield useful insights on spatial patterns beyond simply replicating information about the road network density and spatial configuration of the city.

Nature of Spatial Data, Accuracy, and Validation

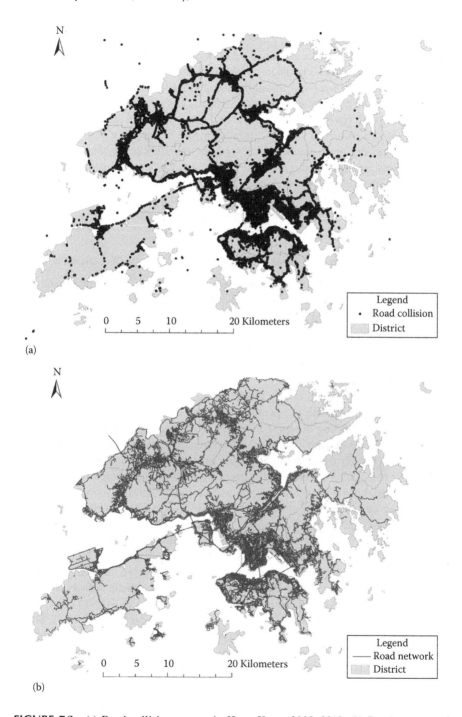

FIGURE 7.2 (a) Road collision pattern in Hong Kong, 2008–2010. (b) Road network of Hong Kong.

7.3 ISSUES INVOLVED WITH COLLISIONS-IN-NETWORKS IN GIS

Spatially, what are so special about networks? Road networks consist of nodes (including starting and end points/terminals and interchanges) and links (essentially continuous stretch of roads for through traffic). While roads differ in terms of various geometric and nongeometric features, they are essentially represented by centerlines in GIS, with road width, type/class, and other attributes stored in a relational database. Nodes in a road network usually involve a slowing down of speed due to either curvature or at junctions/intersections, roundabout, or simple dead ends. In road safety research, nodes are of key interest because traffic conflicts are most abundant at junctions. All network phenomena, therefore, can always be assigned to the nodes or lines where they occur, and be simplified and represented as 1D events. This is sometimes termed the link-attribute approach in spatial analysis.

7.3.1 Requirements of Spatial Accuracy and Precision of Collision Data

Correctly perceiving collisions as network phenomena in GIS raised two fundamental methodological challenges. First, the requirement of spatial accuracy of collision data becomes much higher. In geocoding, correctly assigning a collision to a district (an area) or matching the location of a collision to a nearby junction (a point entity with name) or a street (a line entity with name) is much easier than to plot the spatial reference of a collision (x, y) so that it exactly overlaps or intersects with the centerline of a road network.

In the reviews by Khan et al. (2004), and Loo and Tsui (2007), the police was found to be entrusted with the primary responsibility of collecting road collision data, primarily by filling in report forms using the pen-and-paper method, in both developed and developing countries. Moreover, the major aims are mainly for enforcement (notably prosecution, litigation, and insurance) and administrative (such as keeping road safety records and monitoring problems) purposes rather than for scientific spatial analysis. While the police information is usually very comprehensive yielding up to 99 pieces of information in relation to different characteristics of the road environment, vehicles, and road users, the accuracy, precision, and reliability of the collision data are often inadequate for spatial analysis (Shinar and Treat 1979; Shinar et al. 1983; Ibrahim and Silcock 1992; Austin 1995). The hustle and bustle at the collision scenes, and the need of expert knowledge on specific information like drivers' state and conditions, injury levels, and vehicle defects are major reasons for the discrepancies (Shinar and Treat 1979; Tsui et al. 2009).

Among different types of road collision data, Ibrahim and Silcock (1992), Austin (1993), and Khan et al. (2004) found that collision location was the *least* reliable. In particular, Ibrahim and Silcock (1992, 494) found that "the problem that occurred most frequently is the inaccuracy of the accident location by the grid reference." In places where the collision locations are not identified by the grid reference, the collision locations are usually assigned to some specific road features like kilometer marks or lamp posts. These road features, however, may be sparser or even absent in lower-order roads. For instance, the Belgian Analysis Form for Traffic Accidents uses the road identification number or the kilometer mark (Geurts et al. 2003).

Nonetheless, while regional roads have milestones (markers) every hectometer (100 m), they do not exist at the local roads level in Belgium (Steenberghen et al. 2004). Moreover, other researchers have resorted assigning collisions to the nearest road junctions, which are increasingly being considered as undesirable. In Honolulu of Hawaii, an error range of 0 to about 800 m (0.5 miles) was expected (Levine et al. 1995). The situation is worse in rural areas. In expressways with limited entry and exit points, the assignment of collisions to the nearest junctions is obviously either very imprecise or incorrect altogether.

7.3.2 Concept of Distance in Networks

Correctly perceiving traffic collisions as network phenomena also requires researchers to modify their concept of distance. The distance separating traffic collisions should no longer be seen as straight-line Euclidian distance commonly used in 2D spatial analysis such as nearest neighbor analysis. Instead, network distance should be used. The latter is unlikely to be identical to the former. Consider the case of emergency response; the Euclidian distance between a traffic collision scene and the hospital can be much shorter than the network distance. However, only the shortest network distance but not the Euclidian distance is relevant for ambulance dispatch. Moreover, two collisions located a few street blocks away on the same main road can have much higher relationship than two collisions located at the same location but one on a flyover above another. Hence, a proper measure of spatial proximity is required.

An early work of Moellering (1976) pointed this as the point-in-line versus point-in-area problem. For the former, random points can be generated to compare with the purely clustered and purely dispersed patterns. The distance separating the collisions should be measured quite simply along the line, instead of calculating from the geometric Euclidian distance. In reality, lines intersect to form a network. A network has structure, as measured by hierarchy, connectivity, and complexity. Hence, the seemingly simplified analysis of collisions in 1D becomes complicated again as points-in-network phenomena. Farber et al. (2010) looked at how network topology would affect the spatial proximity matrix and hence the estimation results of network autoregressive models. The relationship among different lines becomes important with the network proximity matrix, whereby line segments not in the right sequence can be neighbors to each other. This network structure becomes important as geographers have to identify spatial clusters in networks.

7.4 GEOVALIDATION BEFORE COLLISION ANALYSIS

The transformation of traffic collision data into spatial attributes with high precision for points-in-network analysis is a challenge in many parts of the world. Generally, the spatial data recorded in the traffic collision databases fall into two categories. The first category directly records the spatial locations with reference to a geographical feature. Depending on the details of the geographical feature, there may be sufficient detail for spatial analysis. The use of a road identification number or the kilometer mark in the Belgium is a case in point (Geurts et al. 2003).

The second category mainly records the collision locations by spatial reference, such as the grid reference (Ibrahim and Silcock 1992; Austin 1993; Kam 2003; Loo 2006) or GPS (Anderson 2009). In other words, traffic collisions are denoted by x- and y-coordinates. Theoretically, the latter would allow more precise collision locations (such as between two lamp posts or kilometer marks or junctions) to be recorded. However, unless the positional accuracy is very high, it is unlikely that these spatial coordinates can be mapped directly on the road centerlines for spatial analysis. Loo (2006), for instance, analyzed the grid reference information of road collisions in Hong Kong from 1993 to 2004 and found that only 0%–0.1% of the collisions identified by the five-figure grid references in the raw collision database could fall directly on or intersect with the centerlines of the road network. Data transformation is therefore necessary to assign collision locations identified by grid references, GPS, or other spatial references based on the Cartesian coordinate system back to the network before spatial analysis. Geovalidation is the process of validating the geocoding of spatial data. The essential concept is to countercheck the validity of the spatial locations using all relevant information in the collision and/or related databases. In most situations, additional information about spatial locations can be found in other locational variables, such as district variables or textual descriptions of the collision scene.

In a network context, there are two major approaches of geovalidation. The first approach is through buffering, which is to assign collisions to the lines as long as they are within the buffers of the lines. Collisions are assigned to road features based on the creation of buffer zones of the road network. This makes sense as many roads are of multiple lanes and the line represented in the spatial analysis is just a centerline of the road. Nonetheless, this method has the problem of assigning collisions that may fall within the buffers of more than one road. This is particularly the case in a highly dense urban road network. Problems of double-counting and ambiguity arise. Moreover, this can give rise to errors of mismatch of road attributes (notably, road name) at a later stage when collisions are assigned to the wrong road. Therefore, further steps of geovalidation are necessary. Notably, Austin (1995) used the highway feature data to validate the spatial locations of 156 road collision records in Humberside of the United Kingdom. In his GIS-validation system, collision locations were first plotted using the five-figure grid references of the collision database (layer 1). Then, based on the highway database, buffer zones of 24 m from either side of the centerlines were generated (layer 2). When a collision fell within the buffer zone of a highway feature, all locational variables of the collision were matched with the corresponding highway feature. The locational variables of the collision database used in the geovalidation include road class, road number, district, speed limit, pedestrian crossing facilities, junction control, junction detail, and carriageway type and markings. Austin's (1995) study suggested that the validity for different types of locational variables varied from more than 90% to less than 80%. Moreover, mismatches in carriage type, junction control, junction detail, and pedestrian crossing facilities were most serious.

The second approach is to assign collisions to the road network features directly by using other information in the collision database and other databases through address matching and other relevant fields. This is called snapping to the link-node system.

Nature of Spatial Data, Accuracy, and Validation 141

Levine et al. (1995) snapped collisions in Honolulu, Hawaii, to the nearest intersections. They then built a standardized dictionary of street names using the "AutoStan" software. The street names in the collision database were then matched with the files of topologically integrated geographic encoding and referencing (TIGER). If the matching was successful, the intersection was assigned as the collision location. This methodology, however, can lead to bias by assigning collisions happening far away from junctions to their nearest junctions. Distinguishing between collisions happening at junction and mid-block locations, Loo (2006) developed a GIS-based spatial data validation system to validate and identify traffic collision locations with the link-node system. Her methodology first makes use of additional information in other node, line, and polygon features collected in the traffic collision database to validate the exact locations of the collisions. The additional information about nodes and lines include road names, intersection names, lamp post numbers, distance to traffic lights, and other textual descriptions of the collision circumstances. The additional information about other polygon features includes district names and region names. The GIS-based spatial data validation system is developed mainly using ArcGIS (version 8.2) and ArcObject from the Environmental Systems Research Institute (ESRI). ArcGIS was used to incorporate and display the spatial data. ArcObject is a module that contains the ArcGIS software component library to customize the functionalities of the software. The module can be accessed and launched by using Microsoft Visual Basic through the Component Object Model (COM). A flowchart of the six-step geovalidation procedures is shown in Figure 7.3. Details of the implementation, together with a case study of Hong Kong, are described in the next section.

7.5 CASE STUDY OF HONG KONG

7.5.1 Database Preparation

Road collision data in Hong Kong, as many cities and countries worldwide, are primarily collected by the police. Specifically, the police filled out the "Traffic Accident Report Booklet (Injury Case)" for traffic collisions involving injuries. The data are processed and computerized by the police and the Transport Department into the collision, vehicle, and casualty databases. These three databases represent the major sources of road collision information in Hong Kong and are collectively known as TRADS. At present, only the collision database of TRADS contains spatial variables. In particular, the five-figure grid references (GRID_N, GRID_E) give the precise location of a road collision. Apart from the grid references, three additional codified locational variables—first street name (ST_NM), second street name (SECND_ST), and district board (DBOARD)—and three textual variables that describe the nearby landmarks (IDEN_FTR), precise locations (PREC_LOCTN), and circumstances (HAPPEN) in the TRADS collision database are found.

Next, the geovalidation involves the building of a link-node system about the road network in Hong Kong. In this respect, the Lands Department maintains the digitalized road network database in Hong Kong. As explained earlier, the creation of standardized buffer zones creates double-counting and false error problems,

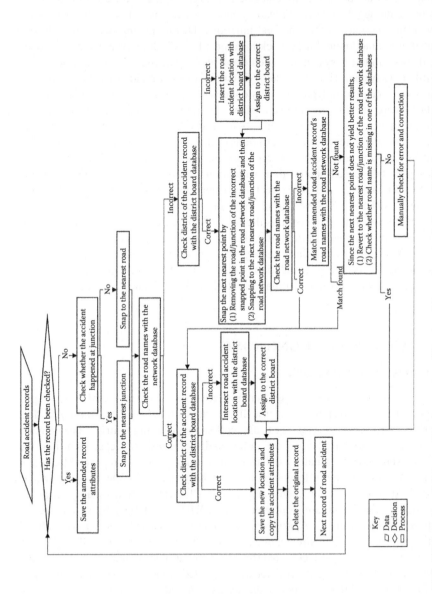

FIGURE 7.3 The GIS-based spatial data validation system. (Reprinted from *Acc. Anal. Prev.*, 38(5), Loo, B.P.Y. Validating crash locations for quantitative spatial analysis: A GIS-based approach, 879–886, Copyright 2006, with permission from Elsevier.)

especially in areas with narrow and dense roads (like urban Hong Kong) and near junctions. The Cartesian coordinates (X and Y) and the road name (ST_ENGNM) in the road network database are later used for matching information in the collision database.

Finally, this spatial data validation system makes use of the district board coverage. District board is chosen as the spatial subunit for validation, because it is the territorial unit not only for local election but also for the compilation and release of most official demographic and socioeconomic statistics in Hong Kong. Locational information of the collision database of TRADS is later validated against the spatial variables (X, Y, and Dist_Abbre) of the district database.

7.5.2 Methodology

Following the above conceptual framework of Figure 7.3, a six-stage GIS-based spatial data validation system is developed. In Loo (2006), the validation results of the police-recorded road collisions in Hong Kong from 1993 to 2004 were presented. In this chapter, updated validation results from 2005 to 2010 are shown in Table 7.1 to illustrate the methodology. Over this period, the total number of police-recorded road collisions in Hong Kong has stayed at around 15,000 per year.

In the first step, road collisions are snapped to the link-node system. To see whether the road collisions intersect with the link-node system, the five-figure grid references are first transformed into a GIS-compatible format (by adding the prefix "8" in this case). Then, a simple intersection process was performed in GIS. To distinguish between collisions happening at junctions and mid-block locations, the information contained in the textual descriptive variables is utilized. Specifically, a standardized library containing the terms used for denoting collisions happening at intersections is compiled. Examples of these terms used in Hong Kong include "intersection," "junction," "J/," and "JW." If the spatial variables contain these keywords, the collisions are snapped to the nearest junctions on the link-node system. Otherwise, the collisions are snapped to the nearest centerline of the road network. From 2005 to 2010, about 40%–45% of the collisions intersected with the link-node system. The shares were much higher than the 1993–2004 period with less than 1% of the collisions (Loo 2006). After snapping the collision locations to the nearest points on the link-node system, it was found that about 90%–95% of the road collisions could be validated as having both correct road names and district boards. Once again, there have been substantial improvements, when compared with 79.2% back in 2004 (Loo 2006).

At the second step, road names of the collision records are matched with the road network database. Since the road names in Hong Kong do not follow any particular system (e.g., the numbering system from north to south or east to west), another library was developed to identify road names in the textual spatial variables. Examples of these terms include "Road," "Street," "Avenue," "Path," "Circuit," "Highway," "Roundabout," "Lane," and their numerous forms (including abbreviations). If the road name matching is successful, the spatial variables of the collision database are then validated against the district board database in the third step. If not, the fourth step is triggered, that is, collisions are snapped to the *next* nearest

TABLE 7.1
Results of the Geovalidation of Traffic Collisions in Hong Kong, 2005–2010

	2005	2006	2007	2008	2009	2010
The raw collision database						
Number of road collisions	15,062	14,849	15,315	14,576	14,316	14,943
On road centerline (%)	44.5	43.5	42.5	42.2	41.9	40.9
Phase One: collisions snapped to the nearest points on the link-node system						
Road names and district matched (%)	46.1	46.3	45.7	48.5	46.0	49.0
Cum. correct (%)	90.6	89.8	88.2	90.7	87.9	89.9
Phase Two: incorrect district board information identified and amended						
Road names and district matched (%)	4.5	4.2	4.5	3.9	4.3	3.5
Road names remained incorrect (%)	1.0	0.8	1.1	0.8	0.5	1.0
Cum. correct (%)	95.1	94.0	92.7	94.6	92.2	93.4
Phase Three: unmatched collisions snapped to the next nearest points on the link-node system						
Road names and district matched (%)	3.0	3.7	5.3	3.5	5.9	4.2
Cum. correct (%)	98.1	97.7	98.0	98.1	98.1	97.6
Phase Four: incorrect district board information identified and amended						
Road names and district matched (%)	0.1	0.1	0.1	0.1	0.1	0.1
Cum. correct (%)	98.2	97.8	98.1	98.2	98.2	97.7
Phase Five: unmatched collisions snapped to the "Identified" road						
Road names and district matched (%)	0.7	0.9	0.8	0.7	0.6	1.1
District further amended (%)	0.1	0.2	0.2	0.1	0.1	0.2
Cum. correct (%)	99.0	98.9	99.1	99.0	98.9	99.0
Phase Six: missing road names and typo-errors identified and amended (manually corrected)						
Missing road names (%)	0.6	0.6	0.5	0.6	0.6	0.6
Wrong spellings or other typo-errors (%)	0.4	0.5	0.4	0.4	0.5	0.4
Cum. correct (%)	100	100	100	100	100	100

road junction or section before a new round of road name matching exercise. If the matching is successful, the district board information is then validated again. This step is necessary because the next nearest point may lie in a different district board. During 2005–2010, about 3.5%–4.5% of the collision records were found to be having correct road names but wrong district board information. Should the focus of validation be put on road names only, 88.2%–90.7% of the collision records could be considered as correct. After step four, the cumulative percentage of validated collision records increased to 97.7%–98.2% in 2005–2010.

For unsuccessful matching at the second round, the fifth step of the system is to identify the road names recorded in the spatial variables of the collision database and then try to find a match in the road network database. This step can be performed by the address matching function of GIS (see, e.g., Levine and Kim 1998). If the matching is successful, the collision is snapped to the nearest road junction or section of the "identified" road. At step five, almost 98.9%–99.0% of the collision records had been geovalidated using the GIS system with computer programs.

Then, the last step of the system checks for problems related to missing road names in either the collision or the road network database. All these and remaining records are then checked manually for wrong spellings and other typographical errors. With about 15,000 collisions per year, it means that only about 150 collision records (1%) had to be checked manually. With the GIS-validation procedures, the task of spatial data validation has been not only improved but also simplified substantially. In particular, a distinction of junctions and road segments is a big breakthrough in assigning collisions correctly. Moreover, the development of a GIS algorithm to geovalidate automatically means an enormous reduction in efforts to improve spatial accuracy. During the study period, the majority of the collision records (14,900 or 99%) had been checked and validated by the computerized GIS-based validation system.

7.6 CONCLUSION

With the advance of GIS, researchers have applied the capabilities to assign and to validate the spatial accuracy of collision locations before spatial analysis, which ranges from simple visualization to generating descriptive spatial statistics and building complex spatial interaction and autocorrelation models. Generally, the smaller the spatial unit of spatial analysis, the higher the precision requirement. Moreover, the more complex the spatial model, the higher the precision requirement. After all, the reliability and validity of the results of a model depends critically on the quality of the data used to develop that model. In spatial analysis, geovalidation is an essential step of safeguarding and ensuring the validity and reliability of spatial data before any scientific spatial analysis.

REFERENCES

Anderson, T. K. 2009. Kernel density estimation and K-means clustering to profile road accident hotspots. *Accident Analysis & Prevention* 41 (3): 359–364.

Austin, K. 1993. The collection and use of additional sources of road safety data in highway authorities. *Traffic Engineering & Control* 34 (11): 540–543.

Austin, K. 1995. The identification of mistakes in road accident records: Part 1, Locational variables. *Accident Analysis & Prevention* 27 (2): 261–276.

Bailey, T. C. 1994. A review of statistical spatial analysis in geographical information systems. In *Spatial Analysis and GIS*, eds. S. Fotheringham and P. Rogerson, pp. 8–25. London, U.K.: Taylor & Francis.

Farber, S., A. Páez, and E. Volz. 2010. Topology, dependency tests and estimation bias in network autoregressive models. In *Progress in Spatial Analysis*, eds. A. Páez, J. Gallo, R. N. Buliung, and S. Dall'erba, pp. 29–57. Berlin, Germany: Springer.

Geurts, K. and G. Wets. 2003. Black spot analysis methods: Literature review. Flemish Research Center for Traffic Safety, Diepenbeek, Belgium.

Ibrahim, K. and D. T. Silcock. 1992. The accuracy of accident data. *Traffic Engineering & Control* 33 (9): 492–496.

Kam, B. H. 2003. A disaggregate approach to crash rate analysis. *Accident Analysis & Prevention* 35 (5): 693–709.

Khan, M. A., A. S. Al Kathairi, and A. M. Grib. 2004. A GIS based traffic accident data collection, referencing and analysis framework for Abu Dhabi. In *Proceeding of CODATU XI: Towards More Attractive Urban Transportation*, Bucharest, Romania. Lyon, France: Association CODATU.

Levine, N. and K. E. Kim. 1998. The location of motor vehicle crashes in Honolulu: A methodology for geocoding intersections. *Computers, Environment and Urban Systems* 22 (6): 557–576.

Levine, N., K. E. Kim, and L. H. Nitz. 1995. Daily fluctuations in Honolulu motor vehicle accidents. *Accident Analysis & Prevention* 27 (6): 785–796.

Loo, B. P. Y. 2006. Validating crash locations for quantitative spatial analysis: A GIS-based approach. *Accident Analysis & Prevention* 38 (5): 879–886.

Loo, B. P. Y. and K. L. Tsui. 2007. Factors affecting the likelihood of reporting road crashes resulting in medical treatment to the police. *Injury Prevention* 13 (3): 186–189.

Moellering, H. 1976. The potential uses of a computer animated film in the analysis of geographical patterns of traffic crashes. *Accident Analysis & Prevention* 8 (4): 215–227.

Shinar, D. and J. R. Treat. 1977. Tri-level study: Modification task 3: Validity assessment of police-reported accident data. U.S. Department of Transportation, National Highway Traffic Safety Administration, Washington, DC.

Shinar, D., J. R. Treat, and S. T. McDonald. 1983. The validity of police reported accident data. *Accident Analysis & Prevention* 15 (3): 175–191.

Steenberghen, T., T. Dufays, I. Thomas, and B. Flahaut. 2004. Intra-urban location and clustering of road accidents using GIS: A Belgian example. *International Journal of Geographical Information Science* 18 (2): 169–181.

Tsui, K. L., F. L. So, N. N. Sze, S. C. Wong, and T. F. Leung. 2009. Misclassification of injury severity among road injuries in police reports. *Accident Analysis & Prevention* 41 (1): 84–89.

8 Collisions in Networks

8.1 INTRODUCTION

A road collision is a point or an occurrence on a map. However, in many modeling and prediction techniques, traffic collisions tend to be assigned either to links on the road network or to administrative units for aggregate analysis. This then leads to a fundamental problem experienced by many geographers, which is determining the most appropriate size and shape of the spatial units used for analysis, since this may heavily influence the visual message of mapping and the outcome of statistical tests. There has been considerable debate in recent years concerning the optimal length of basic spatial units for road collision analysis. Largely this is a size and scale issue well known to geographers, that is, the modifiable areal unit problem (MAUP). Often the choice of the level of aggregation is constrained by the format of available data, because collision data and explanatory variables are often collected by different agencies. This chapter seeks to address to the problem of network segmentation and optimal length of basic spatial unit, and the ways collisions are assigned to segments. Besides, methods for assessing spatial autocorrelation have existed for several decades and stem from the work of Moran (1948). An extension of spatial autocorrelation analysis for assessing departures from randomness in regression residuals for flows on a network has been explored in the last two decades. This chapter also seeks to explore the notion of network autocorrelation, and how it can be measured using global and local variations (e.g., local Moran's I). It will take case studies and examples to expand the theory and statistics.

8.2 MAUP IN NETWORKS

Road safety analysts often require additional training to be capable of conducting meaningful spatial analysis, one reason being the fact that events in space are more often than not aggregated for data analysis. Whenever events are aggregated, the challenge of MAUP exists. In road safety research, it can be demonstrated most clearly with an analysis of traffic collisions as point events directly, that is, before the road network is introduced. Figure 8.1a shows a hypothetical spatial distribution of collisions over space. Suppose the collision pattern is analyzed by superimposing standard cells to measure the central tendency (mean, $\bar{x}$) and variability (standard deviation, sd) of collision counts or density over space in the study area. All key aspects of defining the cells or basic spatial units (BSUs), including their size, shape, and orientation, will affect the statistics obtained. Generally, the higher the spatial resolution (small cells and larger number of cells), the lower the expected mean and standard deviation. To illustrate, the use of 2×2 (grey solid lines) and 4×4 (grey dash lines) grid cells as in Figure 8.1b will yield average cell collision counts of

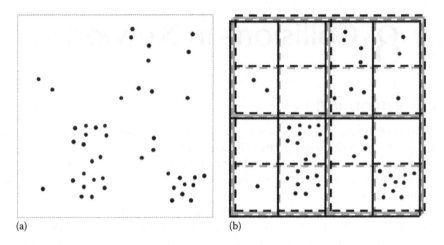

FIGURE 8.1 An illustration of MAUP in 2D point pattern analysis: (a) a hypothetical distribution of 44 traffic collisions in space, (b) alternative ways of dividing the study area into BSUs for spatial analysis.

11 and 2.75, respectively. The standard deviations are 7.07 and 3.53, respectively. However, the variability of the datasets, as measured by the coefficient of variations (cv), actually increases from 0.64 to 1.28 as the spatial resolution increases. The challenge is often called the scale problem. In addition, both the shape and the orientation of spatial unit of analysis will affect the statistics. To demonstrate, the use of rectangles with short edges as width (dark dash lines) as the BSU will lead to a $\bar{x}$ of 5.5, sd of 6.02, and cv of 1.10. With the same size and shape (rectangles), using the rectangles with long edges (dark solid lines) as width, will dramatically reduce sd to 3.85 and cv to 0.70. In reality, a traffic collision may be assigned to a traffic analysis zone, planning units, census tracts, or borough, which may be of a similar spatial scale but irregular shape. The fact that there are more than one way of assigning an object to a BSU is also called the boundary problem.

Following Chapter 7 that traffic collisions are network phenomena, a spatial analysis of traffic collisions is most appropriately conducted using nodes (points) and arcs (lines) rather than polygons (area). As a result, the challenge of MAUP needs to be reexamined carefully. First, if nodes are analyzed directly and independently as points in space, there is no spatial aggregation and the MAUP does not apply. The only key problem is associated with assigning collisions to the "correct" nodes. For instance, should traffic collisions happening with 70 m from a junction be assigned to it? Why not 50 m? 100 m? Do we ignore the nonjunction collisions? These questions, however, are primarily dealt with at the geovalidation stage and do not involve MAUP in defining spatial unit of analysis. Second, when road segments or lines are used as BSUs to analyze traffic collisions, the MAUP applies. Nonetheless, the shape problem is irrelevant as BSUs must take the form of lines. So, the remaining key issue is the scale problem. In other words, what is the optimal BSU length (l) for the spatial analysis of traffic collisions? Taking the entry/exit points of motorways to define road sections, Thomas (1996) considered three types of road segments as

BSUs to analyze the road sections. With the measurement unit as hectometer (hm) or 100 m in Belgium, type A road segments are obtained using 1–49 hm as l of a standard BSU throughout the road sections, ignoring any "leftover bits." Type B road segments are obtained using 3 hm as l of a standard BSU in the middle segment of a road section. For each road section, only the middle segment is considered. Type C road segments have variable l with each BSU having l equal to the length of the road sections. His results confirm that MAUP does exist in networks. Hence, "generalizations made at one level do not necessarily hold at another level, and that conclusions we derived at one scale may be invalid at another" (Haggett 1965, 263). When motorway collisions are analyzed, his analysis suggests that the larger the scale (longer and smaller number of BSUs, ceteris paribus), the closer the statistical distribution of collision counts and ratios approximate a normal distribution. When l is 1 hm, the distribution is close to a Poisson distribution, which is consistent with the vast nonspatial literature on statistical modeling of road collisions. Given the MAUP in network phenomena, it is important to recognize that all relationships (spatial or otherwise) that one establishes are always relative to the ways that the spatial units are defined.

8.3 NETWORK SEGMENTATION

For meaningful spatial analysis of traffic collisions to take place, a BSU must be shorter than the entire road or road network under study. In other words, there is a need to "break down" the road or road network into finer units/BSUs before analysis. This process is generally called network segmentation. In conducting network segmentation, the standard procedures are typically as follows (Loo 2009; Yamada and Thill 2010). First, take all junctions and road ends in the network as nodes. With the nodes defined, road sections of varying length will result. Should the BSU be of a standard length of l, road sections are checked to see whether their lengths exceed l. If so, the road sections are further divided from the starting nodes at intervals of l. In situations whether road junctions are close or where the road section length cannot be divided fully into l, BSUs of lengths shorter than l will result. Under a controlled or one road situation (Black 1991; Thomas 1996; Flahaut et al. 2003), these shorter nonstandard-length BSUs are excluded from the analysis. Nonetheless, the problem cannot be simply ignored should complex empirical road networks are analyzed. In the case of Hong Kong, segmenting the 1,090 km of road in Hong Kong into BSUs of 100 m will yield 14,292 BSUs, of which 45.1%, 24.3%, and 11.5% are shorter than 100, 50, and 25 m, respectively (Loo and Yao 2013).

To reduce the undesirable effect of fragmented road segments, Loo and Yao (2012) proposed a network dissolving algorithm suitable for handling complex empirical road network. Essentially, the raw GIS link-node road network is "dissolved" with a priority sequence before network segmentation. The steps are summarized in Figure 8.2 and described in the following text:

1. Take a link in the raw link-node system. Identify all neighboring links sharing a common end node as the subject link.
2. If there is only one neighbor, dissolve it with the neighbor.

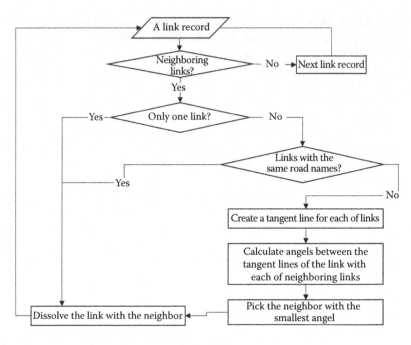

FIGURE 8.2 A flowchart of the network dissolution procedures.

3. If there are two or more neighbors, dissolve the subject link with the one having the same road name. Repeat until there is no neighboring link sharing the same road name.
4. If none of the contiguous links shares the same road name, create a tangent line for each of the contiguous links at the end node of the subject link.
5. Calculate the angle (0°–90°) between the tangent lines of subject link with each of the contiguous links. The one with the smallest angle is picked as the merged segment.
6. Repeat until no links sharing a common end node. Repeat for the start node.

As this kind of dissolving task is highly laborious, a GIS program is developed to accomplish the task. After the procedures of dissolving the network, the number of BSUs for the Hong Kong network reduced from 14,292 to 11,398, of which the shares of BSUs with length less than 100, 50, and 25 m are dramatically reduced to 23.3%, 4.4%, and 2.0%, respectively (Loo and Yao 2013).

8.4 BASIC SPATIAL UNITS IN COLLISION ANALYSIS

Apart from the MAUP, the definition of a BSU needs to be considered very carefully because it represents the smallest unit of analysis. All finer details beyond this scale will not be considered in further analysis. Hence, if the quality of spatial data on collision locations is good, researchers should consider carefully whether the use of longer BSUs will involve substantial trade-offs in details about the road environment.

In reality, the minimum length of a BSU is often dictated by the collision reporting system. For instance, given the data limitations in Belgium, the precise location of a collision within a hectometer is unknown. Hence, Thomas's (1996) paper did not address issues of alternative ways of segmenting the road section (e.g., from the middle of a hectometer to the middle of the next) and the optimal l for the spatial analysis of collisions. The minimum distance increment for reporting collision locations vary substantially worldwide. Within the U.S., it ranged from 0.1 to 1 mile in different states, with 0.1 mile in Virginia, Florida, Idaho, Oklahoma, California, and Connecticut; 0.2 mile in Michigan; 0.4 mile in Alabama; and 0.1–1 mile within North Carolina (Deacon et al. 1975; Black 1991). Unless vigorous geovalidation procedures (which take into account detailed textual descriptions, including street addresses, of collision circumstances) are in place, higher spatial resolution beyond 100 m is either unobtainable or unreliable at the city level or above. An exception is perhaps London, whose traffic police are equipped with GPS to measure collision locations with a 10 m resolution (Anderson 2009).

8.5 ASSIGNING COLLISIONS TO NETWORKS

Next, traffic collisions have to be assigned properly to predefined BSUs of the road networks before scientific spatial analysis. As this step follows the geovalidation process, all traffic collisions should already be on the road network. However, segmenting the road network into BSUs will still give rise to the double-counting issue, because a traffic collision may happen at the meeting point of an end node of a BSU and the start node of the following contiguous BSU. When the traffic collision is counted toward both BSUs, double-counting will arise. While the situation will arise at junctions, it is noteworthy to pinpoint that it will also happen along continuous stretch of highways (with no entry or exit) and mid-block locations (with no junctions) whenever a road section is longer than l of the BSU. To solve this problem, Loo (2009) suggested assigning the collision to one of the BSUs by random selection or by a predefined rule. Typically, GIS stores the location of a BSU by recoding a series of geographic coordinates. The minimum and the maximum x (*xMin* and *xMax*) and y (*yMin* and *yMax*) coordinates of each BSU can be calculated. The collision can then be assigned to one of the BSUs according to their locations, such as the left (smaller *xMin*), the right (larger *xMax*), the upper (higher *yMax*), and the lower (smaller *yMin*). Labeling all intersecting BSUs and drawing random numbers to assign the collision can be an alternative. Unless there are compelling reasons to assign collisions to a predefined rule (for instance, a traffic collision may always be assigned to higher-order road or road with higher traffic volume), ensuring randomness in assigning the collisions should be the general principle.

8.6 SPATIAL AUTOCORRELATION ANALYSIS IN NETWORKS

Next, how do we analyze the spatial pattern of collisions in networks? In road safety, the primary concern in spatial analysis is almost inevitably to consider: Is there a significant clustering of collisions? If so, where are these clusters? To geographers, it means: Does positive spatial autocorrelation exist? "Spatial autocorrelation is the tendency for the level of a variable at one location to influence the level of that variable at sites in proximity to the first location" (Black 1991, 75). In other words, spatial dependency exists between the value of a variable at a location and the value of the same variable at nearby locations. There are two types of spatial autocorrelation. "If positive spatial autocorrelation is present, it results in a spatial clustering of similar variable values" (Black 1991, 75). Conversely, negative spatial autocorrelation suggests spatial dispersion. Approaches to identify positive spatial autocorrelation can be divided into two major groups (Loo and Yao 2013).

8.6.1 Link-Attribute Approaches

When traffic collisions are assigned to the road network, positive spatial autocorrelation can be identified by mapping and building statistical models to analyze the attribute values (such as collision counts and collision density) of the road segments or BSUs. The BSU, a geographic feature, in the network represents the fundamental spatial unit of analysis. Apart from collisions, both geometric (such as road width and gradient) and nongeometric features (such as traffic volume and presence of road markings) of the BSUs can be stored in the relational database of the road network. With the above information, collisions on a BSU can be expressed in many different ways like collision count, collision density per road distance, and collision density per vehicular traffic volume. All these variables related to traffic collisions are termed collision intensity measures. By assigning traffic collisions to the road network, additional information about the traffic collisions, such as the collision type, number of injuries, degree of injury, and number of fatality, can also be visualized and analyzed spatially by BSUs. Furthermore, additional collision intensity measures, such as the number of serious and fatal collisions per traffic volume, can be derived. Different methodologies of analyzing traffic collisions by considering them as attributes of the road features are generally called to be following the link-attribute approach. One of the key advantages of the link-attribute approach is its ability to integrate various key databases, such as the collision database, hospital database, traffic database, and even land use database, in an appropriate network setting.

In building a spatial model to detect spatial autocorrelation, a fundamental step is to build a matrix **W** containing weights W_{ij} that describe the spatial relationship (e.g., contiguity, proximity, or connectivity) between BSU i and j. In a network setting, **W** should be based on the shortest path of network analysis, with each BSU as a topological step. The use of Euclidian distance is inappropriate, as explained in Chapter 7. Once the spatial matrix is derived, some spatial statistics can be derived. The most common ones include Moran's I, Geary's c, and Getis–Ord General G. In these statistical tests, the null hypothesis is "there is spatial randomness." Then, the statistics measure the deviation from spatial randomness. And a level of statistical significance is chosen to reject or do not reject the null hypothesis.

By and large, Moran's I is the most common spatial statistics. As the statistic is at the global level (i.e., considering the entire study area as a whole), it is produced by standardizing the spatial autocovariance by the variance of the data using a measure of the connectivity of the data.

The mathematical formula is shown in the following:

$$I = \frac{N \sum_{i=1}^{N} \sum_{j=1}^{N} W_{ij}(X_i - \bar{X})(X_j - \bar{X})}{\left(\sum_{i \neq j}^{N} \sum_{j=1}^{N} W_{ij}\right)\left(\sum_{i=1}^{N}(X_i - \bar{X})^2\right)} \quad (8.1)$$

where
N is the total number of BSU
W_{ij} is the spatial distance separating BSU i and j
X is the attribute value of a collision intensity measure
$\bar{X}$ is the global mean value calculated as the average of all data
$(X_i - \bar{X})(X_j - \bar{X})$ is the covariance

In road safety, X is usually a collision intensity measure, and W_{ij} is based on network distance. Partly due to the complexity of calculating the network proximity matrix, the spatial weights are usually simplified as a dummy variable indicating whether the two BSUs are contiguous (1) or not (0) only (Loo 2009). The range of possible values of Moran's I is −1 to 1. Positive values indicate a spatial clustering of similar values and negative values indicate a clustering of dissimilar values. Statistical significance test can be conducted to indicate the level of confidence that one can have about whether the difference/pattern is not simply due to chance. For a significance test of I_i, the Z-score derived is calculated by

$$Z(I_i) = \frac{I_i - E(I_i)}{\sqrt{Var(I_i)}} \quad (8.2)$$

where the expected value of I is calculated by

$$E(I) = \frac{-1}{N-1} \quad (8.3)$$

Depending on whether normal approximation or randomization experiment is assumed, the equations for calculating the variance (VAR) differ (Goodchild 1986; Griffith 1987; Odland 1988). Under the assumption of normality,

$$Var(I) = \frac{N^2 S_1 - N S_2 + 3\left(\sum_{i=1}^{N} \sum_{j=1}^{N} W_{ij}\right)^2}{\left(\sum_{i=1}^{N} \sum_{j=1}^{N} W_{ij}\right)^2 (N^2 - 1)} \quad (8.4)$$

where

$$S_1 = \frac{1}{2} \sum_{i=1}^{N} \sum_{j=1}^{N} (W_{ij} + W_{ji})^2 \quad (8.5)$$

and

$$S_2 = \sum_{i=1}^{N}\left(\sum_{i=1}^{N} W_{ij} + \sum_{j=1}^{N} W_{ji}\right) \tag{8.6}$$

Nonetheless, Moran's I is a global measure that provides a single value of spatial autocorrelation for the entire dataset. The statistic does not show where the clusters are or provide quantitative information about where high spatial covariance or dependence is detected. Improvements of the Moran's I index to cover the local scale, that is, local indicator of spatial association (LISA), are usually attributed to Anselin (1995). To detect local clusters, a BSU-specific local Moran's I statistic can be calculated using Equation 8.7. Through plotting LISA in maps, local spatial clusters can be identified.

$$I_i = \frac{(X_i - \bar{X})}{Var} \sum_j W_{ij}(X_j - \bar{X}) \tag{8.7}$$

Statistical significance test can be conducted to indicate the level of confidence that the difference/pattern is not simply due to chance. For a significance test of I_i, the Z-score derived by either normal approximation or randomization experiments is calculated by

$$Z(I_i) = \frac{I_i - E(I_i)}{\sqrt{Var(I_i)}} \tag{8.8}$$

where $E(I_i)$ is the expected value of I and given in the equation:

$$E(I_i) = \frac{-1}{N-1} \sum_{j=1}^{N} W_{ij} \tag{8.9}$$

Next, Geary's c uses the sum of squared differences between pairs of data values as its measure of covariation. Mathematically, it is shown in the equation

$$c = \frac{(N-1)\sum_{i=1}^{N}\sum_{j=1}^{N} W_{ij}(X_i - X_j)^2}{2\sum_{i=1}^{N}\sum_{j=1}^{N} W_{ij}\sum_{i}^{N}(X_i - \bar{X})^2} \tag{8.10}$$

The notations are the same as before. The range of possible values of c is 0–2. A value of c close to 0 means that the distribution of values is clustered; conversely, a value of c close to 2 means that the distribution of values is dispersed. $c=1$ implies that there is no spatial autocorrelation. Similar to Moran's I, a significance test can be conducted by the Z-score method, that is, deducting the observed and expected value of c, and dividing it by the standard deviation of Geary's c.

Nonetheless, both Moran's *I* and Geary's *c* methods indicate clustering of high or low values, but these methods cannot distinguish between the high–high or low–low situations. Other spatial autocorrelation models, including the joint count, semivariance, second-order, and Getis model, place more restrictions on W_{ij} and/or the covariance (Getis 2010).

While the spatial dependency effects may be tested directly (as in the above tests), they may also be entered into regression models, such as the spatial lag model (SLM), which introduces a spatial lag variable (see, e.g., Levine et al. 1995), and the spatial error model (SEM) that incorporates a spatial error term. Moreover, other more advanced spatial interaction models, such as the gravity model, may be built to express the spatial relationship as vectors of attributes related to *i* and *j* and a vector of separation attributes, such as distance or intervening opportunities. Nonetheless, "interactions" among traffic collisions are quite uncommon, and these types of spatial models are not commonly used in the spatial analysis of traffic collisions.

8.6.2 Event-Based Approaches

In event-based analysis, collisions are represented as points. This kind of analysis can be further classified into distance-based methods that examine distances between events and density-based methods that examine the crude density or overall intensity of a point pattern (O'Sullivan and Unwin 2003). Frequently used distance-based methods that directly analyze the distances among collisions as spatial events include the nearest neighbor distance analysis.

The alternative to distance-based methods is density-based measures. Quadrat count methods and density estimation belong to this type. The kernel density estimation (KDE) methods are particularly promising in analyzing collision patterns (Pulugurtha et al. 2007; Delmelle and Thill 2008; Erdogan et al. 2008; Anderson 2009; Yao et al. 2015). The fundamental concept of kernel density estimates is that collisions do not happen at discrete "points" in space only. Instead, they can happen over continuous space, whether continuous over a line (1D) or over space (2D). Let us take the simpler 2D case first. It means that collision is considered as a continuous (rather than a discrete) variable over space. Researchers, therefore, are not dealing with the probability of a collision happening at point *x* out of a finite set of points with the range of *a* and *b*, but a probability density function of a collision happing within a continuous stretch of road, defined by the bandwidth *h*. Therefore, a kernel function *k* satisfies the condition $\int_{-h}^{h} k(x)\,dx = 1$ (Silverman 1986). And the kernel intensity estimator over point *x* is given by

$$\hat{\lambda}(x) = \sum_{i=1}^{n} \frac{1}{h^2} k\left(\frac{x - x_i}{h}\right) \qquad (8.11)$$

(Silverman 1986; Flahaut et al. 2003). Further dividing the kernel intensity estimator by the number of observations, *n* (number of collisions falling within *h* from point *x*), will give the kernel probability density function (Flahaut et al. 2003).

Arguably, bandwidth, also called window width or the smoothing parameter, is the most important criterion for determining the most appropriate density surface (Silverman 1986; Bailey and Gatrell 1995; Fotheringham et al. 2000). The use of mean integrated squared error (MISE), variable h, and adaptive h (adaptive kernel estimation) are some methods of choosing the appropriate h for a study (see Flahaut et al. 2003; Bailey and Gatrell 1995).

There are different ways of specifying the functional form of the k functions. Among them, the quadrat kernel represents a typical choice, which can be shown in Figure 8.3 (reproduction of Bailey and Gatrell (1995). The other commonly used forms are the Gaussian function and the minimum variance function (Schabenberger and Gotway 2005).

When $0 < x - x_i \leq h$, the k intensity as defined using the quadrat function is

$$k\left(\frac{x-x_i}{h}\right) = \frac{1}{\sqrt{2\pi}} \exp\left(-\frac{x-x_i}{2h^2}\right) \qquad (8.12)$$

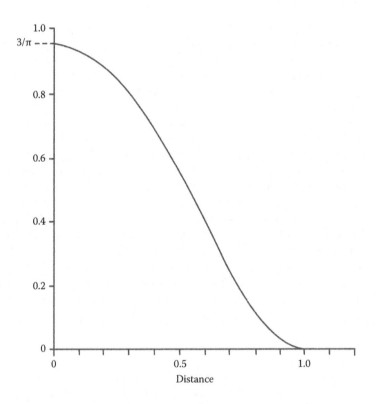

FIGURE 8.3 A cross section of the kernel using the $3/\pi$ quartic function. (Reprinted from Bailey, T.C. and Gatrell, A.C., *Interactive Spatial Data Analysis*, Longman Scientific & Technical, Essex, U.K., 1995, p. 86. With permission from Taylor & Francis Group.)

Using the quartic function, it is

$$k\left(\frac{x-x_i}{h}\right) = \frac{3}{\pi}\left(1 - \frac{(x-x_i)^2}{h^2}\right) \qquad (8.13)$$

or

$$k\left(\frac{x-x_i}{h}\right) = \frac{3}{4}\left(1 - \frac{(x-x_i)^2}{h^2}\right) \qquad (8.14)$$

Using the minimum variance function, it is

$$k\left(\frac{x-x_i}{h}\right) = \frac{3}{8}\left(3 - 5\frac{(x-x_i)^2}{h^2}\right) \qquad (8.15)$$

When $x - x_i > h$, $k((x-x_i)/h) = 0$, regardless of the functions used. Nonetheless, it was recognized that the different functions all yield reasonable estimates and do not affect the kernel density as much as the choice of bandwidth.

Although conventional distance-based methods were originally developed for 2D space, researchers have extended them to 1D, where distances between events are more appropriately calculated using network measurements other than Euclidian distances. An essential raised recently, however, is that traffic collisions as network-constrained phenomena are to consider n not as the number of collisions falling within the radius of h of point x, but as the number of collisions falling within the network distance of h in different directions from point x. Figure 8.4, reproduced from Yamada and Thill (2004), illustrates the concept.

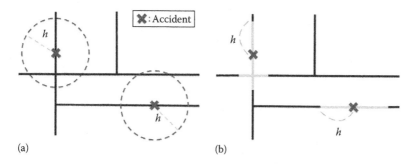

FIGURE 8.4 Planar versus network K-function: (a) the planar K-function with h, (b) the network K-function with h. (Reprinted from *J. Transp. Geogr.*, 12(2), Yamada, I. and Thill, J.-C., Comparison of planar and network K-functions in traffic accident analysis, 149–158, Copyright 2004, with permission from Elsevier.)

The traditional K-function examines the extent to which events occur within a distance of other events. However, for the identification of traffic hot zones, one is not interested in collisions around which other collisions are concentrated, but is more concerned with those *road locations*, RPs, where collisions are clustered. Therefore, Yamada and Thill (2007) suggested using RPs along the network to examine the clustering tendency of traffic collisions. Their local K-function (LK) indicator for RP i is given by

$$LK_i = \sum_{j=1}^{n} f_{ij} \tag{8.16}$$

$$f_{ij} = \begin{cases} 1, & \text{if } d_{ij} \leq h \\ 0, & \text{otherwise} \end{cases} \tag{8.17}$$

where
  n denotes the number of collisions
  d_{ij} is the network distance between RP i and event j
  h (no less than *Int*) is the search distance from RP i

8.7 CONCLUSION

Analyzing traffic collisions over the network is full of methodological challenges. This chapter briefly discussed the two common approaches of the link-attribute approach and the network-constrained event-based approaches in the spatial analysis of traffic collisions. Generally, the link-attribute approaches have been more well developed and researched since the pioneer work of Black and Thomas (1998). Yet, the network-constrained event-based approach was only attracting more academic attention since the mid-2000s with the team efforts of Yamada and Thill (2004, 2007). With the recent methodological advances of the two approaches, Loo and Yao (2013) made a systematic attempt in comparing the two approaches based on simplified hypothetical networks and the empirical collision pattern in Hong Kong. In the future, more efforts should be spent in identifying the relative advantages and disadvantages of the two approaches.

REFERENCES

Anderson, T. K. 2009. Kernel density estimation and K-means clustering to profile road accident hotspots. *Accident Analysis & Prevention* 41 (3): 359–364.
Anselin, L. 1995. Local indicators of spatial association—LISA. *Geographical Analysis* 27 (2): 93–115.
Bailey, T. C. and A. C. Gatrell. 1995. *Interactive Spatial Data Analysis*. Essex, U.K.: Longman Scientific & Technical.
Black, W. R. 1991. Highway accidents: A spatial and temporal analysis. *Transport Research Record* 1318: 75–82.

Black, W. R. and I. Thomas. 1998. Accidents on Belgium's motorway: A network autocorrelation analysis. *Journal of Transport Geography* 6 (1): 23–31.

Deacon, J. A., C. V. Zegeer, and R. C. Deen. 1975. Identification of hazardous rural highway locations. *Transportation Research Record* 543: 16–23.

Delmelle, E. C. and J. C. Thill. 2008. Urban bicyclists: Spatial analysis of adult and youth traffic hazard intensity. *Transportation Research Record* 2074: 31–39.

Erdogan, S., I. Yimaz, T. Baybura, and M. Gullu. 2008. Geographical information systems aided traffic accident analysis system case study: City of Afyonkarahisar. *Accident Analysis & Prevention* 40 (1): 174–181.

Flahaut, B., M. Mouchart, E. S. Martin, and I. Thomas. 2003. The local spatial autocorrelation and the kernel method for identifying black zones—A comparative approach. *Accident Analysis & Prevention* 35 (6): 991–1004.

Fotheringham, A., C. Brubsdon, and M. E. Charlton. 2000. *Quantitative Geography: Perspectives on Spatial Data Analysis*. London, U.K.: Sage Publication.

Getis, A. 2010. Spatial interaction and spatial autocorrelation: A cross-product approach. In *Perspectives on Spatial Data Analysis*, eds. L. Anselin and S. J. Rey. Berlin, Germany: Springer, pp. 23–33.

Goodchild, M. F. 1986. *Spatial Autocorrelation*, Vol. 47 of *Concepts and Techniques in Modern Geography*. Norwich, U.K.: Geo Books.

Griffith, D. A. 1987. *Spatial Autocorrelation: A Primer*. Washington, DC: Association of American Geographers, Resource Publications in Geography.

Haggett, P. 1965. *Locational Analysis in Human Geography*. London, U.K.: Edward Arnold.

Levine, N., K. E. Kim, and L. H. Nitz. 1995. Spatial-analysis of Honolulu motor-vehicle crashes: I. Spatial patterns. *Accident Analysis & Prevention* 27 (5): 663–674.

Loo, B. P. Y. 2009. The identification of hazardous road locations: A comparison of the black-site and hot zone methodologies in Hong Kong. *International Journal of Sustainable Transportation* 3 (3): 187–202.

Loo, B. P. Y. and S. Yao. 2012. Geographic information systems. In *Injury Research: Theories, Methods, and Approaches*, eds. G. Li and S. Baker. New York: Springer.

Loo, B. P. Y. and S. Yao. 2013. The identification of traffic crash hot zones under the link-attribute and event-based approaches in a network-constrained environment. *Computers, Environment and Urban Systems* 41: 249–261.

Moran, P. A. P. 1948. The interpretation of statistical maps. *Journal of the Royal Statistical Society. Series B (Methodological)* 10 (2): 243–251.

Odland, J. 1988. *Spatial Autocorrelation*. Newbury Park, CA: Sage Publications.

O'Sullivan, D. and D. J. Unwin. 2003. *Geographic Information Analysis*. Hoboken, NJ: John Wiley & Sons.

Pulugurtha, S. S., V. K. Krishnakumar, and S. S. Nambisan. 2007. New methods to identify and rank high pedestrian crash zones: An illustration. *Accident Analysis & Prevention* 39 (4): 800–811.

Schabenberger, O. and C. A. Gotway. 2005. *Spatial Methods for Spatial Data Analysis*. New York: Chapman & Hall/CRC.

Silverman, B. W. 1986. *Density Estimation for Statistics and Data Analysis*. New York: Chapman & Hall/CRC Press.

Thomas, I. 1996. Spatial data aggregation: Exploratory analysis of road accidents. *Accident Analysis & Prevention* 28 (2): 251–264.

Yamada, I. and J. C. Thill. 2004. Comparison of planar and network K-functions in traffic accident analysis. *Journal of Transport Geography* 12 (2): 149–158.

Yamada, I. and J. C. Thill. 2007. Local indicators of network-constrained clusters in spatial point patterns. *Geographical Analysis* 39 (3): 268–292.

Yamada, I. and J. C. Thill. 2010. Local indicators of network-constrained clusters in spatial patterns represented by a link attribute. *Annals of the Association of American Geographers* 100 (2): 269–285.

Yao, S., B. P. Y. Loo, and W. W. Y. Lam. 2015. Measures of activity-based pedestrian exposure to the risk of vehicle–pedestrian collisions: Space-time path vs. potential path tree methods. *Accident Analysis & Prevention* 75: 320–332.

9 Cluster Identifications in Networks

9.1 INTRODUCTION

This chapter follows closely to the previous one with regard to outlining network analysis and network autocorrelation. This chapter is concerned with how, once one has established the optimal BSU (basic spatial unit) length or bandwidth, collisions can be measured together to identify hot zones. This issue has been tackled by some researchers in the past, and this chapter seeks to outline how to determine clusters on networks (or spatial contiguity). Many existing geostatistical methods for detecting clusters do not consider the specific nature of the network involved, often leading to biased conclusions. In traffic collision analysis, two techniques are commonly used for determining dangerous locations: the local spatial-autocorrelation method following the link-attribute approach and the kernel method following the event-based approach. Both methods are easily applicable and give comparable results for simplified road segments, such as exclusive highways or hypothetical linear roads, as independent spatial units with no intersections (1D). The analysis of road networks (2D), however, is not so straightforward because the special nature of road networks (notably connectivity) has to be considered. The local autocorrelation method requires the division of the road network into basic statistical units of standard length. There is no unique solution for this task, and it almost inevitably produces a number of statistical units that are often too short and excluded from further analyses. This results in a nonexhaustive coverage of the study area. The use of Euclidian distances in the planar 2D kernel method also disregards the network density: the statistical units are mostly created between, and not across, the intersections. This chapter will be supplemented with case studies from Hong Kong using different approaches to examine traffic collision patterns.

9.2 WHAT ARE HAZARDOUS ROAD LOCATIONS?

Many governments worldwide officially designate hazardous road locations (HRLs) as "black spots" and devote dedicated funding to address them. The AusLink Black Spot Projects of the Australian Government are among the most elaborate in methodology and well-funded by the government (Australian Government 2008, 2009). Similar schemes like "blacksites" and the Priority Investigation Locations (PILs) exist in Hong Kong, New York, and many other administrations (Loo 2009). However, what are HRLs? The fundamental concept is that these areas are having *abnormally* high incidences of traffic collisions involving death and injury than other locations. Taken together, HRLs constitute a small portion of the total network in terms of length but accounted for a much higher share of the traffic injury burden.

In the case of China, it was estimated that HRLs accounted for about 15%–25% of the road network but 40%–69% of all traffic collisions (Guo et al. 2003).

In light of the above, the identification, analysis, and treatment of HRLs is considered as one of the most effective approaches to improve road safety (Deacon et al. 1975; Transportation Research Board 1982; Hoque and Andreassen 1986; Nicholson 1989; Ogden 1996; Elvik 1997). In this chapter, we shall focus on the identification process. The investigation/analysis and treatment of HRLs are dealt with in Chapters 10 through 16. The key methodological aspects of the identification process, in turn, include defining the sites, setting the criteria of "hazardous," considering exposure and other factors as appropriate, and ranking the HRLs.

9.2.1 On the Definition of Sites

Early studies of the identification of HRLs do not follow a spatial algorithm. Typical examples are taking all junctions, or together with their nearby roads, as the unit of analysis. For instance, in the blacksite definition of the Hong Kong SAR Government until 2011, road junctions together with the 70 m of roads nearby had been considered as the unit of analysis for identifying as "blacksites" (Loo 2009). While road junctions can be plotted on maps and have spatial coordinates, junctions of the road network are essentially treated as a nonspatial list in the entire process of HRL identification.

Scientific studies of defining sites generally follow the link-attribute and event-based approaches outlined in the previous chapter. With reference to the link-attribute approach, sites may be considered by dividing the whole road network into BSUs. Then, each BSU is either taken independently or considered together with its contiguous BSUs as "sites." For the former, subsequent HRLs identified are often called hot spots or black spots. For the latter, the HRLs identified are called hot zones or black zones (Thomas 1996; Flahaut et al. 2003; Geurts and Wets 2003; Brijs et al. 2006; Loo 2009; Yao et al. 2015). A noteworthy point is that the distinction of hot spots and hot zones in spatial analysis is not based on the length of HRLs. The difference lies in the methodology. Using the link-attribute approach to illustrate, a hot zone consists of two or more contiguous hazardous BSUs. If each BSU is 100 m long, a hot zone will have a minimum length of 200 m. Depending on the spatial collision pattern, there is no theoretical maximum number of BSUs in a hot zone. However, a hot spot always consists of one BSU only. Its length depends on the length for the standard BSU, which may be much longer (say 500 m or 1 km). Moreover, some hot spots may be clustered or contiguous but network contiguity is not considered in the process of identification. Similarly, using the event-based approach, hot spots are identified without explicit consideration of network contiguity among reference points. The opposite is true for hot zones. Essentially, a hot zone is only found when there are spatially interdependent HRLs at contiguous reference points. Road segments are considered as spatially independent objects in the hot zone methodology. This definition is in contrast to the more traditional and nonspatial analysis of using hot spots to refer to short road segments (0.15 mile/0.24 km for intersection spots, and 0.3 mile/0.48 km for nonintersection spots) and sections or hot zones to refer to longer road segments (typically 3 miles/4.8 km) (Deacon et al. 1975).

Cluster Identifications in Networks

Theoretically, sites may also be defined as areas. However, areas are essentially planar space, including all land occupied by buildings or open space with no or little vehicular traffic. The smaller the spatial scale, the larger the intervening space. Moreover, the smaller the spatial scale, the more heterogeneous the land uses and other socioeconomic environment, and the concentration of traffic collisions may not be attributable to one or several common causes applicable to that whole area. Hence, though the identification of HRLs can be conducted at the regional zonal level (e.g., Erdogan 2009), the most fruitful analysis is always at the local network level.

9.2.2 SETTING THE CRITERIA

What is hazardous? Following Elvik (2007), there are three common groups of definitions, that is, simple numerical, statistical, and model based.

9.2.2.1 Magic Figures

Simple numerical definitions are overwhelmingly popular among road safety administrations worldwide (Elvik 2006). In Norway, any site with a maximum length of 100 m where at least four injury collisions have been recorded during the last 5 years is considered as an HRL (Statens vegvesen 2006). Similar conclusion was made by Elvik (2008) after systematically surveying how HRLs were identified in eight European countries: Austria, Denmark, Flanders, Germany, Hungary, Norway, Portugal, and Switzerland. In Kentucky of the United States, HRLs were considered as road segments of 0.1 mile (0.16 km) having three or more accidents in a 12-month period (Deacon et al. 1975). Similar method was adopted in the state of Arizona to identify HRLs by the Arizona Local Government Safety Project (ALGSP) Model (Carey 2001).

Using numerical definitions, "hazardous" is defined by collision frequency or count, sometimes taking into account injury severity of the traffic collision victims, rather than the collision potential based on risk and exposure. The use of observed collision frequency (O_i), compared to a predetermined critical number (CN), which may be the observed average of counts of comparison, is the most common. Location i is identified as unsafe if O_i exceeds the magic figure of CN. To illustrate, the magic figure approach identifies a road location i as an HRL, in the following manner:

$$HRL_i = \begin{cases} 1, & \text{if } O_i > CN \\ 0, & \text{otherwise} \end{cases} \quad (9.1)$$

The value of CN, in turn, depends a lot on the absolute "tolerable" levels of traffic collisions in the society and the resources available to the road safety administrations, because administrations have to have sufficient resources for tackling the HRLs identified.

9.2.2.2 Statistical Definitions

Statistical definitions recognize that traffic collisions are random events. In the 1970s, Hakkert and Mahalel (1978) proposed that hot spots should be defined as those sites whose collision frequency is significantly higher than expected at some prescribed level of significance. In other words, *CN* is no longer a magic figure, but it involves generating the descriptive statistics of the empirical collision pattern, specifying the statistical significance level, and calculating the confidence interval at the specified significance level accordingly. Moreover, the rate or quality control method can be applied with collisions per some exposure measures defined by statistical methods (Deacon et al. 1975). Often, the observed collision frequency O_i is first divided by some exposure factors, say the traffic volume, $AADT_i$, to calculate the observed collision rate ($R_i = O_i/AADT_i$), before comparing it with a predetermined critical collision rate (*CR*) for identifying the HRLs. *CR* is defined statistically and refers to the "normal level of safety" expected for a road location (Elvik 2008).

The statistical approach will identify road location i with O_i (or R_i) outside the confidence intervals at a specified significance level. The essential idea is to ascertain whether an HRL's poor collision records is or is not due to chance. The null hypothesis (H_0) is therefore whether the difference of $D_i = O_i - E_i$ is due to chance. At 95% confidence level, H_0 is rejected when O_i lies outside the confidence interval of $\bar{x} \pm 1.96 SE_x$, where $\bar{x}$ is the mean of all O_i and SE_x is the standard error of mean (Elvik 1988). It is worthwhile to highlight that road safety researchers are not so much interested when D_i is negative, that is, O_i being lower than E_i, and to test whether the good road safety record of i is due to chance. However, when D_i is positive, one is interested to know whether the high record of i is or is not due to chance. Hence, a one-tail test is appropriate. When H_0 is rejected, an HRL is identified

$$HRL_i = \begin{cases} 1, & \text{if } O_i > \bar{x} + 2.54 SE_x \\ 0, & \text{otherwise} \end{cases} \quad (9.2)$$

Typically, $\bar{x}$ is the global mean of all O_is (count data) and is same for all is within the road network. SE_x is the standard error, which in turn is the standard deviation of all O_i divided by the total sample size or the total number of BSUs of that road network, *N*. When the rate-quality method is used, $\bar{x}$ is the global mean of all R_is (ratio data). SE_x is the standard error of the rate data, which in turn is the standard deviation of all R_i divided by *N*. Conceptually, the statistical approach modifies the rationale of identifying HRLs as screening for sites with high collision records to sites with high collision intensity records over a certain level of statistical significance.

9.2.2.3 Model-Based Definitions

The final major group of model-based definitions defines *CR* based on some forms of collision prediction models. Moreover, the values of *CR* for different road segments (CR_i) in the same road network need not be the same. CR_i is defined by models based on risk levels of other sites. These other sites can be further defined as comparable sites. With model-based definitions, the major aim of identifying HRLs further changes from screening road locations for high collision frequency or rate to those of screening road locations for high potential of collision reduction. As a result, some of the HRL identification literature with the model-based definitions prefer to use terms like "sites with promise" (Hauer 1996; Hauer et al. 2002).

McGuigan (1981, 1982) was among the earliest to propose the use of potential for collision reduction (PCR) as the difference between the observed and expected number of collisions at a site given exposure. In comparison with statistical definitions, the collision rate is more complicated than taking O_i over some exposure factor like $AADT_i$, but is based on a more sophisticated understanding of many more relevant factors that pose road hazards, and the existence of that factor at the specific road location i.

$$E_i = \sum_{j=1}^{F} \left(K_j * PO_{i,j} \right) \tag{9.3}$$

where
E_i is the expected collision frequency at the ith road segment
F is the total number of risk factors considered
K_j is the increased collision frequency per unit increase of the jth risk factor
$PO_{i,j}$ is the level of exposure of the jth risk factor at the ith road segment

Many more risk factors (such as the junction type or the presence of steep gradient) beyond road length and traffic volume are often considered. In situations where a local risk factor is not applicable, the risk exposure level is zero ($PO_{i,j}=0$). The difference between the observed and expected collision frequency (D_i) is no longer simply used in the hypothesis testing of statistical significance. With statistical definitions, D_i is mainly analyzed to reject or do not reject the null hypothesis that the difference is or is not due to chance. With model-based definitions, the difference becomes a direct measure of PCR at the specific road location (PCR_i):

$$PCR_i = O_i - E_i \tag{9.4}$$

PCR_i is treated as an indicator of collision risk reduction potential and/or in ranking HRLs (Mahalel et al. 1982; Maher and Mountain 1988).

More recently, the Empirical Bayes (EB) approach has become more popular. It takes E_i as a variable not depending directly on "theoretical" risk factors and exposure, but the "empirical" collision records of road locations belonging to the same relatively homogenous road type. Roadway elements are generally classified in different types ($g = 1, 2, \ldots, G$). Each roadway element type will then have its own expected collision counts (E_g), which may simply be the average of all collision counts of that roadway type, or be modeled to depend on specific relevant risk factors, or be modeled to depend on previous collision records termed *prior information*. Among all model-based definitions, the EB methods are considered as one of the most promising and preferred by statisticians and road safety researchers. Cheng and Washington (2005) used experimentally derived data to compare three hot spot identification methods—simple ranking, confidence interval, and EB. They considered EB to be much better but also much more complicated.

9.3 RANKING ISSUES, FALSE POSITIVES, AND FALSE NEGATIVES

Regardless of the definitions used, researchers cannot completely remove the "randomness" of traffic collision events. Table 9.1 illustrates the problems. The false positive problem arises when a safe site is being wrongly identified as hazardous or having high risk. It was due to the inability of the identification process in differentiating between sites that are truly hazardous and those that are actually safe but were having a random surge in collision records during the study period (Cheng and Washington 2005). The existence of many false positive HRLs is undesirable because it leads to wasted resources in site investigations, which may be used to investigate and to treat truly hazardous road locations. The HRL identification process is therefore inefficient. Furthermore, the false positive problem is not expected to lessen over time. Different false positives will exist (like noise) among the pool of identified HRLs every year. Nonetheless, the researchers cannot tell the exact number and locations of these false negatives. Conversely, the false negative problem arises when a high-risk site is not being identified as an HRL and, hence, not further considered for road safety improvements. These road locations with true high collision risk were having random "down" fluctuations of collision records during the study period. All efforts and resources used to improve road safety by addressing HRLs will have no effects on these locations. The seriousness of this problem depends on whether the HRLs successfully identified are more dangerous than these unidentified HRLs. Furthermore, when the HRL identification process is continuous, the problem of false negatives is likely to be less significant. If these true HRLs persist, it is highly unlikely that they will experience random "down" fluctuations continuously for an extended period of time. While the false positive and negative problems exist regardless of the criteria used, researchers found that the extent of these problems varies with the definitions used.

It follows that the identified HRLs inevitably include both true positives and false positives, as they are by definition not distinguishable. Next, which one(s) of the identified HRLs should be treated and in what order? There needs to be a follow-up ranking exercise within the identification process because site investigation, data analysis, and treatment require substantial time and other resources. Generally, the larger is the pool of identified HRLs and/or the smaller are the resources available for follow-up actions, the more important is the ranking exercise in ensuring that HRLs posing different levels of road hazards are treated with the correct priority.

TABLE 9.1
Problems of False Positives and False Negatives Illustrated

Decision	True State — Safe	True State — Not Safe
HRL	Incorrect, false positive	Correct
Not HRL	Correct	Incorrect, false negative

Following magic figure definitions, the ranking of HRLs usually follows the simple ranking (SR) method. Most notably, the set of HRLs is ranked in descending order based on their observed collision frequency (O_i). A good example of using the SR technique in compiling the collision count profile can be found in Nicholson (1989). Despite its easy implementation, the SR method is found to suffer from problems of producing large numbers of false positives caused by the random annual fluctuations of collisions (Hauer and Persaud 1984; Persaud 1986; Hauer 1997). In Kentucky, HRLs identified by the magic figure approach were screened monthly, basically following the SR technique. Approximately 10% were selected for thorough field investigation by traffic engineers, maintenance engineers, and police personnel. Improvements recommended were then implemented. However, through this approach, "in as much as approximately 35% of the locations investigated in the field do not warrant improvement" (Deacon et al. 1975, 16). For the same reason of high random fluctuations of annual collision frequency at any specific location, the SR method also suffers from producing excessive number of false negatives and, hence, allowing truly hazardous locations to escape identification and result in inefficient use of resources.

The scale of false positives and negatives seems to be clearly specified with statistical definitions because the yardsticks are based on classical statistical confidence intervals (*CI*). Typical statistical significance chosen is 0.95 or 0.99. The Type I error, which corresponds to the false positive error in road safety, is therefore 0.05 or 0.01, respectively. Through increasing the statistical significance chosen, the number of HRLs that pass the statistical test will reduce. While it is not possible to say for certain (i.e., confidence level of 100%), it is at least possible to specify the level of confidence that the researchers have on the results. The problem, however, is that the statistical significance can only be specified with respect to an assumed underlying statistical distribution. In most situations, the normal distribution is assumed (i.e., $z = 2.54$) (Oppe 1979; Ceder and Livneh 1982). Nonetheless, traffic collisions happening at a specific road segment over a year are really rare events, which follow the Poisson or negative binomial distribution rather than the normal distribution (Cheng and Washington 2005). Various statistical distributions, such as the generalized Poisson (Kemp 1973), logarithmic models (Andreassen and Hoque 1986), Poisson log-linear regression (Blower et al. 1993), and the negative binomial (Persaud 1990; Hauer 1997; Abdel-Aty and Radwan 2000) models, have been used to address this statistical drawback (Anderson 2009).

In the late 1980s, Maher and Mountain (1988) introduced the simulation-based approach for the ranking exercise. Over time, the Monte Carlo simulation has been the most widely used for the purpose of defining statistically meaningful threshold levels (*TL*) for identifying and ranking HRLs, independent of the theoretical underlying statistical distribution/form of traffic collisions (Yamada and Thill 2007, 2010). The general procedures are to simulate sufficient number of randomly distributed collision patterns so as to establish the statistical significance. In each simulation, the total number of collisions are distributed randomly with equal chance over the entire road network. Following the event-based approach, it is not possible to allocate collisions randomly to the theoretically infinite number of points on the network that a collision may happen. Hence, GIS can be used to identify representative points with an equal interval along the road network, similar to the logic of the Geographical Analysis Machine (GAM) (Openshaw et al. 1987). Following the link-attribute approach, all collisions $\left(\sum_{i=1}^{N} O_i\right)$ are randomly assigned to one of the N BSUs of the road network in each simulation. After each simulation, the simulated collision frequency, S_i, can be obtained. When the simulation is repeated 1000 times, the 10th largest value of S_i can be used as the threshold level TL_i at the significance level of 0.01. The larger the number of repeated randomization, the more stable the resulting estimates and the more reliable the pseudo-significance levels (Yamada and Thill 2004; Loo and Yao 2013). Following the EB methods, the statistical significance is usually established by specifying the upper percentiles of the distribution of EB estimates of safety specific to the roadway element ($g = 1, 2, \ldots, G$) (Elvik 2008).

After the statistical tests are passed (either by making assumptions about the statistical distributions or the simulation approach), the ranking of HRLs may simply follow the SR method or the rate-quality method by controlling certain key exposure factors. More sophisticated yet data-intensive ranking exercise may follow the benefit–cost method. The rationale of ranking HRLs is

$$\text{Max}(B_i - C_i) \qquad (9.5)$$

or

$$\text{Max}\left(\frac{B_i}{C_i}\right) \qquad (9.6)$$

where
B_i is the potential benefits of improving HRL_i
C_i is the estimated costs of improving HRL_i

B_i/C_i is the well-known benefit–cost ratio. Historically, $B_i = f(PCR_i)$. In other words, the benefits of improving HRL_i are directly dependent on PCR_i, and multiplied by the average saving of preventing a collision. More detailed ranking exercise further weighs collisions by types, such as property damage only, collisions causing slight injury only, and collisions causing serious injury and fatality. Nonetheless, all these collision-based estimates do not consider the fact that the number, injury severity levels, and health outcomes of persons injured or killed in a collision can vary substantially. A collision involving buses, for instance, can involve more than a hundred persons killed or injured. Another collision may involve a slightly injured passenger only. Hence, the use of an average saving of preventing a collision, whether further classified by types or not, is inadequate. Hence, Loo et al. (2013) proposed the use of the person-based rather than collision-based approach in the identification of HRLs, so that potential benefits of addressing HRLs can be more accurately reflected and human based. Their method, put simply, is to analyze the number of persons injured or killed in traffic collisions directly (PE_i), rather than considering the collision frequency (O_i) or rate (R_i) indirectly. Next, the cost of improving HRL_i needs to be estimated. As the ranking exercise aims to screen HRLs for more expensive and detailed site investigation and analysis, C_i is usually estimated using ballpark figures from standard improvement measures. Theoretically, C_i can be obtained by combining the individual cost of a bundle of road safety measures known to be effective for addressing road hazards based on risk factors. Practically, a manual of standard costs, such as installation of pedestrian railings, is available in the more advanced road safety administrations. With the EB approach, past records of expenses in improving HRLs of roadway element g or typical improvement scheme costs for roadway element g are used to estimate C_i (Geurts and Wets 2003).

9.4 HRL IDENTIFICATION USING SPATIAL ANALYSIS

Suppose a road safety analyst is well aware of the pertinent issues of HRL identification described earlier, he/she will be ready to take the additional step of applying spatial analysis in HRL identification. To do so, there are three major stages (Loo 2009; Loo and Yao 2013). Stage I is geovalidation. Stage II is to define the spatial unit of analysis and calculate the collision statistics. Stage III is the HRL identification. HRLs can broadly be identified using either the hot spot methodology or the hot zone methodology. As we have covered geovalidation in Chapter 7, this chapter assumes that the researchers are satisfied with the quality of the spatial data and proceeds to discuss the major methodological issues in Stages II and III.

9.4.1 DEFINING THE SPATIAL UNIT OF ANALYSIS AND CALCULATING COLLISION STATISTICS

Following the link-attribute approach, Stage II will first involve cutting up the entire road network into small road segments as BSUs having a standard length, l. As far as possible, l should be equal for all BSUs. Following the event-based approach, this step involves determining the number and positions of reference points (RPs), for calculating the kernel density with a standard window width, h. Strictly speaking, the spacing or interval (Int) of RPs is independent of h. The value of h determines the width of the kernels placed over individual collisions (see Chapter 8). An RP is simply an "accounting" point for summarizing the total height of the kernels at a location. Ideally, Int should be equal, resulting in RPs at regular intervals covering the entire road network. As BSUs are also of standard length, one may simply take the starting point, mid-point, or random point r within l of each BSU to generate the RPs (Loo and Yao 2013). The details are described in Chapter 8.

9.4.2 HOT ZONE IDENTIFICATION

At Stage III, HRLs need to be identified based on the collision statistics of individual spatial units (BSUs or RPs). For sake of illustration, we shall define HRL sites as hot zones rather than hot spots. To recall, hot zones explicitly take into consideration spatial interdependency among neighboring spatial units. Moreover, the hot zone methodology is more complex and it can be easily modified to become a hot spot methodology (by setting the network proximity weighs all to zeros), if desired (e.g., for comparison purpose). Hence, we define HRL sites as hot zones in the illustrations later in the chapter. For the setting of criteria, we shall use the statistical definition with repeated randomization. The major consideration is to keep the procedures reasonably simple, compact, and comparable using both the link-attribute and the event-based approaches. The simple numerical definitions are not used because they are known to suffer from serious false positive and negative problems. The model-based definitions are highly heterogeneous and much more data intensive. Many of them, such as the EB method, require detailed discussion on the model-building process and are elaborated in other parts of the book.

9.4.2.1 Link-Attribute Approach

At Stage III of HRL identification, researchers need to consider both the network connectivity and the statistical significance of collision records at the same time. These concerns raise methodological challenges. To properly consider spatial connectivity of BSUs in HRL identification, Loo (2009) proposes an index, called the hot zone index $I_{(HZ)}$, on the basis of the local Moran's I method (Anselin 1995). Depending on the spatial relationships among BSUs, $I_{(HZ)i}$ for BSU i is defined as:

$$I_{(HZ)i} = z_i \sum_{j=1, j \neq i}^{N} W_{ij} z_j \qquad (9.7)$$

$$z_i = \begin{cases} 1, & \text{if } O_i \geq t_i \\ 0, & \text{otherwise} \end{cases} \qquad (9.8)$$

where

N is the number of BSUs
t_i is the threshold collision rate of BSU i
O_i is the observed collision rate at the ith BSU
W_{ij} is the network proximity matrix

In spatial analysis, matrices are widely used for representing spatial concepts such as distance, adjacency, interaction, and neighborhood. For hot zone identification, we focus on those contiguous BSUs with relatively high risks. Generally, most collision patterns do not strongly exist beyond the first degree of spatial proximity (Flahaut et al. 2003; Flahaut 2004). Thus, W_{ij} is denoted as a contiguity (0,1) matrix whose elements are only ones or zeros. Nonetheless, researchers may use other distance-based proximity matrix, such as d_{ij}^{-2}, when more than one degree of neighbor is considered.

To establish the statistical significance of the hot zone results, t_i is defined statistically using the simulation approach and the Monte Carlo method. The introduction of statistical definitions as critical thresholds for detecting link-attribute traffic hot zones is first presented in Loo and Yao (2013). In each simulation, the procedures are to randomly allocating the total number of road collisions over the BSUs and obtain $z_{i(sim)}$ for each BSU. Then, simulations can be repeated 100, 500, 1000, or more times. Using the value of $z_{i(sim)}$ of the top 1% or 5% of all $z_{i(sim)}$, the pseudo-significance level of 95% or 99% can be obtained, respectively. These cutoff values are then substituted into t_i to define the threshold value and compute $I_{(HZ)i}$ in Equations 9.7 and 9.8.

Cluster Identifications in Networks

After determining the key parameters and methods used in hot zone identification, the implementation procedures involved are summarized in Figure 9.1. Details of the GIS-based algorithm are reported in Loo (2009). To begin with, the first BSU record is examined to see whether its observed collision frequency O_i is greater than or equal to its threshold value t_i. If the answer is positive, a new working table is

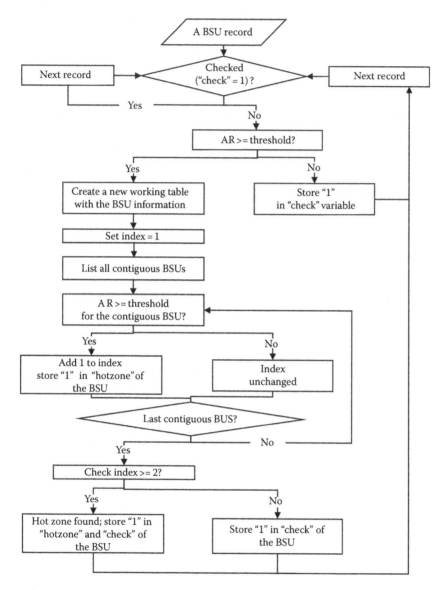

FIGURE 9.1 A flowchart showing the steps of hot zone identification. (Reprinted from Loo, B.P.Y., *Int. J. Sustain. Transp.*, 3(3), 187, 2009. With permission from Taylor & Francis Group Ltd., http://www.tandfonline.com/.)

created with an index number equal to one. All contiguous BSUs are checked by GIS and listed out in this working table. Then, each contiguous BSU is analyzed. $I_{(HZ)i}$ is computed with the assistance of GIS, and the result is recorded as a variable in the attribute table of the BSU dataset. Whenever the observed collision frequency of any one of the contiguous BSUs is also greater than or equal to the respective t_i, the index number $I_{(HZ)i}$ increases by one. The checking will continue until all contiguous BSUs in the working table have been checked. When this is done, the index number is examined. If the index number is greater than one, a hot zone has been identified and the "hot zone" variable of the BSU in the main table (default=0) is updated. The entire process repeats until all BSUs in the road network have been checked. Mathematically, the value of $I_{(HZ)}$ is either positive ($I_{(HZ)} = 1, 2, ..., N-1$) or equal to zero. A positive value of $I_{(HZ)i}$ indicates that the observed collision rates of BSU i and at least one of its neighboring BSUs are no less than their threshold values, and a hot zone is detected. The spatial pattern of HRLs can then be visualized and analyzed by plotting the O_i or other attributes of all BSUs that form part of a hot zone (BSUs with $I_{(HZ)} \geq 1$) in a road network map of an appropriate spatial scale.

9.4.2.2 Event-Based Approach

How to consider the network proximity of RPs properly under the event-based approach? By drawing reference to $I_{(HZ)}$, Loo and Yao (2013) introduce an event-based hot zone indicator $LK_{(HZ)}$ based on the KLINCS approach of Yamada and Thill (2007):

$$LK_{(HZ)i} = Z_i \sum_{j=1, j \neq i}^{m} f_{(HZ)ij} Z_j \quad (9.9)$$

$$Z_i = \begin{cases} 1, & \text{if } LK_i \geq t_i \\ 0, & \text{otherwise} \end{cases} \quad (9.10)$$

where
LK_i is the local network-constrained K-function index for the RP i
t_i is the threshold value at RP i

LK_i can be calculated following Equations 9.9 and 9.10. Similarly, t_i can be defined statistically by Monte Carlo simulations rather than an arbitrary number (see preceding text). $f_{(HZ)ij}$ is a binary variable indicating whether or not RP i and j are contiguous. It is measured by

$$f_{(HZ)ij} = \begin{cases} 1, & \text{if } d_{(HZ)ij} \leq Int \\ 0, & \text{otherwise} \end{cases} \quad (9.11)$$

where $d_{(HZ)ij}$ is the network distance between RP i and j. The value of $LK_{(HZ)}$ is also either positive or equal to zero. Once again, the identification of hot zones only focuses on contiguous RPs with positive $LK_{(HZ)}$. For each of the hot zones identified, the profile of LK_i can be further analyzed and compared.

Cluster Identifications in Networks 173

9.5 SOME ADDITIONAL METHODOLOGICAL REMARKS

Despite the various key methodological challenges discussed earlier, we found that a few remarks are necessary before we draw this chapter to a close. These issues will be faced by administrators, researchers, consultants, students, or indeed anyone who is interested in identifying HRLs.

9.5.1 STUDY PERIOD

The first remark is about the study period for considering traffic collision patterns, whether spatial or nonspatial. As traffic collisions are rare events, annual fluctuations of collision frequency especially at specific locations are likely to be high. Hence, most scholars have recommended that the collision data be pooled over a longer study period than a few months or a year before data analysis. However, how long should researchers pool collision statistics for HRL identification? A short study period is problematic because it increases the false positive and negative problems. A long study period is also problematic because it may violate the assumptions that road conditions and other relevant risk factors are relatively homogenous within the study period. Moreover, it may mean doing nothing in the short term and, hence, miss the opportunity of improving the situation before it gets worse. May (1964) first discussed about the issue of how many years of collision data should be analyzed when determining HRLs. He considered study periods from 1 to 13 years. His results suggest that the marginal benefit of having longer study period to improve estimates declines as the study period extends beyond 3 years. Nonetheless, Deacon et al. (1975) considered the use of 1- and 2-year intervals for consolidating collision statistics to be "desirable." With a better understanding of the false positive problem in the scientific community, more studies have adopted the pool data approach of 3 years or above when conducting collision analysis (Mueller et al. 1988; Cheng and Washington 2005). In particular, Cheng and Washington (2005, 870) remarked that "three years of collision history appears, in general, to provide an appropriate crash history duration." More recently, Elvik (2008) suggested 3–5 years as "suitable" for collision analysis. Generally, a 3-year period is the most preferred (May 1964; Mueller et al. 1988; Cheng and Washington 2005; Elvik 2008). However, the final decision depends a lot on the local circumstances as well.

9.5.2 DEGREE OF INJURY

As the primary aim of road safety research is to reduce death and human suffering resulting from traffic collisions, many researchers and road safety administrators consider it desirable to consider the severity of traffic collision victims in the HRL identification process. Apart from considering this factor as the potential benefits of improving an HRL in the ranking exercise, other approaches are also commonly adopted. One of the ways is to identify HRL for specific collision type only. For the sake of illustration, collision types may be classified into fatal, serious injury, slight injury, and property damaged only. HRLs involving fatal collisions, for instance, can be identified for priority investigation and treatment. A second way is to identify

HRLs based on multiple criteria with respect to different collision types or severity. Typically, the criteria for more serious collision types are set at lower levels. An HRL is, therefore, identified if the sites had z collisions involving deaths or serious injury, or y collisions involving slight injury or property damaged only. The value of y is usually much higher than that of z. A third way is to identify HRLs not based on the observed collision frequency O_i or collision rate R_i directly but some composite road collision indices RS_i. Once again, different weighing (w_γ) are applied to different collision types ($\gamma = 1, 2, \ldots, Y$).

$$RS_i = \sum_{\gamma=1}^{Y} w_\gamma O_{i\gamma} \qquad (9.12)$$

where
RS_i is the composite collision score of the ith road segment
γ refers to the collision type
Y is the number of collision types considered
$O_{i\gamma}$ is the number of observed collision frequency at the ith road segment of the γth collision type

In the survey of Elvik (2008), collision severity was considered in the HRL identification process in four out of the eight European case studies. In Flanders, the weighting of deaths, serious injury, and slight injury are in the ratio of 5:3:1. Moreover, RS is called the score of priority (S). In Portugal, RS takes the form of a Severity Index. The weighting of deaths, serious injury, and slight injury are in the ratio of 100:10:1. Moreover, the weighting is based on the number of traffic collision victims (individuals) rather than collisions (events).

9.6 CONCLUSION

This chapter outlines the key issues in identifying HRLs. Various methodological issues are addressed. Moreover, there does not seem to be a best way of HRL identification. Many different approaches and methods are available for researchers to consider when analyzing a specific situation. Different methods have different strengths and weaknesses. In making choices about the HRL identification process, many factors need to be considered. People working in road safety administrations are most commonly faced with making choices about the HRL identification process. Understandably, striking a balance is important. Moreover, practical considerations and the administrations road safety policy or priority will have a strong influence in the entire process. For instance, drunk-driving may represent a priority area for legislation and enforcement in a society. Hence, HRLs involving drunk-driving can be conducted to inform the decision makers (Tarko et al. 2012).

Another more drastic approach is to conduct road safety audits for all roads in an administration. Generally, road safety audits are based on a comprehensive checklist of ensuring that road safety standards are followed. They are most fruitfully applied at the infrastructure planning and construction stage. However, to what extent have these guidelines been already incorporated in standard engineering manuals? If not, can they be incorporated? This seems to be a more rational approach than to implement separate road safety audits as a separate process. For treating HRLs, the road safety audits again can help identifying local risk factors and provide useful information about the potential areas of improvements (Robert and Veeraragavan 2004). Nonetheless, improvements in a retrospective manner following audit guidelines may not be feasible. Another key consideration is that research has shown that poor road engineering design is not the most important or the sole factor in contributing to HRLs. Resources need to be put on investigating nearby land uses, environment, road user types, and road user behavior so that road safety records can really be improved. These are some of the issues that we shall look into in the coming chapters.

REFERENCES

Abdel-Aty, M. A. and A. E. Radwan. 2000. Modeling traffic accident occurrence and involvement. *Accident Analysis & Prevention* 32 (5): 633–642.

Anderson, T. K. 2009. Kernel density estimation and K-means clustering to profile road accident hotspots. *Accident Analysis & Prevention* 41 (3): 359–364.

Andreassen, D. C. and M. M. Hoque. 1986. LATM and road safety: Accidents in road classes in Melbourne. Paper presented at the *13th Australian Road Research Board (ARRB) Conference*, Adelaide, Australia.

Anselin, L. 1995. Local indicators of spatial association—LISA. *Geographical Analysis* 27 (2): 93–115.

Australian Government. 2015. Department of Infrastructure and Regional Development website. http://investment.infrastructure.gov.au/funding/blackspots/index.aspx (accessed March 4, 2015).

Blower, D., K. L. Campbell, and P. E. Green. 1993. Accident rates for heavy truck-tractors in Michigan. *Accident Analysis & Prevention* 25 (3): 307–321.

Brijs, T., F. V. D. Bossche, G. Wets, and D. Karlis. 2006. A model for identifying and ranking dangerous accident locations: A case study in Flanders. *Statistica Neerlandica* 60 (4): 457–476.

Carey, J., 2001. Arizona local government safety project (LGSP) analysis model. Final Report 504. Arizona Department of Transportation, Phoenix, AZ.

Ceder, A. and M. Livneh. 1982. Relationships between road accidents and hourly traffic flow—I: Analyses and interpretation. *Accident Analysis & Prevention* 14 (1): 19–34.

Cheng, W. and S. P. Washington. 2005. Experimental evaluation of hotspot identification methods. *Accident Analysis & Prevention* 37 (5): 870–881.

Deacon, J. A., C. V. Zegeer, and R. C. Deen. 1975. Identification of hazardous rural highway locations. *Transportation Research Record* 543: 16–33.

Elvik, R. 1988. Some difficulties in defining populations of entities for estimating the expected number of accidents. *Accident Analysis & Prevention* 20 (4): 261–275.

Elvik, R. 1997. Evaluations of road accident blackspot treatment: A case of the iron law of evaluation studies? *Accident Analysis & Prevention* 29 (2): 191–199.

Elvik, R. 2006. New approach to accident analysis for hazardous road locations. *Transportation Research Record* 1953: 50–55.

Elvik, R. 2007. State-of-the-art approaches to road accident black spot management and safety analysis of road networks. Institute of Transport Economics, Oslo, Norway.

Elvik, R. 2008. A survey of operational definitions of hazardous road locations in some European countries. *Accident Analysis & Prevention* 40 (6): 1830–1835.

Erdogan, S. 2009. Explorative spatial analysis of traffic accident statistics and road mortality among the provinces of Turkey. *Journal of Safety Research* 40 (5): 341–351.
Flahaut, B. 2004. Impact of infrastructure and local environment on road unsafety: Logistic modeling with spatial autocorrelation. *Accident Analysis & Prevention* 36 (6): 1055–1066.
Flahaut, B., M. Mouchart, E. S. Martin, and I. Thomas. 2003. The local spatial autocorrelation and the kernel method for identifying black zones—A comparative approach. *Accident Analysis & Prevention* 35 (6): 991–1004.
Geurts, K. and G. Wets. 2003. Black spot analysis methods: Literature review. Flemish Research Center for Traffic Safety, Diepenbeek, Belgium.
Guo, Z., J. Gao, and L. Kong. 2003. The road safety situation investigation and characteristics analysis of black spots of arterials highways. *Advances in Transportation Studies* 1 (1): 9–20.
Hauer, E. 1996. Identification of sites with promise. *Transportation Research Record* 1542: 54–60.
Hauer, E. 1997. *Observational Before-After Studies in Road Safety: Estimating the Effect of Highway and Traffic Engineering Measures on Road Safety*. Oxford, U.K.: Pergamon.
Hauer, E., J. Kononov, B. Allery, and M. S. Griffith. 2002. Screening the road network for sites with promise. *Transportation Research Record* 1784: 27–32.
Hauer, E. and B. N. Persaud. 1984. Problem of identifying hazardous locations using accident data. *Transportation Research Record* 975: 36–43.
Hakkert, A. S. and D. Mahalel. 1978. Estimating the number of accidents at intersections from a knowledge of the traffic flows on the approaches. *Accident Analysis & Prevention* 10 (1): 69–79.
Hoque, M. M. and D. C. Andreassen. 1986. Pedestrian accidents: An examination by road class with special reference to accident 'cluster'. *Traffic Engineering & Control* 27 (7–8): 391–397.
Kemp, C. D. 1973. An elementary ambiguity in accident theory. *Accident Analysis & Prevention* 5 (4): 371–373.
Loo, B. P. Y. 2009. The identification of hazardous road locations: A comparison of the blacksite and hot zone methodologies in Hong Kong. *International Journal of Sustainable Transportation* 3 (3): 187–202.
Loo, B. P. Y., C. B. Chow, M. Leung, T. H. J. Kwong, S. F. A. Lai, and Y. H. Chau. 2013. Multidisciplinary efforts toward sustained road safety benefits: Integrating place-based and people-based safety analyses. *Injury Prevention* 19: 58–63.
Loo, B. P. Y. and S. Yao. 2013. The identification of traffic crash hot zones under the link-attribute and event-based approaches in a network-constrained environment. *Computers, Environment and Urban Systems* 41: 249–261.
Mahalel, D., A. S. Hakkert, and J. N. Prashker. 1982. A system for the allocation of safety resources on a road network. *Accident Analysis & Prevention* 14 (1): 45–56.
Maher, M. J. and L. J. Mountain. 1988. The identification of accident blackspots: A comparison of current methods. *Accident Analysis & Prevention* 20 (2): 143–151.
May, J. F. 1964. A determination of accident prone location. *Traffic Engineering* 34 (5): 21–27.
McGuigan, D. R. D. 1981. The use of relationships between road accidents and traffic flow in "black-spot" identification. *Traffic Engineering and Control* 22 (8–9): 448–451.
McGuigan, D. R. D. 1982. Non-junction accident rates and their use in "black-spot" identification. *Traffic Engineering and Control* 23 (2): 60–65.
Mueller, B. A., F. P. Rivara, and A. B. Bergman. 1988. Urban-rural location and the risk of dying in a pedestrian-vehicle collision. *Journal of Trauma & Acute Care Surgery* 28 (1): 91–94.
Nicholson, A. J. 1989. Accident clustering: Some simple measures. *Traffic Engineering & Control* 30 (5): 241–246.

Ogden, K. W. 1996. *Safer Roads: A Guide to Road Safety Engineering.* Aldershot, U.K.: Avebury Technical.

Openshaw, S., M. Charlton, C. Wymer, and A. Craft. 1987. A mark 1 geographical analysis machine for the automated analysis of point data sets. *International Journal of Geographical Information Systems* 1 (4): 335–358.

Oppe, S. 1979. The use of multiplicative models for analysis of road safety data. *Accident Analysis & Prevention* 11 (2): 101–115.

Persaud, B. N. 1986. Safety migration, the influence of traffic volumes, and other issues in evaluating safety effectiveness—Some findings on conversion of intersections to multi-way stop control. *Transportation Research Record* 1068: 108–114.

Persaud, B. N. 1990. Blackspot identification and treatment evaluation. Ontario Ministry of Transportation, Ottawa, Ontario, Canada.

Robert, R. V. and A. Veeraragavan. 2004. Hazard-rating scores for prioritization of accident-prone sections on highways. *Transportation Research Record* 1878: 143–151.

Statens vegvesen. 2006. Håndbok 115: Analyse av ulykkessteder. Draft dated October 19, Vegdirektoratet, Oslo, Norway.

Tarko, A. P., S. M. Azam, J. Thomaz, and M. Romero. 2012. Identifying traffic safety needs—A systematic approach: Research report and user manual. Publication FHWA/IN/JTRP-2012/02. Joint Transportation Research Program, Indiana Department of Transportation and Purdue University, West Lafayette, IN.

Thomas, I. 1996. Spatial data aggregation: Exploratory analysis of road accidents. *Accident Analysis & Prevention* 28 (2): 251–264.

Transportation Research Board. 1982. Highway accident analysis systems. Transportation Research Board, Washington, DC.

Yamada, I. and J. C. Thill. 2004. Comparison of planar and network K-functions in traffic accident analysis. *Journal of Transport Geography* 12: 149–158.

Yamada, I. and J. C. Thill. 2007. Local indicators of network-constrained clusters in spatial point patterns. *Geographical Analysis* 39 (3): 268–292.

Yamada, I. and J. C. Thill. 2010. Local indicators of network-constrained clusters in spatial patterns represented by a link attribute. *Annals of the Association of American Geographers* 100 (2): 269–285.

Yao, S., B. P. Y. Loo, and W. W. Y. Lam. 2015. Measures of activity-based pedestrian exposure to the risk of vehicle–pedestrian collisions: Space-time path vs. potential path tree methods. *Accident Analysis & Prevention* 75: 320–332.

10 Exposure Factor 1
Traffic Volume

10.1 INTRODUCTION

Statistical models have been developed to analyze different types of collisions at intersections, and on sections on urban roads, rural roads, carriageways, and motorways. Empirical tests about the relationship between collisions and independent variables (including traffic flows) can be based on different statistical techniques, all having their own limitations. Besides, different data definitions (such as time and spatial units) are used in different applications in the literature; careful definition and selection of the data are rare. Collision data consist of counts and thus are considered as Poisson or negative binomial distribution. This chapter explores the nature of the underlying distribution and methods, which are fundamentally linked to traffic volume, notably average annual daily traffic (AADT). There has been much discussion concerning the aggregation of road segments and the application of Poisson-based methods to analyze larger segments of road collisions. Poisson regression is a commonly used technique; however, research has shown that it is the most effective with lower levels of aggregation.

10.2 RELATIONSHIP BETWEEN TRAFFIC FLOW AND COLLISIONS

Perhaps one of the most fundamental concepts in road safety research is the concept of risk, which we have touched upon in Chapter 9 in deriving the expected collision frequency at a specific road location. To recall,

$$E_i = \sum_{j=1}^{F} \left(K_j * PO_{i,j} \right) \quad (10.1)$$

where
E_i is the expected collision frequency at the ith road segment
F is the total number of risk factors considered
K_j is the increased collision frequency per unit increase of the jth risk factor
$PO_{i,j}$ is the level of exposure of the jth risk factor at the ith road segment

In reality, how is K_j derived? How much is the expected increase in collision frequency per unit increase of the jth risk factor? This chapter attempts to get to know better about the relationship between empirical collision frequency, O_i, and one of the most fundamental risk factors, whose data are relatively easy to obtain and of reasonable validity and reliability in most road safety administrations worldwide—vehicular traffic volume. If we take vehicular traffic volume as the *j*th risk factor, then the exposure level ($PO_{i,j}$) of road segment *i* to the vehicular traffic risk can be estimated. For the sake of simplicity, take the exposure level to be the average vehicle-kilometers traveled (VKT) and the study period to be 3 years. Then,

$$PO_{i,j} = l_i \sum_{t=1}^{3} AADT_{i,t} * 365 \tag{10.2}$$

where
l_i is the length of the road location *i*
$AADT_i$ is the average annual daily traffic of *i*
t is the time period in years

The risk associated with per unit increase in VKT at *i* is often estimated by

$$K_{i,j} = \frac{O_i}{PO_{i,j}} \tag{10.3}$$

where
$K_{i,j}$ is a quantified risk measure of location *i* for the *j*th risk factor
O_i is the total number of collisions happening at location *i* over the study period
$PO_{i,j}$ is a measure of the exposure of location *i* to the *j*th risk factor

A simple comparison of $K_{i,j}$ in Equation 10.3 (the empirical risk measure estimated at location *i* for the *j*th risk factor) with K_j in Equation 10.1 (the theoretical risk measure of the *j*th risk factor in general) shows that the former will lead to an overestimation of the risk measure of the *j*th risk factors, because O_i is actually the result of the sum of all relevant risk factors present at road location *i* (and a random factor, which we ignore for now) rather than that of the *j*th risk factor only. Theoretically, K_j should not vary across *i*. To avoid confusion and make this distinction clear in the following discussion, we replace $K_{i,j}$ and simply call it a *collision–exposure ratio* ($\Gamma_{i,j}$), which does not measure the effects of traffic exposure per se but all other factors (including the random factor and other risk factors) that may present at *i*. In addition, traffic volume is only one of the many measures to quantify vehicular traffic exposure (elaborated later in this chapter). Hence, we also replace $PO_{i,j}$ with X_i, where X_i is a *traffic volume index*. Traffic volume is defined as "the rate of flow of traffic on a facility, aggregated over time, for example, vehicles per day or vehicles per hour (vph)" (Ivan 2004, 134). Equation 10.3 therefore becomes

$$\Gamma_{i,j} = \frac{O_i}{X_i} \tag{10.4}$$

In this way, the collision–exposure ratios of different risk factors are not much different from traditional collision rates, such as collisions per kilometer of roads and collisions per million vehicle-kilometers traveled. Using different exposure measures will give us different perspectives about the road safety situation than analyzing collision frequency per se. For instance, in the international comparisons of road safety performance of seven administrations, including Australia, California, Great Britain, Hong Kong, Japan, New Zealand, and Sweden (Loo et al. 2005, 2007), Hong Kong ranked the safest in terms of collisions per population, but it was the second worst in terms of collisions per kilometer of roads. As shown in Table 10.1, the situation did not change much over time. One of the reasons lies in the importance of off-road transport modes, particularly railways, in Hong Kong's local transport system, which reduces people's exposure to on-road collision risks in a high-density environment.

10.3 TRAFFIC VOLUME

In this section, we explore various issues related to measuring the traffic volume index, X_i. On a road, exposure is generally perceived as the number of opportunities for a traffic collision to happen. The opportunities are, in turn, related to the number of vehicles on the road. This seems intuitively simple because there will not be any traffic collision with no vehicle on the road. Nonetheless, as the number of vehicles increases, does the collision risk increase proportionately and in a linear manner? This question is not as simple and straightforward. Generally, traffic exposure is an aggregation of individual traffic volumes observed during the period of analysis. Traffic volume may be further disaggregated into density and speed. The choice of the best traffic volume index is often determined by the significance and explanatory power of the statistical model, which specifies the functional form of the collision-traffic exposure relationship.

To begin with, if location i is a road segment without any intersection, the traffic volume index can often be calculated in a relatively straightforward manner using VKT, or AADT multiplied by the number of days and segment length (Equation 10.2) (Jorgensen 1972). When junctions are considered, the estimation of X_i becomes quite different. Moreover, it was argued that collision risks at junctions were so different from road sections that they had to be analyzed separately (Smeed 1955; Mathewson and Brenner 1957; Breuning and Bone 1960; Hakkert and Mahalel 1978; McGuigan 1981). In particular, the junction type, traffic directions and flows could have major effects on resulting collision risk at junctions. The first group of studies uses the sum of entering traffic flow as the traffic flow index X_i (Babkov et al. 1970; Schaechterle et al. 1970; Tamburri and Smith 1970; McGuigan 1981). The concept is essentially the same as traffic throughput of a junction, which is an estimate of annual traffic flow entering the junction in thousands or millions of vehicles.

However, the sum measure does not explicitly recognize that traffic collision risk is higher when two vehicles cross, rather than when traffic diverges or runs in the same direction. Therefore, the second major group defines X_i as the product of the traffic volume of the first (X_1) and second (X_2) approaches ($X=X_1X_2$). In other words, X_i is measured as the product of traffic flows rather than the sum of traffic flows. Tanner (1953) and McDonald (1953) were among the first to propose the product

TABLE 10.1
Road Collision Fatality Numbers and Rates in Seven Administrations

Fatality	Australia	California	Great Britain	Hong Kong	Japan	New Zealand	Sweden
Absolute figures							
2001	1737	3956	3450	173	10,071	455	583
2002	1715	4088	3431	171	9,645	405	560
2003	1621	4224	3508	202	8,944	461	529
2004	1583	4120	3221	166	8,561	435	480
2005	1627	4333	3201	151	7,990	405	440
2006	1598	4240	3172	144	7,326	393	445
2007	1603	3995	2946	160	6,681	421	471
2008	1437	3434	2538	162	6,067	366	397
2009	1491	3090	2222	139	5,831	385	358
2010	1353	2720	1850	117	5,806	375	266
Per million population							
2001	89.48	113.62	60.08	25.77	79.21	117.25	65.54
2002	87.27	116.84	59.50	25.36	75.68	102.57	62.75
2003	81.48	116.87	60.55	30.01	70.03	114.47	59.05
2004	78.65	113.60	55.31	24.47	67.01	106.42	53.37
2005	79.78	105.48	54.54	22.16	62.53	97.97	48.73
2006	77.21	112.58	53.69	21.00	57.34	93.92	49.01
2007	76.97	105.99	49.46	23.13	52.29	99.57	51.49
2008	67.63	89.46	42.27	23.28	47.51	85.74	43.06
2009	68.74	79.95	36.75	19.93	45.71	89.21	38.50
2010	61.41	73.01	30.35	16.66	45.55	85.86	28.36
Ranking in 2010	5	6	3	1	4	7	2
Per 100,000 vehicles							
2001	13.92	13.52	11.60	32.93	11.56	16.45	7.77
2002	13.38	13.56	11.23	32.92	11.02	14.25	7.37
2003	12.31	13.71	11.24	32.62	10.21	15.73	6.85
2004	11.70	12.86	9.98	37.91	9.74	14.29	6.11
2005	11.69	13.07	9.73	30.70	8.99	12.86	5.49
2006	11.13	12.50	9.59	27.31	8.24	12.18	5.47
2007	10.85	11.51	8.75	25.48	7.54	12.73	5.63
2008	9.39	10.03	7.49	27.82	6.87	10.92	4.70
2009	9.51	8.78	6.54	27.74	6.63	11.38	4.22
2010	8.42	8.56	5.42	22.87	6.61	11.04	3.05
Ranking in 2010	4	5	2	7	3	6	1

(Continued)

TABLE 10.1 (Continued)
Road Collision Fatality Numbers and Rates in Seven Administrations

Fatality	Australia	California	Great Britain	Hong Kong	Japan	New Zealand	Sweden
Per 1000 km of road							
2001	2.15	14.56	8.83	90.53	8.59	4.93	1.39
2002	2.12	15.13	8.76	88.88	8.19	4.38	1.33
2003	2.00	15.48	8.94	104.45	7.56	4.98	1.25
2004	1.95	15.08	8.31	85.43	7.21	4.69	1.13
2005	N/A	15.85	8.25	77.24	6.38	4.35	1.03
2006	N/A	15.47	8.04	72.58	5.83	4.21	1.04
2007	1.97	14.50	7.46	79.64	5.30	4.50	1.10
2008	1.76	12.37	6.44	79.41	4.80	3.90	0.69
2009	1.81	11.17	5.63	67.80	4.60	4.10	0.62
2010	1.64	9.82	4.69	56.36	4.58	3.99	0.46
Ranking in 2010	2	6	5	7	4	3	1
Per 100 million vehicle-kilometers							
2001	0.89	1.27	0.73	1.50	1.03	1.27	0.84
2002	0.86	1.27	0.71	1.48	0.99	1.10	0.80
2003	0.79	1.31	0.72	1.81	0.92	1.21	0.74
2004	0.74	1.25	0.65	1.50	0.90	1.11	0.66
2005	0.76	1.32	0.65	1.35	0.80	1.03	0.60
2006	0.74	1.29	0.63	1.25	0.74	1.00	0.59
2007	0.73	1.22	0.58	1.34	0.67	1.05	0.61
2008	0.65	1.05	0.51	1.35	0.63	0.92	0.51
2009	0.68	0.95	0.45	1.18	0.41	0.96	0.47
2010	0.61	0.84	0.38	0.97	0.50	0.94	0.35
Ranking in 2010	4	5	2	7	3	6	1

Sources: Data from Australian Bureau of Statistics, Year Book Australia, 2001–2010, http://www.abs.gov.au/ausstats/abs@.nsf/mf/1301.0; Organisation for Economic Co-operation and Development, OECD database, 2001–2010, http://stats.oecd.org/index.aspx; National Highway Traffic Safety Administration, Fatality analysis reporting system, 2001–2010, http://www-fars.nhtsa.dot.gov/States/StatesFatalitiesFatalityRates.aspx; Statistics Japan, Japan Statistical Yearbook, 2001–2010, http://www.stat.go.jp/english/data/nenkan/; Statistics New Zealand, Yearbook collection, 2001–2010, http://www.stats.govt.nz/yearbooks; Statistics Sweden, Finding statistics, 2001–2010, http://www.scb.se/en_/; Statistics U.K., Statistics, 2001–2010, https://www.gov.uk/government/statistics; Transport Analysis, Statistics, 2001–2010, http://www.trafa.se/en/; Transport Department, Publications, 2002–2010, http://www.td.gov.hk/en/publications_and_press_releases/publications/free_publications/index.html; Census and Statistics Department, Hong Kong statistics, 2001–2010, http://www.censtatd.gov.hk/home.html; World Bank, Data, 2001–2010, http://data.worldbank.org/ (all accessed November 3, 2014).

Key: N/A = Data not available.

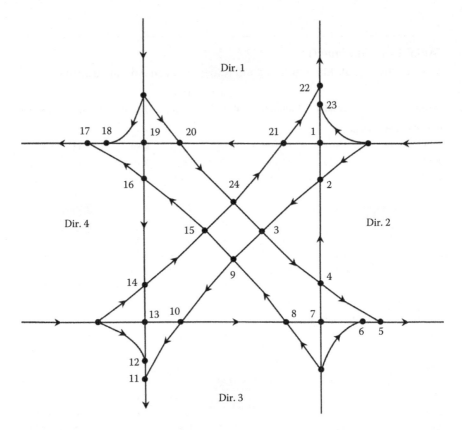

FIGURE 10.1 Conflict points in a four-arm two-way junction. (Reprinted from *Accid. Anal. Prev.*, 10(1), Shalom Hakkert, A. and Mahalel, D., Estimating the number of accidents at intersections from a knowledge of the traffic flows on the approaches, 69–79. Copyright 1978, with permission from Elsevier.)

measure. Taking a step further, Grossman (1954) defines exposure as the sum of flows at junction crossing points. *Crossing points* are being defined as those points in a junction where two streams of traffic cross each other. Hakkert and Mahalel (1978) systematically illustrated these crossing points in a diagram. As shown in Figure 10.1, there are altogether 24 conflict points in a 4-legged 2-way junction. It was argued that "exposure at intersections may be defined as the number of opportunities of being involved in accidents," and these opportunities are related to vehicle maneuvers through the intersection (Hakkert and Mahalel 1978, 72). Hence, they proposed using the sum of the products of the two traffic streams involved at all 24 conflict points that vehicles merge or cross within a junction as the exposure measure. The Hakkert and Mahalel's index, however, was considered to be intuitively wrong by McGuigan (1981). For instance, right-turning traffic does not appear in X_i and that the left-turning traffic is double counted.

10.4 METHODS

Next, researchers have to quantify the *relationship* between X_i and O_i. The aim is to better understand collision frequency after taking out the vehicular traffic exposure, which is often *not* considered as a "treatment variable" in safety analysis.

10.4.1 Simple Ratios

A simple way to take traffic exposure into account in understanding the collision-exposure relationship is to use X_i as the denominator in analyzing collision frequency. This approach has been used in Austria for identifying hazardous road locations (HRLs) (Elvik 2007). When plotted in a diagram with collision frequency on the y-axis and traffic volume on the x-axis, the relationship becomes a positively sloping straight line from the origin with the slope equal to the collision–exposure ratio (Γ). In other words, this approach assumes that the collision–exposure relationship is not changing over different ranges of traffic volumes. Moreover, it is a positive and directly proportional relationship ($O_i = \Gamma \times X_i$). In reality, while there is some good correlation between collision frequency and vehicle exposure, the relationship is not a simple directly proportional one.

10.4.2 Simple Exponents

Tanner (1953) argues that the collision–exposure relationship is best described by the square root of the product flow of the two crossing roads at junctions. The relationship is specified as

$$O_i = \sqrt{X_1 X_2} \tag{10.5}$$

McDonald (1953), working along a similar line, found that the traffic volume of different approaches to the junctions needs to be duly considered, because the potential number of conflicts will be very different. The equation becomes

$$\hat{O} = \alpha X_1^{\beta_1} X_2^{\beta_2} \tag{10.6}$$

where
 X_1 specifies the traffic volume of the main road
 X_2 is the traffic volume of the minor road
 β_1 and β_2 are coefficients that can be estimated by the maximum likelihood method

In a linear form, it becomes

$$\ln(\hat{O}) = \ln(\alpha) + \beta_1 \ln(X_1) + \beta_2 \ln(X_2) \tag{10.7}$$

McDonald's (1953) estimates suggest β_1 and β_2 to be 0.455 and 0.633. However, Leong (1973) found that the estimates of β_1 and β_2 were not significantly different for 234 junctions studied, and suggested that they need not be estimated separately.

McGuigan (1981) found that the square root of the cross flow product at junctions did not provide a better model fit in most cases, except in priority T-junctions.

10.4.3 Linear Regression Models

McGuigan (1981) was among the first to build a linear regression model to quantify the relationship between collision frequency and traffic volume with

$$O_i = \alpha + \beta X_i \qquad (10.8)$$

where X_i is the VKT or traffic throughput at a junction. After establishing traffic throughput as the best estimate of X_i for a junction, McGuigan (1981) analyzed the collision records of 3211 junctions in Lothian and estimated that α is 0.054 and β is 0.107. The relationship, though statistically significant ($p<0.001$), only has a low explanatory power ($R^2=0.138$). For links (nonjunctions), using 118 randomly selected single-carriageway, two-way road links, he estimated α to be 0.078 and β to be 0.737, with slightly higher explanatory power ($R^2=0.406$). In both situations, categorization or breakdown of junctions and links by type and location (urban versus rural) improves the ability to predict collision totals. As the categories are very flexible, regression-based models have been further extended and applied widely to examine the relationships between collision frequency and traffic exposure (Zegeer et al. 1990). Nonetheless, the assumptions of linearity of the relationship and that the error term is independent, normally distributed, and with constant variance (homoscedacity) do not hold for the traffic collision phenomenon (Hair et al. 1995). The challenges to theoretical justification, model assumptions, and model fit when using empirical data have been documented in Jovanis and Chang (1986), Joshua and Gerber (1990), and Miaou et al. (1992).

10.4.4 Poisson Regressions

Collisions happening at a specific location over a given period of time are essentially count data, which are discrete and strictly nonnegative. Like a statistical experiment of flipping a coin with "head" defined as a success, the occurrence of each collision at location i during a time period can be counted as "a success" in a Bernoulli experiment. With repeated Bernoulli trials, the probability of success follows the Poisson distribution, because the true/underlying probability of success (λ) is extremely small that the expected collision frequency at a location i, E_i, is close to zero. The probability that a road segment i having n collisions during the study period is given by

$$P_i(n) = \frac{e^{-\lambda_i}\lambda_i^n}{n!} \qquad (10.9)$$

where e is the base of the natural logarithm or 2.71828. λ_i is unknown and it can be perceived as the underlying or true collision rate of road segment i. (*Note*: When restricting the analysis to a single road element [e.g., in before-and-after analysis], i is often taken to refer to the ith time interval in the time dimension. λ_i becomes the long-term average collision rate. The resulting models are different. Hence, care should be taken in interpreting the results.)

In reality, the Poisson parameter λ_i can be specified as

$$\lambda_i = e^{X_i \beta} \qquad (10.10)$$

where

X_i is a vector of other explanatory variables, such as road geometric design variables
β is a vector of estimable coefficients

With Poisson regression, either linear additive or multiplicative form can be specified (Jovanis and Chang 1986). For instance, should a multiplicative form of three explanatory variable be specified, then

$$\lambda_i = \beta_0 X_1^{\beta_1} X_2^{\beta_2} X_3^{\beta_3} \qquad (10.11)$$

Let the number of explanatory variables be M and the total number of ith set of road segments (set of observations) be G. The Poisson regression (as defined by Equations 10.9 and 10.10) can be estimated by the standard maximum likelihood methods:

$$L(\beta) = \prod_{i=1}^{G} \prod_{n=0}^{\infty} P_i(n)^{D_{in}} \qquad (10.12)$$

With the logarithm transformation of Equation 10.12, the log-likelihood value can be obtained by

$$LL(\hat{\beta}) = \sum_{i=1}^{G} \sum_{n=0}^{\infty} D_{in} \log[P_i(n)] \qquad (10.13)$$

$LL(0)$ is defined as the log-likelihood value of the model in which only the constant term is used. The value of $2[LL(\hat{\beta}) - LL(0)]$ follows a Chi-squared distribution with $M - 1$ degrees of freedom. It is a statistic for testing the significance of all explanatory variables included in the model. ρ^2, defined as $1 - (LL(\hat{\beta})/LL(0))$, is an informal goodness-of-fit measure, similar to R^2 used in regression (Jovanis and Chang 1986). Further refinements (such as the Empirical Bayes approach explained in Chapter 11) are possible. The two other less popular estimation methods are weighted least squares and minimum Chi-squared estimation (Jovanis and Chang 1986).

Statistically, the Poisson distribution is only defined by one parameter, λ_i, because the variance, $V(N)$, is equal to the mean, $E(N)$, where

$$E(N) = \sum_{n=0}^{\infty} nP_i(n) = \lambda_i \qquad (10.14)$$

$$V(N) = \sum_{n=0}^{\infty} [n - E(N)]P_i(n) = \lambda_i \qquad (10.15)$$

If $V(N)$ is bigger than λ_i, the data are said to suffer from overdispersion (see Burt and Barber 1996), and the overall statistical significance of the estimated coefficients will be overestimated (Miaou et al. 1990).

When applied to empirical data, the number of road segments having zero collisions often exceeds that assumed in a Poisson regression. To account for the higher probability of "spike" (additional) zeros, P_{0i} is used to represent the "additional" probability of segment i to have no collisions. $1 - P_{0i}$ represents the probability that segment i follows the Poisson distribution. With a Poisson distribution, the probability that segment i has no collision is $e^{-\lambda i}$. The total probability of observing zero collisions, therefore, is the sum of the two probabilities of having no collision together. The entire probability distribution is called the zero-inflated Poisson distribution, and the model is called the zero-inflated Poisson (ZIP) model (Lambert 1992; Miaou 1994; Shankar et al. 1997; Vogt and Bared 1998). The probability density function is given as follows:

$$P(n_i) = \begin{cases} P_i + (1 - P_{0i})e^{-\lambda i} & (n_i = 0) \\ (1 - P_{0i}) \dfrac{e^{-\lambda i} \lambda_i^{ni}}{n_i!} & \text{(otherwise)} \end{cases} \qquad (10.16)$$

Given the flexibility of the Poisson distribution, researchers often consider other road geometric features other than traffic volume in modeling. However, only findings related to traffic volume is reported in this chapter. In the 1980s, Jovanis and Chang (1986) used the Poisson regression to examine the relationship between collisions, vehicle-miles of travel (VMT), and environmental data. They found that the collision frequency did increase as travel mileage increased. In the pooled model, the coefficients of VMT automobiles (in millions) and trucks (in millions) were estimated to be 0.255 and 0.299, respectively. Subsequently, Miaou et al. (1992) estimated a Poisson model of truck collision frequency with AADT per lane, horizontal curvature, and vertical gradient. They found that there was a significant relationship between truck collisions, traffic, and highway geometric design variables. Ivan and O'Mara (1997) also used the Poisson regression model to analyze collisions happening in Connecticut. AADT and posted speed limit were again found to be critical explanatory variables for collision frequency.

Following the exposure concept, Ivan (2004) considered traffic volume and segment length to be reflecting *exposure* and that the characteristics of other geometric variables to be the *collision risk separately* influencing O_i. As there is no confusion, subscript i for individual road segment is dropped for simplicity. Hence, the Poisson regression looks like

$$\lambda = X^{\beta_X} L^{\beta_L} e^{(\beta_0 + \beta_W W + \beta_S S)} \quad (10.17)$$

where
λ is the estimated mean number of collisions per year
X is AADT (in thousands of vehicles)
L is the length of the road segment in miles
W is the width of both lanes and paved shoulders in feet
S is the speed limit in miles per hour
β are the parameters to be estimated

Focusing on β_X, the results obtained by Ivan (2004) are shown in Table 10.2.

In other words, the coefficient is generally less than 1 for single-vehicle collisions, and very close to or greater than 1 for all others. The results suggest that "for roads with the same geometry (pavement width and speed limit), one would expect to find a lower rate of single-vehicle collisions per vehicle-mile traveled on the roads with higher AADT than on the roads with lower AADT. For the other collision types, one would expect the opposite effect" (Ivan 2004, 136).

Qin et al. (2004) used a similar ZIP model to estimate the effects of AADT and segment length on two-lane rural highways in Michigan without major intersections. The ZIP model has a functional form as specified in Equation 10.16. The collision data of each year from 1994 to 1997 were used to estimate the exponentials separately. The results are summarized in Table 10.3. Their hypothesis that the relationship between AADT and collision frequency is linear (i.e., $\beta_X = 1$) has been rejected; similarly, the relationship between segment length and collision frequency as linear (i.e., $\beta_L = 1$) has also been rejected. The collision frequency increases nonlinearly. In most situations, β_X (or β_L) are less than 1, suggesting that the increase in collision

TABLE 10.2
Estimated Parameters β_X

	Single-Vehicle Collisions	Same-Direction Collisions	Opposite-Direction Collisions	Intersecting-Direction Collisions
Michigan	0.397	1.422	1.203	1.123
California	0.685	1.263	1.091	0.915
Washington	0.788	1.000	0.944	0.877
Illinois	0.795	1.740	1.326	0.948

Source: Data from Ivan, J.N., *Transp. Res. Rec.*, 1897, 134, 2004.

TABLE 10.3
Estimated β_X and β_L

	Single-Vehicle Collisions	Multi-Vehicle Same Direction Collisions	Multi-Vehicle Opposite Direction Collisions	Multi-Vehicle Intersecting-Direction Collisions
β_X	0.254–0.363	1.063–1.270	0.551–1.126	0.650–0.807
β_L	0.725–0.792	0.407–0.442	0.354–0.748	0.245–0.191

Source: Data from Qin, X. et al., *Accid. Anal. Prev.*, 36(2), 183, 2004.

frequency is lower as AADT (or segment length) increases. Moreover, the hypothesis that $\beta_X = \beta_L$ cannot be rejected.

10.4.5 NEGATIVE BINOMIAL METHODS

While the specification of the Poisson distribution is attractive because only one parameter needs to be estimated, the overdispersion problem is commonly encountered when examining empirical collision patterns (Jovanis and Chang 1986; Joshua and Garber 1990; Jones et al. 1991; Shankar et al. 1995). Maycock and Hall (1984) and Miaou and Lum (1993) suggested using the more general negative binomial (NB) model instead. When each collision happening at road location i over a study period is considered as "a success," the probability of having n successes in Z trials is given by

$$P_i(n) = C_n^Z p^n (1-p)^{Z-n} \tag{10.18}$$

where
  p is the probability of success
  q is the probability of failure ($q = 1 - p$, as they are mutually complimentary events)
  C is the combination of n out of Z

The latter defines the number of ways that the total of n can be obtained out of Z trials regardless of the order that the successes appear. This is also called the binomial distribution. In a binomial distribution, Z and p are the key parameters. However, there is no specified Z trials or maximum number of "successes" in the case of road safety. Hence, Z is not fixed in advance. Instead, the sequence of Bernoulli trials will continue until a certain number, say r, of "successes" occurs. If $r = 1$, there will be $Z - 1$ failures before the first "success" that occurs. In other words, Z has a geometric distribution (Hogg and Tanis 1997). The negative binomial distribution is applicable:

$$P_i(n) = C_{r-1}^{n-1} p^r (1-p)^{n-r}, \quad n = r, r+1, r+2\ldots \tag{10.19}$$

Mean is

$$\mu = E(n) = \frac{r}{p} \tag{10.20}$$

and variance is

$$\sigma^2 = \frac{r(1-p)}{p^2} \tag{10.21}$$

In road safety analysis, take the number of observed collisions at segment i as n_i. λ_i is the expected number of collisions at segment i, which is not governed by a single underlying probability of success, p, but many collision risk factors $\mathbf{X}_i$. The NB model is

$$\lambda_i = e^{\beta X_i} + \varepsilon_i \tag{10.22}$$

where $\exp(\varepsilon_i)$ is a gamma-distributed error term with mean equal to 1 and a variance of v. For the purpose of the maximum likelihood estimation, the equation becomes

$$P(n_i) = \frac{\Gamma(\theta + n_i)}{[\Gamma(\theta) n_i!]} \lambda_i^\theta (1 - \lambda_i)^{n_i} \tag{10.23}$$

or

$$P(n_i) = \frac{(\theta + n_i - 1)!}{(\theta - 1)! n_i!} \lambda_i^\theta (1 - \lambda_i)^{n_i} \tag{10.24}$$

where
$\theta = 1/v$
$\lambda_i = \theta(\theta + n_i)$

The variable v, to recall, is the variance of the error term. This variance allows the overdispersion problem to be taken into account as

$$var(n_i) = E(n_i)[1 + v E(n_i)] \tag{10.25}$$

If v is not significantly different from zero, $var(n_i) = E(n_i)$, and the Poisson distribution is appropriate (Poch and Mannering 1996; Abdel-Aty and Radwan 2000). In particular, the elasticity from NB model is useful in defining collision risk,

$$a_{x_{ik}}^{\lambda_i} = \frac{\partial \lambda_i}{\partial x_{ik}} * \frac{x_{ik}}{\lambda_i} \tag{10.26}$$

where
λ_i is the expected collision frequency at i
x_{ik} is the value of the explanatory variable k at i

Differentiating Equation 10.22 and applying Equation 10.26 give

$$a_{x_{ik}}^{\lambda_i} = \beta_k x_{ik} \qquad (10.27)$$

Average elasticities calculated based on the above equation over all road segments will give elasticity estimates of the explanatory variables, which can be readily interpreted as the percentage change in collision frequency, should there be a 1% increase in the explanatory variable.

With the flexibility of NB models, numerous road features such as median width (Knuiman et al. 1993) and horizontal curves (McGee et al. 1995) have been considered in understanding collision frequency. Among the NB models, which focus on traffic volume, Abdel-Aty and Radwan (2000) have used negative binominal regression models. Taking State Road 50 (SR50) in Central Florida, they divided it into 566 highway segments with any detectable change in the geometric and/or roadway variables (including AADT, degree of horizontal curvature, shoulder type, divided/undivided, rural/urban classification, posted speed limit, number of lanes, road surface and shoulder types, and lane, median, and shoulder widths). The aim is to have relatively homogenous/uniform highway segments. Then, the 3-year collision records (1992–1994) (totaling 166) were assigned to the highway segments to examine the relationship of collision frequency with the geometric and traffic characteristics of the segment. Among the different variables included, two exposure variables were found to be significant, that is, the log of the segment length ($a=0.33$) and the log of AADT per lane ($a=0.62$). Their results confirm that both the traffic volume and segment length exposure effect are nonlinear. Moreover, among other variables included in the model, AADT per lane had the greatest relative effect on collision frequency.

The NB models of Poch and Mannering (1996) focus on intersections and approach conditions. Using 7 years of collision data (1987–1993) from 63 urban intersections in Bellevue, Washington, their NB models reveal that collision frequency was elastic with respect to left-turn traffic volume (AADT in thousands) ($a=2.28$) and total opposing approach volume (AADT in thousands) ($a=2.95$), and inelastic to the right-turn volume ($a=0.92$). Speed limit was another significant variable. Moreover, a finer categorization of collisions into rear-end, angle, and approach-turn collisions led to better model fit and more specific diagnosis of the collision explanatory variables.

10.5 IMPLICATIONS ON INTERVENTIONS

10.5.1 Collision Count versus Collision Rate in Road Safety Analysis

The above discussion has reconfirmed a few important points. First, traffic exposure should be considered separately from other risk factors (such as number of lanes, curvature, other behavioral factors). Second, the relationship between traffic exposure and that of collision counts is not a linear one. Hence, the use of collision rate in safety analysis (e.g., Miaou et al. 1992) is not scientific and recommended, because it implicitly assumes that collision counts have a direct linear relationship with exposure. The collision counts should be used in road safety analysis.

10.5.2 "Regression-to-Mean" Problems

Another important point coming out from the above analysis is that there is a certain random element in collision counts. An "inherently safe" road segment i can have high collision counts simply due to chance. When this happens, it is highly likely that the collision counts of road segment i will fall in subsequent periods even without any treatment. This tendency for road segments having high collision counts as a result of "chance" and "naturally" falling in subsequent periods back to the underlying collision risk is called the "regression-to-mean" problem. The problem is particularly important for evaluating treatments, which we shall turn to in Chapter 13. For now, it is important to recognize that a high observed collision frequency of a road segment i at any point in time can be the result of real hazards (as reflected in the risk factors present at the road segment), purely random element (chance), or a combination of both. If the randomness is not taken into account, any gain from improvements resulting from the road safety measures will be overestimated (Elvik 1997; Maher and Mountain 1988; Geurts and Wets 2003).

10.6 CONCLUSION

Taking traffic volume as an exposure factor implicitly implies that road safety administrations are not considering traffic volume as other collision risk variables for road safety improvement measures. In the short term, it is generally true that the traffic volume of a specific road segment cannot be easily changed. However, major traffic diversions or even modal change (e.g., with the building of metro) particularly at road junctions or segments with consistently high traffic volume and high collision frequencies should be actively considered in a road safety strategy.

REFERENCES

Abdel-Aty, M. A. and A. E. Radwan. 2000. Modeling traffic accident occurrence and involvement. *Accident Analysis & Prevention* 32 (5): 633–642.

Babkov, V. F., E. M. Lobanov, and B. M. Lebedev. 1970. Accident risks and capacity of single level intersections. Theme 2, *10th International Study Week in Traffic & Safety Engineering*, Rotterdam, the Netherlands.

Breuning, S. M. and A. J. Bone. 1960. Interchange accident exposure. *Highway Research Board Bulletin* 240: 44–52.

Burt, J. E. and G. M. Barber. 1996. *Elementary Statistics for Geographers*. New York: The Guilford Press.

Elvik, R. 1997. Evaluations of road accident blackspot treatment: A case of the iron law of evaluation studies? *Accident Analysis & Prevention* 29 (2): 191–199.

Elvik, R. 2007. State-of-the-art approaches to road accident black spot management and safety analysis of road networks. Institute of Transport Economics, Oslo, Norway.

Geurts, K. and G. Wets. 2003. Black spot analysis methods: Literature review. Flemish Research Center for Traffic Safety, Diepenbeek, Belgium.

Grossman, L. 1954. Accident-exposure index. *Highway Research Board Proceedings* 33: 129–138.
Hakkert, A. S. and D. Mahalel. 1978. Estimating the number of accidents at intersections from a knowledge of the traffic flows on the approaches. *Accident Analysis & Prevention* 10 (1): 69–79.
Hair Jr., J. F., R. E. Anderson, R. L. Tatham, and W. C. Black. 1995. *Multivariate Data Analysis: With Readings*, 4th edn. Upper Saddle River, NJ: Prentice-Hall.
Hogg, R. V. and E. A. Tanis. 1997. *Sampling Distribution Theory in Probability and Statistical Inference*. Upper Saddle River, NJ: Prentice-Hall.
Ivan, J. N. 2004. New approach for including traffic volumes in crash rate analysis and forecasting. *Transportation Research Record* 1897: 134–141.
Ivan, J. N. and P. J. O'Mara. 1997. Prediction of traffic accident rates using Poisson regression. Paper presented at the *Transportation Research Board Annual Meeting*, Washington, DC.
Jones, T. A., J. Y. Zou, S. W. Cowan, and M. Kjeldgaard. 1991. Improved methods for building protein models in electron density maps and the location of errors in these models. *Acta Crystallographica Section A: Foundations of Crystallography* 47 (2): 110–119.
Jorgensen, N. O. 1972. Statistical detection of accident blackspots. In *11th International Study Week in Transportation and Safety*, Brussels, Belgium.
Joshua, S. C. and N. J. Garber. 1990. Estimating truck accident rate and involvements using linear and Poisson regression models. *Transportation Planning and Technology* 15 (1): 41–58.
Jovanis, P. P. and H. L. Chang. 1986. Modeling the relationship of accidents to miles traveled. *Transportation Research Record* 1068: 42–51.
Knuiman, M. W., F. M. Council, and D. W. Reinfurt. 1993. Association of median width and highway accident rates. *Transportation Research Record* 1401: 70–82.
Lambert, D. 1992. Zero-inflated Poisson regression, with an application to defects in manufacturing. *Technometrics* 34 (1): 1–14.
Leong, H. J. W. 1973. Relationship between accidents and traffic volumes at urban intersections. *Australian Road Research* 5 (3): 72–90.
Loo, B. P. Y., W. T. Hung, H. K. Lo, and S. C. Wong. 2005. Road safety strategies: A comparative framework and case studies. *Transport Reviews* 25 (5): 613–639.
Loo, B. P. Y., S. C. Wong, W. T. Hung, and H. K. Lo. 2007. A review of the road safety strategy in Hong Kong. *Journal of Advanced Transportation* 41 (1): 3–37.
Maher, M. J. and L. J. Mountain. 1988. The identification of accident blackspots: A comparison of current methods. *Accident Analysis & Prevention* 20 (2): 143–151.
Mathewson, J. H. and R. Brenner. 1957. Indexes of motor vehicle accident likelihood. *Highway Research Board Bulletin* 161: 1–8.
Maycock, G. and R. D. Hall. 1984. Crashes at four-arm roundabouts. Transport Research Laboratory, Crowthorne, U.K.
McDonald, J. W. 1953. Relation between number of accidents and traffic volume at divided-highway intersections. *Highway Research Board Bulletin* 74: 7–17.
McGee, H. W., W. E. Hughes, and K. Daily. 1995. Effect of highway standards on safety. Transportation Research Board, Washington, DC.
McGuigan, D. R. D. 1981. The use of relationships between road accidents and traffic flow in "black-spot" identification. *Traffic Engineering and Control* 22 (8–9): 448–451.
Miaou, S. P. 1994. The relationship between truck accidents and geometric design of road sections: Poisson versus negative binomial regressions. *Accident Analysis & Prevention* 26 (4): 471–482.
Miaou, S. P., P. S. Hu, T. Wright, A. K. Rathi, and S. C. Davis. 1992. Relationship between truck accidents and highway geometric design: A Poisson regression approach. *Transportation Research Record* 1376: 10–18.

Miaou, S. P. and H. Lum. 1993. Modeling vehicle accidents and highway geometric design relationships. *Accident Analysis & Prevention* 25 (6): 689–709.

Miaou, S. P., A. K. Rathi, F. Southworth, and D. L. Greene. 1990. Forecasting urban highway travel for year 2005. Oak Ridge National Laboratory, Oak Ridge, TN.

Poch, M. and F. Mannering. 1996. Negative binomial analysis of intersection-accident frequencies. *Journal of Transportation Engineering* 122 (2): 105–113.

Qin, X., J. N. Ivan, and N. Ravishanker. 2004. Selecting exposure measures in crash rate prediction for two-lane highway segments. *Accident Analysis & Prevention* 36 (2): 183–191.

Schaechterle, K., H. Kurzak, K. Pfund, and W. Mensebach. 1970. Risk of accidents and capacity at at-grade road junctions. Theme 2, *10th International Study Week in Traffic & Safety Engineering*, Rotterdam, the Netherlands.

Shankar, V., F. Mannering, and W. Barfield. 1995. Effect of roadway geometrics and environmental factors on rural freeway accident frequencies. *Accident Analysis & Prevention* 27 (3): 371–389.

Shankar, V., J. Milton, and F. Mannering. 1997. Modeling accident frequencies as zero-altered probability processes: An empirical inquiry. *Accident Analysis & Prevention* 29 (6): 829–837.

Smeed, R. J. 1955. Accident rates. *International Road Safety & Traffic Revue* 3 (2): 30–34.

Tamburri, T. N. and R. N. Smith. 1970. The safety index: A method of evaluating and rating safety benefits. *Highway Research Record* 332: 28–39.

Tanner, J. C. 1953. Accidents at rural 3-way junctions. *Journal of the Institute of Highway Engineers* 2 (11): 56–67.

Vogt, A. and J. Bared. 1998. Accident models for two-lane rural segments and intersections. *Transportation Research Record* 1635: 18–29.

Zegeer, C., R. Steward, D. Reinfut, F. Council, T. Neuman, E. Hamilton, T. Miller, and W. Hunter. 1990. *Cost Effective Geometric Improvements for Safety Upgrading of Horizontal Curves.* Chapel Hill, NC: University of North Carolina.

11 Exposure Factor 2
Road Environment

11.1 INTRODUCTION

The Empirical Bayes (EB) approach to road safety estimation has developed gradually during the past 30 years. It is now firmly established and recommended as the state-of-the-art approach to the estimation of the expected number of collisions. The EB approach was originally developed for the purpose of controlling for the "regression-to-mean" problem in before-and-after studies evaluating the effects of road safety measures. This has remained an important area of application, but EB methods are now also used to identify hazardous road locations. This chapter outlines several versions of EB that exist now. Developing good collision prediction models is a complex process, and there are many analytic choices. A case study on the usage of motorcycle helmets in a Chinese city will be included to illustrate the methods.

11.2 RELATIONSHIP BETWEEN ROAD ENVIRONMENT AND COLLISIONS

Following the conceptual framework that road collisions are the results of road environment, vehicle, and human factors, variability in each of three major groups of factors will affect the likelihood of collisions happening. It follows that when collisions are assigned and analyzed geographically, the physical locations become the primary unit of analysis. Given that the road environment of different physical locations is so variable, Elvik (2008) has proposed the concept of site-specific safety. Most importantly, the safety of a road location should not be directly compared with some aggregate averages, whether world, country, or city figures. It is only meaningful to make comparisons of a road location's safety with similar sites. However, what are similar sites? Similar sites can be and are most often defined by road environment conditions, such as lighting, road geometry, road surface, and speed. In order to conduct fair comparisons, statistical methods are used to take into account of variability in road environment conditions.

11.2.1 Intersections and Mid-Block Locations

One of the key road environment factors that has an impact on road safety is whether the location is an intersection or a mid-block location. At intersections, traffic flows in different directions and, hence, are often in direct conflict with each other. More details are provided in Chapter 9. In contrast, vehicles in mid-block locations are traveling in parallel and often within specific lanes. Hence, the potential conflicts are minimal, except with lane-changing activities. Hence, it is typical to classify road locations into the two key categories of intersections and mid-block locations before comparisons of their collision records are made.

11.2.2 OTHER GEOMETRIC FEATURES

The concept of fair comparisons of road locations with similar road conditions can arguably be extended to all different road features that have significant bearings on road safety. For instance, mid-block locations with different speed limits, such as those on a rural road and an expressway, are arguably incomparable. Other road geometric features having noticeable effects on safety include speed, curvature, slope, and road surface.

11.3 METHODS

Some other statistical methods, such as the generalized linear model (GLM) (Geurts and Wets 2003), are also used to take into account the different road conditions. Many of these studies can be grouped under the category of collision prediction models. In other words, the aim is to primarily predict the collision occurrence (and/or severity) rather than to make fair comparisons or evaluate road safety measures.

11.3.1 LOGISTICAL REGRESSION

One of the methods of controlling for variability in road environment factors is the use of logistical regression. Though the method does not really single out or group observations into groups, it allows the impact of different road environment factors on road safety to be estimated by the regression coefficients and the relative impacts of different groups to be quantified and easily conceptualized as the odds ratios. For instance, with the outcome variable as road collisions causing serious or fatal injury, or not (binary logistical regression), and the mid-block locations as the reference group for the independent variable of road location, an odds ratio of 2 for intersections (supposedly only two groups for road locations) will mean that the chance of having a road collision causing serious or fatal injury will be twice as high for intersections than mid-block locations, ceteris paribus. The disadvantage is that there will be many different reference groups for different road environment variables (e.g., of speed and road surface) and the results become more difficult to relate to empirical records with more independent variables. For instance, a mid-block location with different speed limits and road surface will not be twice as dangerous. In addition, the outcome variable must be in categories and the collision rate or collision frequency is not used directly in the analysis. Hence, some information will be lost in most situations.

11.3.2 GEOGRAPHICALLY WEIGHTED REGRESSION

Apart from spatial autocorrelation, spatial heterogeneity has also attracted much attention in recent years. Geographically weighted regression (GWR) is an attempt at exploring spatial heterogeneity (Brunsdon et al. 1996), which examines relationships among variables varying from location to location. Erdogan (2009) modeled collision and death rates by GWR under the assumption that there was a nonstationary spatial relationship between variables and found that the GWR model significantly improved model fitting over the ordinary least squares (OLS) model. In the GWR model of Erdogan et al. (2008), the coordinates of province centroids are used as reference points for building a spatially weighted least square regression. Mathematically, the relationship between the variables may be expressed as

$$Y(u,v) = \beta_0(u,v) + \beta_1(u,v)x_1 + \beta_2(u,v)x_2 + \cdots + \beta_n(u,v)x_n + \varepsilon(u,v) \quad (11.1)$$

where
  β indicates that the parameters are to be estimated at a location (u, v)
  ε is the random error term

GWR softwares, such as GWR4, typically provide a set of local parameter estimates for each relationship, which may in turn be mapped to visualize the nature of the variation within the study area. Yet, how useful are these local parameter estimates beyond visualization? Another limitation of this method is that the analysis must be area based. Hence, collision density rather than collision count or collision frequency in the networks is analyzed. Moreover, as an area-based technique, the parameter estimation depends on the choice of weighting function and the kernel used (Fotheringham et al. 2002).

11.3.3 Empirical Bayes Methods

According to Geurts and Wets (2003), the applications of EB methods in road safety only originated in the 1980s. The major characteristics of the EB method are that it compares the collision frequency of road segment i with *similar sites,* whose true collision risk is believed to be the same as road segment i and the historical collision records of that site. Sites that are grouped as "similar sites" with the same true underlying collision risk are labeled the reference population. Generally, the "true" underlying collision frequency of the reference population can be estimated by the method of moments and statistical models. Suppose the study area is a city and the study period is 3 years. All collisions happening on all roads of that city over the 3-year period are identified. Next, the road segments (link based) are grouped into relatively distinct reference populations. Each road segment will belong to one and only one of the reference populations. For the sake of simplicity, the three groups are junctions, local roads, and highways. The collision frequency for each reference population group, O_{cat}, is calculated as

$$O_{cat} = \frac{\sum_{g=1}^{G} O_g}{G} \quad (11.2)$$

$$s^2 = \frac{\sum (O_c - O_{cat})^2}{N-1} \tag{11.3}$$

where
O_g is the collision frequency of a site in group G
G is the total number of sites in category G

The EB adjusted collision frequency at each site,

$$O_{EBi} = O_i + \frac{O_{cat}}{s^2}(O_{cat} - O_i) \tag{11.4}$$

As early as the mid-1980s, Hauer (1986) and Hauer and Persaud (1987) have used the EB approach to estimate the expected number of collisions at specific sites and to compare the performance of different identification procedures of hazardous locations. Hauer et al. (1988) and Belanger (1994) applied EB methods to estimate the safety at signalized intersections. Persaud evaluated the collision potential of Ontario road sections (Persaud 1991) and ranked sites for potential safety improvements (Persaud 1999). Higle and Witkowski (1988) presented a supplemental EB technique that makes use of collision rates. Cheng and Washington (2005, 872) commented that "by accounting for both crash history and expected crashes for similar sites, EB methods have been shown to offer improved ability to identify 'high-risk' sites." Extensive studies have been reported in safety research such as analyzing spatial–temporal patterns of motor vehicle collisions and ranking sites for safety improvements (Miaou et al. 2003; Miaou and Song 2005; Aguero-Valverde and Jovanis 2006, 2008; Li et al. 2007; Quddus 2008). Moreover, though "most of these research studies yielded favorable results in terms of identifying hotspots, but the range of conditions were quite small within studies" (Cheng and Washington 2005, 872).

The EB methods are in general data intensive and depends critically on the correct specification of the reference groups. Moreover, there are two assumptions. First, collision occurrence at a given location obeys the Poisson probability law so that

$$P(x|\lambda) = \frac{e^{-\lambda} \lambda^x}{x!} \tag{11.5}$$

where $P(x|\lambda)$ is the probability of recording x collisions at a location with a long-term expected collision number of λ. Second, "the probability distribution of the λ's of the population of sites is gamma distributed, where $g(\lambda)$ is denoted as the gamma probability density function, and is typically modeled as a function of site covariates. On the basis of the above assumptions, the probability that a site selected randomly records x collisions is approximated by the negative binomial (NB) probability distribution" (Cheng and Washington 2005, 872).

11.3.4 HIERARCHICAL BAYES METHODS

Based on the above logic, more complicated techniques have been developed to allow more precise modeling of the sites to be associated with more than one level of reference groups, for example, four-legged intersections within intersections, or signalized intersections within intersections. Some notable examples include Christiansen et al. (1992), Schlüter et al. (1997), Davis and Yang (2001), Tunaru (2002), Geurts and Wets (2003), and Brijs et al. (2006).

11.4 INTERVENTION

Collision prediction models arguably have no meanings unless proactive and effective measures can be taken to improve the situation. Otherwise, it will just be knowing that x collisions will inevitably occur at location i. Some vehicles (not knowing in advance which vehicles) will be involved, and someone (but not knowing in advance who) will be hurt or killed "inevitably." It is precisely this mentality of "randomness" and "inevitability" that the term "accident" is increasingly avoided in medical and public health research. With a belief that "accidents" can be avoided and are caused by systematic factors, notably human factors, results of collision prediction models using the EB approach should be looked at carefully to identify only sites that are having much higher than expected collision number than its reference group. More systematic road safety measures targeted to address road safety hazards can be introduced.

11.5 EVALUATION

Once a road safety improvement measure is identified and implemented, the next key question then becomes "Is the measure effective?". In order to answer this question, the road safety records of the location before and after the implementation of the road safety measure need to be compared. However, a direct comparison of the before and after scenarios suffers from two major problems, the first one being the "regression-to-mean" problem. While road collisions are not simply random events, the collision records of a specific location will have a random element, that is, the empirical collision frequency will not be the same, despite a true mean (an expected long-term collision frequency). If a location of high collision frequency due to randomness is selected for the treatment, it is highly likely that the collision frequency will fall and return to its mean value even if no treatment is made. In this way, attributing the reduction in collision frequency to the treatment will be problematic. The second problem being that road collisions are not laboratory experiments, where other independent variables can be controlled and kept constant when testing the effects of a change of one independent variable, such as the road safety measure. Other changes in the society, including the overall economic situation, the opening of a new railway line, or weather changes, will not stop as the road safety measure is introduced. Hence, not all changes happening between the before and after scenarios can be attributable to the road safety measure. It is noteworthy to mention that these other changes may also happen in opposite directions leading to an overestimation or an underestimation of the true effects of the road safety measure. Events that lead to a reduction of road traffic collisions, such as an economic slum, may lead to an overestimation of the effects of the road safety measure. The co-occurrence of events that lead to increases in road collisions, such as worsening drink-driving problems, will lead to an underestimation of the effectiveness of the road safety measure. In general, the longer is the period between the before and after scenarios, the more likely that the second problem will loom large. Given these two pitfalls, a direct comparison of the before and after scenarios in road safety improvement evaluation is often considered unscientific and naïve before-and-after studies.

The EB method, however, may also be applied to overcome the above problems. In particular, "uncontrollable" and "unobservable" favorable and unfavorable factors to road safety in the after period will affect not just the subject location but also locations in the same reference group. Hence, the EB methods can provide a more accurate estimation of the effectiveness of a road safety measure. The logic is illustrated with a case study in the following text.

In a study by Wu and Loo (2013), the naïve before-and-after studies and the EB method have been applied to examine the effects of helmet law enforcement in Maoming, southern China. While more than 160 countries in the world have laws requiring the use of motorcycle helmets (World Health Organization 2009), the lack of law enforcement, especially in developing countries, has greatly jeopardized the effectiveness of these laws in protecting motorcyclists. In China, there has not been much research on the effects of law enforcement on increasing helmet use. Transport authorities are often uncertain about how long a law enforcement should last and how often they should conduct law enforcement activities. They typically rely on the number of penalty tickets issued or the number of motorcycles they have temporarily seized in enforcement episodes. These data obviously are not good indicators of the effectiveness of the enforcement actions in achieving the ultimate aim, that is, increasing the helmet usage rate.

Taking advantage of an enforcement action of the Maoming Transport Authority at an intersection (location not known in advance to the authors) in the summer of 2012, Wu and Loo (2013) conducted an evaluation of the effectiveness of the enforcement action on the helmet usage rate in the city. Other than the naïve before-and-after comparisons, the method of sample moments following the EB approach was used to estimate the mean and variance of the rates of helmet use in the reference population, which are the other 20 signalized intersections on the main road network in the urban area of Maoming equipped with video cameras. Equation 11.6 is used to estimate the rates of helmet use (π) at the enforcement intersection had the helmet law not been enforced. For simplicity sake, this will be called the de facto rates.

$$E\{k|K\} = \alpha E\{k\} + (1-\alpha)K \qquad (11.6)$$

with

$$\alpha = \frac{E\{k\}}{[E\{k\} + VAR\{k\}]} \qquad (11.7)$$

$E\{k\}$ and $VAR\{k\}$ represent the mean and the variance of the k (the rates of helmet use) in the reference population. The variable α is a function of $E\{k\}$ and $VAR\{k\}$, and its value is between 0 and 1. In Equations 11.8 and 11.9, $\bar{K}$ is the mean of the rates of helmet use, and S^2 is the variance of the rates of helmet use. Therefore, $\hat{E}\{k\}$ and $\widehat{VAR}\{k\}$ are used to obtain α by Equation 11.7. Next, the de facto rate, $\hat{E}\{k\,|\,K\}$ (it equals to π), is estimated using α, $\hat{E}\{k\}$, and K (it equals to λ) in Equation 11.6.

$$\hat{E}\{k\} = \bar{K} \tag{11.8}$$

$$\widehat{VAR}\{k\} = S^2 - \bar{K} \tag{11.9}$$

Therefore, the effects of the helmet law enforcement on helmet use during the three periods (as shown in Table 11.1) can be evaluated. In Table 11.1, π is the rate of helmet use at the subject intersection from 9:00 to 10:30 a.m. had the helmet law not been enforced, whereas λ is the estimated de facto rate of helmet use for the same time period. Therefore, $\lambda - \pi$ is the effect of the helmet law enforcement on helmet use.

A comparison of the results from the naïve before-and-after approach and the EB approach is shown in Table 11.2. The results indicate that the former approach exaggerates the effectiveness of the helmet law enforcement in all three periods. A similar conclusion is echoed by several other studies (Hauer 1997; Wong et al. 2006;

TABLE 11.1
Rates (%) of Helmet Use among Motorcyclists at the Intersection with Helmet Law Enforcement Action

Period	Date	π (%)	λ (%)	$\lambda - \pi$ (%)
During the law enforcement	July 5 (Thursday)	38.48	61.45	22.97
Weekend shortly after the law enforcement	July 7 (Saturday) and 8 (Sunday)	44.38	54.63	10.25
Weekday shortly after the law enforcement	July 9 (Monday)	38.45	58.35	19.90

TABLE 11.2
Increases in the Rate (%) of Helmet Use as the Effect of the Helmet Law Enforcement, Gauged by the Naïve Before-and-After Approach and the EB Approach, Respectively

Period	Date	Naïve Before-and-After Approach (%)	EB Approach (%)
During the law enforcement	July 5 (Thursday)	26.06	22.97
Weekend shortly after the law enforcement	July 7 (Saturday) and 8 (Sunday)	10.33	10.25
Weekday shortly after the law enforcement	July 9 (Monday)	22.96	19.90

TABLE 11.3
Rates (%) of Helmet Use among Motorcyclists, Gauged by the Naïve Before-and-After Approach and the EB Approach (Method of Sample Moments)

Approach	Baseline without the Law Enforcement (%)	During the Law Enforcement (%)	After the Law Enforcement (%)	Z-Test (Sig. at 95% CL?) Baseline vs. During	Z-Test (Sig. at 95% CL?) Baseline vs. After
Naïve before-and-after approach	38.85	59.30	56.00	−14.84 (Yes)	−13.64 (Yes)
EB approach	39.59	42.96	42.41	−2.43 (Yes)	−2.23 (Yes)

Cairney et al. 2012), indicating that the naïve before-and-after approach often provides an overly optimistic view of how safety has been improved by various treatments, such as law enforcement.

The results of the Z-test in Table 11.3 suggest that the same conclusion can be drawn using either the naïve before-and-after approach or the EB approach: a statistically significant difference exists between the helmet use rates in "baseline without the law enforcement" and "during the law enforcement," or between the rates in "baseline without the law enforcement" and "after the law enforcement." Therefore, comparisons between these two different approaches in Tables 11.2 and 11.3 indicate that the naïve before-and-after approach results in the same conclusion on the effect of the law enforcement on helmet use, but it gives overly optimistic results for the effectiveness of law enforcement. Conversely, the EB approach requires more resources (i.e., helmet use rates at 20 signalized intersections as the reference population) to reach the same conclusion with more precise results. For certain studies with limited resources (i.e., financial budget or official support) or a low precision requirement for results, the naïve before-and-after approach remains one of the preliminary approaches.

11.6 CONCLUSION

This chapter highlights that comparisons of road safety records, whether of the same location at different points in time or of different road locations, should be conducted carefully. The purpose of making comparisons should be clear to the researchers in the research design. Different research methodologies, with different assumptions, strengths, and weaknesses, are available. However, it should be recognized that no two road locations will be the same in terms of all road environment factors. Arguably, there are differences between two junctions nearby with different land use, or between the same road at peak hours and at midnight. Judgments about fair comparisons, therefore, will have to be made by the researchers based on the research problem at hand, the data available, the research time frame, and making reference to the research findings of the literature.

REFERENCES

Aguero-Valverde, J. and P. P. Jovanis. 2006. Spatial analysis of fatal and injury crashes in Pennsylvania. *Accident Analysis & Prevention* 38 (3): 618–625.
Aguero-Valverde, J. and P. P. Jovanis. 2008. Analysis of road crash frequency with spatial models. *Transportation Research Record* 2061 (1): 55–63.
Belanger, C. 1994. Estimation of safety of four-legged unsignalized intersections. *Transportation Research Record* 1467: 23–29.
Brijs, T., F. V. D. Bossche, G. Wets, and D. Karlis. 2006. A model for identifying and ranking dangerous accident locations: A case study in Flanders. *Statistica Neerlandica* 60 (4): 457–476.
Brunsdon, C., A. S. Fotheringham, and M. E. Charlton. 1996. Geographically weighted regression: A method for exploring spatial nonstationarity. *Geographical Analysis* 28 (4): 281–298.
Cairney, P., B. Turner, and L. Steinmetz. 2012. An introductory guide for evaluating effectiveness of road safety treatments. Report no. AP-R421/12. Austroads, Sydney, New South Wales, Australia.
Cheng, W. and S. P. Washington. 2005. Experimental evaluation of hotspot identification methods. *Accident Analysis & Prevention* 37 (5): 870–881.
Christiansen, C. L., C. N. Morris, and O. J. Pendleton. 1992. A hierarchical Poisson model with beta adjustments for traffic accident analysis. Center for Statistical Sciences Technical Report 103, University of Texas at Austin, Austin, TX.
Davis, G. A. and Yang, S. 2001. Bayesian identification of high-risk intersections for older drivers via Gibbs sampling. *Transportation Research Record* 1746: 84–89.
Elvik, R. 2008. A survey of operational definitions of hazardous road locations in some European countries. *Accident Analysis & Prevention* 40 (6): 1830–1835.
Erdogan, S. 2009. Explorative spatial analysis of traffic accident statistics and road mortality among the provinces of Turkey. *Journal of Safety Research* 40 (5): 341–351.
Erdogan, S., I. Yimaz, T. Baybura, and M. Gullu. 2008. Geographical information systems aided traffic accident analysis system case study: City of Afyonkarahisar. *Accident Analysis & Prevention* 40 (1): 174–181.
Fotheringham, A. S, C. Brunsdon, and M. E. Charlton. 2002. *Geographically Weighted Regression: The Analysis of Spatially Varying Relationships*. Chichester, U.K.: John Wiley & Sons.
Geurts, K. and G. Wets. 2003. Black spot analysis methods: Literature review. Flemish Research Center for Traffic Safety, Diepenbeek, Belgium.
Hauer, E. 1986. On the estimation of the expected number of accidents. *Accident Analysis & Prevention* 18 (1): 1–12.
Hauer, E. 1997. *Observational Before-After Studies in Road Safety: Estimating the Effect of Highway and Traffic Engineering Measures on Road Safety*. Oxford, U.K.: Pergamon.
Hauer, E., J. C. N. Ng, and J. Lovell. 1988. Estimation of safety at signalized intersections. *Transportation Research Record* 1185: 48–61.
Hauer, E. and B. N. Persaud. 1987. How to estimate the safety of rail-highway grade crossing and the effects of warning devices. *Transportation Research Record* 1114: 131–140.
Higle, J. L. and J. M. Witkowski. 1988. Bayesian identification of hazardous locations. *Transportation Research Record* 1185: 24–36.
Li, L., L. Zhu, and D. Z. Sui. 2007. A GIS-based Bayesian approach for analyzing spatial-temporal patterns of intra-city motor vehicle crashes. *Journal of Transport Geography* 15: 274–285.
Miaou, S. P. and Song, J. 2005. Bayesian ranking of sites for engineering safety improvements: Decision parameter, treatability concept, statistical criterion, and spatial dependence. *Accident Analysis & Prevention* 37 (4): 699–720.

Miaou, S. P., J. Song, and B. K. Mallick. 2003. Roadway traffic crash mapping: A space-time modeling approach. *Journal of Transportation & Statistic* 6 (1): 33–57.
Persaud, B. N. 1991. Estimating accident potential of Ontario road sections. *Transportation Research Record* 1327: 47–53.
Persaud, B. N. 1999. Empirical Bayes procedure for ranking sites for safety investigation by potential for safety improvement. *Transportation Research Record* 1665: 7–12.
Quddus, M. A. 2008. Modelling area-wide count outcomes with spatial correlation and heterogeneity: An analysis of London crash data. *Accident Analysis & Prevention* 40: 1486–1497.
Schlüter, P. J., J. J. Deely, and A. J. Nicholson. 1997. Ranking and selecting motor vehicle accident sites by using a hierarchical Bayesian model. *The Statistician* 46: 293–316.
Tunaru, R. 2002. Hierarchical Bayesian models for multiple count data. *Austrian Journal of Statistics* 31: 221–229.
Wong, S. C., N. N. Sze, H. F. Yip, B. P. Y. Loo, W. T. Hung, and H. K. Lo. 2006. Association between setting quantified road safety targets and road fatality reduction. *Accident Analysis & Prevention* 38: 997–1005.
World Health Organization. 2009. Global status report on road safety. Geneva, Switzerland: WHO.
Wu, C. and B. P. Y. Loo. 2013. Effects of helmet law enforcement on helmet use: A case study in Maoming, China. In *Travel Behaviour & Society*, eds. W. W. Y. Lam and B. P. Y. Loo, pp. 399–406. Hong Kong: HKSTS.

12 Exposure Factor 3
Distance Traveled

12.1 INTRODUCTION

Time geography has its origins from Torsten Hägerstrand's "space-time model" (Hägerstrand 1970). Generally, it studies the space-time behavior of human individuals; in their daily life, people follow a space-time trajectory. The space-time model is an attempt to understand under what basic condition linkages like collisions develop. Transportation and therefore road collisions are fundamentally space-time oriented, where the human population is conceived as forming a web of paths that flow through a set of space-time locations (Carlstein et al. 1978). Each path, or for the sake of this research, we shall call journey, has a life span (a journey time); however, these paths are not isolated. In terms of transportation, they coexist along the road network. The temporal importance of road collisions is very prominent in affecting the risk of a collision. The movement that drivers take through space or, in other words, from A to B, takes up time and yet has constraints surrounding it, for example attitude to speeding, type of vehicle, and weather. These factors all contribute to the changing temporal dimension of road collisions. This chapter emphasizes the potential applications of intelligent transport systems (ITSs) and three-dimensional (3D) GIS in understanding and analyzing road collisions.

12.2 METHODS

12.2.1 Road Collision per Population and per Vehicle Registered

Typical road safety records, such as road collision per population and road collision per vehicle registered in a society, are not true rates. However, these collision rates are often used in cross-sectional and international comparisons. In a society, vehicles may travel outside of the administrative boundary, and nonlocal vehicles may constitute a large share of the total vehicle fleet traveling on local roads. Moreover, some commercial vehicles, such as buses and taxis, are used much more intensively. In contrast, there are registered vehicles that are seldom used over the year or only at a specific time of the day. Similarly, road collision per population is a true rate only when the local population is a good measure of people's exposure to road traffic and, hence, collisions. However, the local population do travel outside of the administrations and that there can be a substantial number of nonlocal population. In some local contexts like that in continental Europe, where the barriers to movements of people among European Union (EU) countries are substantially removed, road collision per population may not be very meaningful. In other local contexts, such as North Korea, where the inflow and outflow of people are strictly controlled, the use of population as an exposure measure of people to road collisions is reasonable.

In general, the smaller is the jurisdiction, the higher is the likelihood that the assumption of the local population being a good measure of people's exposure to road traffic does not hold. Similarly, the lower are the barriers for the movement of people across jurisdictions, the higher is the likelihood that this assumption does not hold.

12.2.2 Road Collision per Vehicle- and Passenger-km

In order to get a more accurate estimate of road safety risk on the roads of a society, the use of vehicle-km as the exposure value of vehicles is often used, especially in comparison among administrative units. The aggregate people-based road safety indicator fails to consider mobility (Erdogan 2009). The modal split of the society, for example, may have major implications on people's exposure to road traffic as well. A high share of off-road transport modes, such as railways and ferries, can make the comparison of the road collision per population on the road network quite biased. The high share of metro, nearly 70%, in Hong Kong, means that its population-based collision frequency does not truly affect the collision risk that the population is facing when traveling on the road. To take into account people's mobility, passenger-kilometers or more appropriately, on-road passenger-kilometers will be more appropriate. Nonetheless, even the earlier improved exposure indicators of collision per passenger-km and vehicle-km only give an overall *average* impression. Most of them are annual indicators. Understandably, vehicular traffic in a city is highly uneven over time and space. The variability of road collision risk within a year or across space within an administration cannot be captured in these summary safety indicators.

12.2.3 Time-Space Measures

Focusing on mobility suggests that the road safety records should get closer to people-based indicators and recognize that traffic risk cannot be separated from the concept of exposure, and it follows that scientific road safety analysis has to base on even better exposure measures. In some administrations where traffic surveys are well developed, traffic volume data are arguably already collected continuously over the day. These data, however, are often not stored or used at all for calculating time-specific traffic exposure values for road safety analysis. Typically, data from traditional traffic surveys are aggregate and anonymous.

Nowadays, the use of ITS can allow disaggregate and even interactive data to be collected for understanding individual collision risk and even relating collision risk to potential activities that people are undertaking. The increasing popularity of portable car GPS navigation system and GPS-enabled mobile phones also allows dynamic real-time travel and location information to be collected.

With the earlier detailed information, it is possible to measure collision risk with respect to both the spatial intensity (concentration) and the temporal distribution (duration) of the exposure factor. Using the concepts of time geography of Hägerstrand, an individual's movement over time can be visualized with space-time trajectories or space-time path (STP), with the 2D space represented on the x- and y-axes, and time as the z-axis. Given the unique feature of traffic volume as network phenomena, the movement from A and B can be represented as a line on the transport network, and the slope of the line on the z-axis will depend on the travel speed. The higher is the speed, the sharper is the climb. An individual staying at the same place will be represented as a vertical line going up on the z-scale at the activity location. With relational databases storing individual, activity, and trip information, the STP of an individual over a day can be visualized and examined in detail with regard to the individual's life cycle, style, and other specific circumstances. A similar presentation technique is the space-time prism, which encompasses all possible locations for the STP (Miller 2005).

While a detailed microanalysis of an individual may not be very meaningful at the society level, STPs of groups of individuals can be aggregated for the visualization or analysis to highlight time and locations of high concentrations of people. Differences among subgroups, such as gender differences, in their command and autonomy of space or the sharing of household and child care responsibilities, can also be meaningfully analyzed (Loo and Lam 2013). Apart from examining STPs, the space-time data may be extracted to develop into indicators like the potential path area (PPA). PPA is defined as the area (on the 2D space) within reach of an individual given his/her specific time-space constraints (Lam et al. 2014). More recently, these space-time indicators and 3D GIS visualization techniques have been used to analyze not just accessibility and related equity issues but also road safety issues, primarily in developing better exposure measures (Yao et al. 2015).

12.3 INTERVENTION

The relationship between collision frequency and an exposure measure can be described by a safety performance function, with the y-axis showing the former and the x-axis showing the latter. In many cases, the safety performance functions are not strictly linear functions. For instance, when the passenger-km of a society doubles, collision frequency does not double. Often, safety performance functions are curvilinear, suggesting that the increase in collision rates will have a sharp increase at the lower range but will flatten at the higher range (Bauer and Harwood 2000; Hadayeghi et al. 2003; Miaou and Lord 2003; Cheng and Washington 2005).

In the following discussion, an example of using people-based exposure measure in understanding the problem of pedestrian safety and to identify hidden hazardous road locations (HRLs) is shown. The essential point is that unless the exposure factor is properly taken into account, effective measures to address road safety hazards, especially for specific vulnerable subgroups and at specific high-risk locations, will not be possible. The identification of locations with high absolute level of collision frequency may not be amenable to improvement measures, especially if the high collision frequency is simply related to the high volume of road users there. When collision risk is related to individuals, it should be recognized that individuals may not just be vehicle occupants (drivers or passengers) but also cyclists or pedestrians. While the vehicular traffic data are collected in most administrations for traffic management purposes (though with different coverage and levels of details), flow data of cyclists or pedestrians are often completely lacking. As cycling is getting increasingly popular as a sustainable transport mode (especially in Europe), the lack of cycle exposure measure is becoming a big issue for better understanding the safety records of cycling and to make scientific comparisons with that of other vehicular traffic. Walking is an important means of transportation in all societies, and there is an increasing scientific evidence to show that it is not only environmentally friendly but also beneficial to physical and mental health. Yet, no country in the world has a comprehensive database of pedestrian flow volumes on its pedestrian network. As a result, proxies are used. Population-based methods based on population density or walking working population (Qin and Ivan 2001; Ernst 2004; Wier et al. 2009; Chakravarthy et al. 2010; Cottrill and Thakuriah 2010) and trip-based methods based on estimates such as distance traveled, time spent walking, number of roads crossed, and number of trips taken (Keall 1995; Roberts et al. 1997; Beck et al. 2007) are the most often used.

The potential of using mobile phones to track all pedestrians and to derive comprehensive pedestrian flow data is huge and promising. However, privacy concerns (of mobile phone users) and the lack of commercial interests (of mobile phone providers) have made this option formidable in most societies. Moreover, governments (which are responsible for road safety) are most reluctant to intervene for possible accusations of jeopardizing citizens' privacy and upsetting the level commercial field. Fortunately, many societies still conduct rather large-scale travel characteristic surveys (which may also be called travel activity diaries). Against the earlier background, Lam et al. (2014) develop a time-space framework to estimate pedestrian exposure measures using data from multiple sources. Focusing on the elderly (aged 65 or above), the travel activity and vehicle-pedestrian collision records in a selected district in Hong Kong were extracted from the Travel Characteristics Survey and Traffic Road Accident Database System, respectively, for the analysis.

Three pedestrian exposure measures are developed and compared. The first is the population-/place-based methods of pedestrian exposure using population density (POP). The second is the STP method based on the deterministic shortest path assumption (Lam et al. 2013; Loo and Yao 2013). The third is the potential path tree (PPT) method that captures not just the shortest path but also all potential paths, given the specific space-time constraints for a pedestrian (Miller 2007). Given the stochastic properties of pedestrian movement, weights can be assigned to specific paths to reflect the probability that the feasible paths are actually chosen. The PPT method takes into account that an individual may not choose the shortest path due to imperfect information or other considerations, such as avoiding the main streets.

Among the three exposure measures, the models using space-time methods were found to be having better modal fit and higher interpretation power. The STP and PPT exposure variables were statistically significant ($p < 0.01$), but the POP exposure variable is not. The higher is the pedestrian concentration, the more likely that pedestrian-vehicle collisions would occur. With space-time pedestrian exposure factors, the main roads were also found to be more dangerous for pedestrians. This factor was not having as strong an effect when the aggregate exposure measure (POP) was used.

In addition, Lam et al. (2013) uses the Comap approach (Brunsdon 2001; Corcoran et al. 2007) to show that the locations of HRLs can be dramatically different should a pedestrian exposure value be included. Again, their study focused on elderly pedestrian in a selected district in Hong Kong. However, the methodology is applicable to other road users and contexts. Using the 1- and 4-hourly space-time

Exposure Factor 3   211

slices, it was shown that a better collision risk profile can be obtained with the pedestrian exposure factor. Figure 12.1 shows that HRLs identified based on the collision frequency involving elderly. Figure 12.2 shows that HRLs identified after taking into account the exposure measure based on STP. With the insights based on exposure, it is clear that the elderly pedestrians are not facing the highest collision risk during the morning peak (08:00–11:59) at the locations as shown in panel 3, but the midnight (00:00–03:59) and the late afternoon (16:00–19:59) period at the locations shown in panels 1 and 5. In general, these locations are highly concentrated especially along the main roads in the district's town center. Corresponding road safety investigations should be done to identify possible improvement measures.

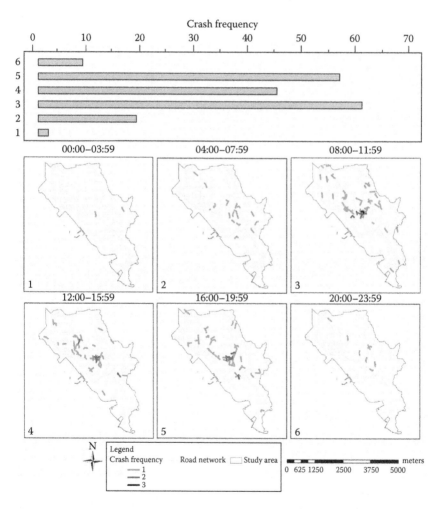

FIGURE 12.1 Comaps showing collision frequency, conditional upon six ST-slices. (Reprinted from Lam, W.W.Y. et al., *Asian Geogr.*, 30(2), 121, 2013, http://www.tandfonline.com/doi/full/10.1080/10225706.2012.735436. With permission from Taylor & Francis Group Ltd.)

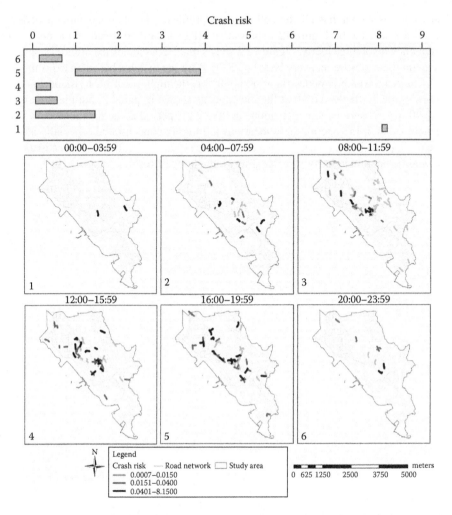

FIGURE 12.2 Comaps showing collision risk, conditional upon six ST-slices. (Reprinted from Lam, W.W.Y. et al., *Asian Geogr.*, 30(2), 122, 2013, http://www.tandfonline.com/doi/ful l/10.1080/10225706.2012.735436. With permission from Taylor & Francis Group Ltd.)

12.4 CONCLUSION

With technological advancement, the application of a disaggregate GIS-based 3D time-space framework in relating individual travel activities with collision patterns over time is becoming possible. Road safety analysis that properly takes into account disaggregated exposure data for analyzing collision data as points-in-networks can help identify HRLs for targeted road safety initiatives. After all, asking vulnerable road user groups like the elderly pedestrians to avoid traveling or to avoid areas with high pedestrian volumes can be contradictory to promoting their mobility and maintaining/enhancing their quality of life.

REFERENCES

Bauer, K. M. and D. Harwood. 2000. Statistical models of at-grade intersection accidents—Addendum. No. FHWA-RD-99-094. Federal Highway Administration, McLean, VA.

Beck, L. F., A. M. Dellinger, and M. E. O'Neil. 2007. Motor vehicle crash injury rates by mode of travel, United States: Using exposure-based methods to quantify differences. *American Journal of Epidemiology* 166 (2): 212–218.

Brunsdon, C. 2001. The Comap: Exploring spatial pattern via conditional distributions. *Computers, Environment and Urban Systems* 25 (1): 53–68.

Carlstein, T., D. Parkes, and N. Thrift, eds. 1978. *Timing Space and Spacing Time*, Vol. 2: *Human Activity and Time Geography*. New York: John Wiley & Sons.

Chakravarthy, B., C. L. Anderson, J. Ludlow, S. Lotfipour, and F. E. Vaca. 2010. The relationship of pedestrian injuries to socioeconomic characteristics in a large Southern California county. *Traffic Injury Prevention* 5: 508–513.

Cheng, W. and S. P. Washington. 2005. Experimental evaluation of hotspot identification methods. *Accident Analysis & Prevention* 37 (5): 870–881.

Corcoran, J., G. Higgs, C. Brunsdon, and A. Ware. 2007. The use of Comaps to explore the spatial and temporal dynamics of fire incidents: A case study in South Wales, United Kingdom. *Professional Geographer* 59 (4): 521–536.

Cottrill, C. D. and P. Thakuriah. 2010. Evaluating pedestrian crashes in areas with high low-income or minority populations. *Accident Analysis & Prevention* 6: 1718–1728.

Erdogan, S. 2009. Explorative spatial analysis of traffic accident statistics and road mortality among the provinces of Turkey. *Journal of Safety Research* 40 (5): 341–351.

Ernst, M. 2004. Mean streets 2004: How far have we come? Pedestrian safety, Washington, DC: Surface Transportation Policy Project, 1994–2003. Surface Transportation Policy Project. http://transact.org/wp-content/uploads/2014/04/Mean_Streets_2004.pdf (accessed January 14, 2012).

Hadayeghi, A., A. S. Shalaby, and B. N. Persaud. 2003. Macrolevel accident prediction models for evaluating safety of urban transportation systems. *Transportation Research Record* 1840: 87–95.

Hägerstrand, T. 1970. What about people in regional science? *Papers in Regional Science* 24 (1): 7–24.

Keall, M. D. 1995. Pedestrian exposure to risk of road accident in New Zealand. *Accident Analysis & Prevention* 27 (5): 729–740.

Lam, W. W. Y., B. P. Y. Loo, and S. Yao. 2013. Towards exposure-based time-space pedestrian crash analysis in facing the challenges of ageing societies in Asia. *Asian Geographer* 30 (2): 105–125.

Lam, W. W. Y., S. Yao, and B. P. Y. Loo. 2014. Pedestrian exposure measures: A time-space framework. *Travel Behaviour and Society* 1: 22–30.

Loo, B. P. Y. and W. W. Y. Lam. 2013. A multilevel investigation of differential individual mobility of working couples with children: A case study of Hong Kong. *Transportmetrica A: Transport Science* 9 (7): 629–652.

Loo, B. P. Y. and S. Yao. 2013. The identification of traffic crash hot zones under the link-attribute and event-based approaches in a network-constrained environment. *Computers, Environment and Urban Systems* 41: 249–261.

Miaou, S. and D. Lord. 2003. Modeling traffic crash-flow relationships for intersections: Dispersion parameter, function form, and Bayes versus empirical Bayes methods. *Transportation Research Record* 1840: 31–40.

Miller, H. J. 2005. A measurement theory for time geography. *Geographical Analysis* 37 (1): 17–45.

Miller, H. J. 2007. Place-based versus people-based geographic information science. *Geography Compass* 3: 503–535.

Qin, X. and J. N. Ivan. 2001. Estimating pedestrian exposure prediction model in rural areas. *Transportation Research Record* 1773 (1): 89–96.

Roberts, I., J. Carlin, C. Bennett, E. Bergstrom, B. Guyer, T. Nolan, R. Norton, I. B. Pless, R. Rao, and M. Stevenson. 1997. An international study of the exposure of children to traffic. *Injury Prevention* 3 (2): 89–93.

Wier, M., J. Weintraub, E. H. Humphreys, E. Seto, and R. Bhatia. 2009. An area-level model of vehicle–pedestrian injury collisions with implications for land use and transportation planning. *Accident Analysis & Prevention* 1: 137–145.

Yao, S., B. P. Y. Loo, and W. W. Y. Lam. 2015. Measures of activity-based pedestrian exposure to the risk of vehicle–pedestrian collisions: Space-time path vs. potential path tree methods. *Accident Analysis & Prevention* 75: 320–332.

13 Enforcement

13.1 INTRODUCTION

Enforcement is part of the three E's that are commonly known as enforcement, education, and engineering. Each has its own unique and intertwined role to play in reducing and preventing road collisions. Often the lines become blurred between the three as they intertwined with each other. First, it is important to make the distinction between automated and manual enforcement. While this chapter deals with both types of enforcement, it is important to differentiate between the two. Automated enforcement uses technology to reduce the need for human intervention in the enforcement process. In many instances, the use of automated enforcement technologies does not completely eliminate human intervention, as the equipment must be set up and monitored during its operation by qualified persons to meet legal or operational requirements. However, there are a number of key characteristics of automated enforcement that distinguish it from manual enforcement. These include the ability to detect a large number of offences per hour of operation, delayed notification of the offence or issuance of infringement notice to the offender, and a lack of human interaction with the driver at the time the offence is committed. Examples of automated enforcement include fixed position and mobile speed cameras. In contrast, manual enforcement, as the name suggests, requires human intervention throughout the enforcement process. In general, manual enforcement will involve the operator identifying that an offence has occurred, intercepting the offender, and immediately issuing the penalty notice where the offence attracts one (Leggett 1997).

In order to discuss the analytic issues of enforcement in more depth, we need to understand that every year collisions kill the equivalent of a highly populated city and describe the deaths as *accidents* something random and unavoidable. What we argue is that they are not *accidents*—they are collisions with a cause and effect; the causes are clearly defined in many countries' collision data. More than 80% of identified contributing factors are either driver errors of commission or omission. Despite the involvement of police in collision management and enforcement, the causes of most collisions cannot truly be considered crimes except in the sense that negligence driving *without due care and attention* or breaking road rules is a crime. Nonetheless, the methods and opportunities for road collision prevention and crime prevention—in particular, situational crime prevention (Clarke 1980)—are similar. Both involve the notion that coordinated action to make breaking the law more difficult or risky can achieve general reductions in the volume of lawbreaking (Clarke 1992).

Effective traffic enforcement is therefore essential for road safety. The main objective of traffic enforcement is the safe and efficient flow of traffic, achieved by essentially persuasion, prevention, and punishment. Safe behavior in traffic does not necessarily come naturally for most people, but with appropriate laws in place, behavior can be modified by traffic law enforcement. Typical offences relate to speeding, drinking and driving, drug taking, nonuse of seat belt or child restraints, and not wearing a helmet. All of these relate to well-known risk factors, where research has shown that limiting noncompliance will reduce the frequency and severity of road collisions. Targeted and appropriate legislation that is consistently enforced and well understood by the public is a critical component of successful enforcement.

Alongside traffic enforcement, most motorized countries adopt an appropriate penalty system. This can vary considerably from country to country. Often fixed penalties can be issued with a written infringement or violation handed out on the spot, requiring the offending driver or rider to pay a fine by a specified date. Confiscation of licenses or of vehicles can be applied for serious offences. Demerit or black-point systems seek to deter drivers from continuing to re-offend for a range of traffic law–related offences. To operate a penalty system effectively, a computerized database is generally needed to record all offences and driver records. There has been research to suggest that the penalty system inherently acts as a deterrent in order to prevent road users from offending. Most motorized countries enforce some level of penalty points system, and there has been some research that has looked at the influence of the penalty system and its link to road collision reduction. De Paola et al. (2010) looked at the introduction of such a penalty scheme in Italy and found that the system (controlling for various variables) led to an overall reduction of 10% of all road collisions and a 25% reduction in traffic fatalities.

13.2 MANAGING SPEEDS

Speeding is seen as one of the leading causes of all road traffic collisions. By increasing the speed, the driver's ability to control and ultimately stop the car decreases, therefore increasing the risk of a collision. In the United States, in 2007, 31% of all the fatal collisions were speed related, resulting in 13,040 fatalities (NHTSA 2009). Research has told us that excessive or inappropriate speed is a common contributing factor in many of these collisions (Graham 1996; Barker et al. 1998; Quimby et al. 1999; Taylor et al. 2000; DETR 2001; Taylor 2001; Spek et al. 2006).

The UK STATS19 data show us that of the injury collisions attended by the police in 2009, only 13% did the police officer record *exceeding the speed limit* or *traveling too fast for the conditions* as a contributory factor. However, this increased to 15% for collisions involving serious injury and 26% for fatal collisions. It is important to note that these factors are recorded only when the officer can be sufficiently sure about them to be able to justify if they are subsequently disclosed in court, often some times after the event. Other factors that are recorded in similar or higher percentages of collisions like failure of judgment, loss of control or being careless, and reckless or in a hurry can also be linked to the choice of speed. So the percentages 13, 15, and 26 just quoted are cautious minimum indications of the involvement of excessive or inappropriate speed in injury collisions, serious injury collisions, and fatal collisions, respectively.

Most drivers want to get to their destination in the quickest time possible, with a reasonable feeling of comfort and safety while they are traveling. Most drivers will trade off travel time against safety for themselves and other road users. Many drivers have the expectation of being able to control their car at much higher speeds than the posted speed limits. This is especially true of young and inexperienced drivers who have a tendency to underestimate traffic hazards (Rumar 1985). If car drivers could chose to travel at any speed, it would be much higher than the speed limits enforced. In practice, there are major differences in the speed at which different drivers drive given the external conditions. High speeds and variations in speed increase the probability of road collisions and serious injury, because the demand on the road user's observation and reactions increase and braking distance increases proportionally with the square of speed.

In most European countries, unrestricted speeds on the entire part of the road network existed from the 1930s. It was only around 1970–1975 that permanent speed limits became common on the road network. In Hong Kong, the speed limit is structured into three bands: low, middle, and high. In general, the 50 km/h is the standard speed limit for all built-up areas in Hong Kong and Kowloon. An American study showed that following a revision of the speed limit of the interstate highways from 88 to 104 km/h, there was an increase of 3–6 km/h in the mean rural interstate speed, which has resulted in an increase of 19%–34% in fatalities in traffic collisions (Garber and Graham 1990). This suggested that for every 1.6 km/h change in the mean traffic speed, there is an associated change of 8% or 9% in the number of fatalities.

Various studies have demonstrated that collision and casualty risk increases with increasing speed. These include analyses of collision risk at given speeds using case control studies (Research Triangle Institute 1970; Kloeden et al. 1997) and studies examining the effect of speed limit changes on resultant changes in speeds and collision rates (e.g., Nilsson 1982; Wagenaar and Reason 1990). While there is further discussion on automated speed enforcement in this chapter, it is noted that actual physical policing has its own disadvantages and advantages. One of the main advantages and deterrents is that traffic violators are stopped immediately by the police. The violator is given immediate feedback and allocation of fine or apprehension. One of the main disadvantages is the intensive labor requirement of policing, and compared with speed cameras, policing does not reach the same level of coverage.

13.2.1 Speed Limits

With the increased engineering and technology of vehicles comes the increased speed at which these vehicles can travel. This increased the culture of speed that speeding is considered to be normal, and it is acceptable to break the speed limits, especially when traveling on motorways (arguably motorways are the safest roads in terms of road collisions). There is a culture in our (developed and motorized) society today that speeding is socially acceptable, accepted by our peers, and something they have little chance of being apprehended for by the police or even causing a collision (Holland and Conner 1996). Evidence suggests that the speed at which drivers chose to drive directly affects both the severity and the number of road collisions (Quimby et al. 1999). A report by Taylor et al. (2000) by the Transport Research Laboratory in the United Kingdom highlighted that driving at slower speeds is effective in reducing road collision frequency. However, Morrison et al. (2003) found that just lowering the statutory speed limit is not effective, because drivers tend to respond poorly to such interventions and the importance of speed enforcement cannot be underestimated.

Speed limit enforcement is essentially the action taken by authorities to monitor that road vehicles are complying with the speed limit in force. In the United States, it is estimated that speeding is a factor in about one-third of all fatal collisions. Speed limits intended to control top speeds often are ignored, and vehicle speed capabilities far exceed posted speed limits, which therefore makes enforcement a necessity. One of the most common approaches to enforcing speed before speed cameras was deploying police officers using radar equipment in patrol cars to identify and arrest violators. This conventional approach arguably has limitations including that it is resource intensive, often inconsistent in application, and actually does little to slow motorists. It may also be difficult to observe speeds at the worst places and times, and police officers may be diverted to other duties considered more important. Over a period of 5 years, researchers monitored motorist response to speed limits at 227 different locations around the United States. First, motorist speeds were measured at all the locations. Next, the speed limits were raised on some roads and lowered on others while yet others remained the same. The results were that speeds did not change. People continued to drive at speeds that they felt were comfortable and safe. This study also measured the relationship of speed limit changes and collision frequency. As one might expect, if speeds did not change much, neither did collision rates. However, in those instances where speed limits were raised, there was a slight reduction in collisions. The U.S. federal and state studies have repeatedly shown that people most likely to be involved in a collision are the ones driving at speeds significantly below the average speed of traffic. In fact, the safest motorists, in terms of avoiding collisions, are those who are driving 5–10 mph above the average speed of traffic.

The relationship between speeding and collision rate, and speeding and injury rate has been examined in many studies and reviewed by the U.S. Transportation Research Board (TRB 1998). The relationship between speeding and the likelihood of collision should be logical: increasing speed increases the reaction distance (the distance traveled while the driver is reacting to a situation) and the braking distance. In reality, the relationship between speeding and collision rate is not simple, but it is consistent across studies. In two car collisions, the greater the deviation in speed from the average, the higher the rate of collisions. This relationship is thought to be due to increased interactions between vehicles when traveling at different speeds. In single-vehicle collision, the higher the speed, the greater the risk of collision. While studies demonstrate that motorists traveling at both significantly higher and significantly lower than average speeds have a greater collision risk, it has also been highlighted that the relationship with lower speeds was due mainly to collisions involving turning and slowing vehicles (TRB 1998; Wegman and Aarts 2006). The relationship between speeding and injury rate is straightforward: the faster the vehicle is traveling, the greater the energy absorbed by the occupants during the rapid change in velocity that occurs during a collision.

Speed limits on roads are used to regulate traffic speed and thus promote road safety by establishing an upper limit on speed and by reducing the variance (*dispersion*) of the speed of vehicles. As injury severity increases nonlinearly in relationship

TABLE 13.1
Factors Considered in the Setting of Speed Limits

Criterion	Factors
Road environment	Road classification
	Undivided or divided road
	Number of lanes and lane widths
	Clearance to roadside obstacles
	Vertical and horizontal alignment
Abutting development	Number and density of abutting developments
	Type and extend of traffic generated
	Land use (schools, houses, apartments, shops, etc.)
Road users and their movements	Cars
	Trucks
	Buses
	Cyclists and pedestrians
	Parked vehicles
	Peak hour traffic
	Recreational traffic
Existing speeds	Average speeds
	85 percentile speeds
Road collision history	To give an indication of speed-related safety problems
Adjacent speed zones	To be consistent
	Minimum lengths for buffer zones are specified
Other factors	Intersections
	Schools
	Pedestrian crossings
	Road alignment

Source: Data from VicRoads, Traffic engineering manual, Volume 1: Chapter 7—Speed zoning guidelines, 5th edn., November 2013, pp. 7–8, VicRoads, Kew, Victoria, Australia.

to speed, curbing *top-end* speeders should also reduce the number of deaths and severe injuries in those collisions that do occur. Speed limits are usually assigned by category, type, and design of the road (Chin 1999; see also Table 13.1). The main purpose of speed limits is to regulate driving speeds to achieve an appropriate balance between travel time and risk (TRB 1998). Many countries provide some type of enforcement to ensure that drivers obey the posted speed limits.

In Australia, during 2002, there were 1715 fatal collisions, of which 562 were identified as involving excessive speed (defined as driving faster than the posted limit or too fast for the prevailing conditions), using data obtained from the Australian Transport Safety Bureau (ATSB 2003). These findings are consistent with a 2004 report on fatality data from the United States, which suggests that excessive speed likewise defined is implicated in about 30% of all fatal collisions there (NHTSA 2004). It is predicted that if the number of drivers who are speeding is reduced, both the likelihood and the severity of a collision will be lowered (Pilkington 2002).

The enforcement of speed limits must be sufficient to ensure that drivers believe if they speed, they will be caught. Police cannot be present on all roads at all times, and therefore, in many countries, there is an increasing use of automatic speed enforcement, using detection devices (speed cameras) that may be manned or unmanned, mobile or fixed, as well as overt or covert.

In the United Kingdom, traffic calming was built into the UK 1865 Locomotive Act, which set a speed limit of 2 mph (3.2 km/h) in towns and 4 mph (6.4 km/h) out of town, by requiring a man with a red flag to walk 60 yards (55 m) ahead of qualifying powered vehicles. The distance ahead of the pedestrian crew member was reduced to 20 yards (18 m) in 1878, and the vehicles were required to stop on the sight of a horse. The speed limit being effectively redundant as vehicle speeds could not exceed the speed at which a person could walk. By 1895, some drivers of early lightweight steam-powered autocars assumed that these would be legally classed as a horseless *carriage* and would therefore be exempt from the need for a preceding pedestrian. A test case was brought by motoring pioneer John Henry Knight, who was subsequently convicted with using a locomotive without a license.

In 1905, the Automobile Association (AA) was formed to actually assist motorists in avoiding police speed traps in the United Kingdom. Even as far back as 1907, a Royal Commission on "Motorcars" raised concern about the manner in which speed traps were being used to raise revenue in rural areas rather than protect lives in towns. One of the key ways in which speed limits are enforced is by manually *clocking* vehicles traveling through *speed traps* defined between two fixed landmarks along a roadway that were a known distance apart; the vehicle's average speed was then determined by dividing the distance traveled by the time taken to travel it.

The Road Traffic Act 1934 introduced a speed limit of 30 mph (48 km/h) in built-up areas for cars and motorcycles. The definition of a built-up area was based on the presence of street lighting. The reintroduction of a speed limit for cars was in response to concern at increased road injuries. The number of fatalities had increased to 7343. Half of the deaths were of pedestrians, and three-quarters of these occurred in built-up areas. Between 1935 and 1940, the number of annual road fatalities increased from 6502 to 8609 (Department for Transport 2004).

13.2.2 Methods of Speed Enforcement

The primary method of speed enforcement has been the use of speed limits. The next section outlines the nature of speed limits and their controversy and future. Latterly, we have seen a shift to see methods to enforce the speed limits themselves. This need has evolved as car manufacturers are designing car with increasing speeds, and drivers themselves are acutely taking more risks than they were, say, 50 years ago. Methods to enforce speed limits have predominantly focused on speed cameras as well as direct police enforcement. Speed limit enforcement also uses traffic calming measures, which we discuss in Chapter 14. This involves fixed-site engineering methods such as speed humps and traffic lights.

The use of speed camera enforcement has seen much controversy over the past 15 years with the common issue being that it perceived as a money-making scheme.

With the wide range of traffic regulations and the limited amount of traffic police resources available, traffic enforcement efforts must be prioritized to suit local problems. A recommended order of offenses for consideration is as follows:

1. *Safety*: Offenses that could lead to a road collision, that is, speeding, signal violations, drink-driving violations, and pedestrian crossing violations by drivers.
2. *Traffic management*: Offenses that, while not inherently dangerous, do not facilitate smooth movement of traffic, that is, illegal parking, or buses loading and unloading passengers within a junction.
3. *Equipment*: Offenses such as lighting or tire defects that could contribute to a road collision, but have a much lower correlation with collisions than the safety offenses.
4. *Administrative*: Paperwork offenses such as improper vehicle registration or transfer of ownership.

Safety violations should be targeted to focus enforcement efforts at actions most closely connected with road collisions. High-risk collision sites should also be targeted. As traffic regulations usually specify the maximum fine for each violation, safety violations should incur the maximum fine to highlight the seriousness of the offense.

13.2.2.1 Controversy

Speed limits and their enforcement have been opposed by various groups and for various reasons since their inception. Historically, the AA was formed in 1905, initially to warn members about speed traps. In more recent times, some advocacy groups seek to have certain speed limits as well as other measures removed. For example, automated camera enforcement has been criticized by motoring advocacy groups like the Association of British Drivers, the North American National Motorists Association, and the German Auto Club.

Arguments used by those advocating a relaxation of speed limits or their removal include the following:

- A 1994 peer-reviewed paper by Charles A. Lave et al. titled "Did the 65 mph speed limit save lives?" stated evidence that a higher speed limit may be positive on a system wide in the United States by shifting more traffic to these safer roads.
- A 1998 report in the *Wall Street Journal* titled "Highways are safe at any speed" stated that when speed limits are set artificially low, tailgating, weaving, and speed variance (the problem of some cars traveling significantly faster than others) make roads less safe (Peters 1998).
- In 2010, German Auto Club (a major motoring organization) concluded that an autobahn speed limit was unnecessary, because numerous countries with a general highway speed limit had worse safety records than Germany, for example, Denmark, Belgium, Austria, and the United States.
- In 2008, the German Automobile Manufacturer's Association called general limits "patronizing," arguing instead for variable speed limits.

The association also noted that "raising the speed limits in Denmark (in 2004 from 110 to 130 km/h) and Italy (2003 increase on six-lane highways from 130 to 150 km/h) had no negative impact on traffic safety. The number of accidental deaths even declined" (Wikipedia 2012).
- Safe Speed, a UK advocacy organization, campaigns for higher speed limits and to scrap speed cameras on the basis that the benefits were exaggerated and that they may actually increase casualty levels. Their e-petition to the UK government in 2007 calling for speed cameras to be scrapped received over 25,000 signatures.

Various other advocacy groups press for stricter limits and better enforcement. Historically, the Pedestrians' Association and the AA were opposed in the early years of UK motoring legislation. The Pedestrians' Association was formed in the United Kingdom in 1929 to protect the interests of the pedestrian. Their president published a critique of motoring legislation and the influence of motoring groups in 1947 titled "Murder most foul," which laid out in an emotional but detailed way the situation as they saw it and called for tighter speed limits (Dean 1929). More recently, RoadPeace was founded in 1991 with an aim to stop road victims being "treated by the economy as acceptable, by the judicial system as trivial and by society as accidents" (RoadPeace 2012) and called for a default 20 mph speed limit in residential areas. Vision Zero is "a philosophy of road safety that eventually no one will be killed or seriously injured within the road transport system" (Tingvall and Haworth 1999, 13).

13.2.2.2 Future

The future of speed limits is complex. In ever-growing urban areas, the mix of road users becomes more interlinked, and the speeds are actually decreasing in urban areas due to the increase in traffic, people tend to become more frustrated and use speed as a defense mechanism. The use of speed limits has another dimension as well as we enter into a sustainable environmental focus. The use of fuel and the increasing cost of fuel have actually decreased speed limits. Studies have shown in United Kingdom that since the rapid increase in petrol prices, speeds have actually decreased. In 2008, Leonard Evans approached the question "Do increases in the cost of fuel actually reduces traffic fatalities?"

According to a 2004 report from the World Health Organization, a total of 22% of all "injury mortality" deaths worldwide were from road traffic injuries in 2002. It states that without "increased efforts and new initiatives," road traffic casualty rates would increase by 65% between 2000 and 2020. The report highlighted that the speed of vehicles was *at the core of the problem* and said that speed limits should be set appropriately for the road function, designed along with physical measures related to the road and the vehicle, and effective enforcement by the police (Peden et al. 2004). Road collisions are said to be the leading cause of deaths among children 10–19 years of age (260,000 children die a year; 10 million are injured). They are also occasionally set to reduce vehicle emissions or fuel use. This has been the subject of debate in recent years with the continued rise of fuel costs. It is hard to

Enforcement

separate whether people are driving slower and safer because they want to conserve fuel. Maximum speed limits place an upper limit on speed choice and, if obeyed, can reduce the differences in vehicle speeds by drivers using the same road at the same time. Road safety professional indicate that the likelihood of a collision happening is significantly higher if vehicles are traveling at speeds faster or slower than the mean speed of traffic. When severity is taken into account, the risk is lowest for those traveling at or below the median speed and "increases exponentially for motorists traveling much faster" (Stuster et al. 1998, 2).

13.2.3 Speed Cameras

Speed cameras are a sensitive issue for many groups of society (general public, politicians, road safety professionals), as they have been the most controversial speed management policy in recent years. People believe the idea of them as a mechanism for raising revenue; however, research has shown they do save lives. It is estimated that if all the speed cameras in the United Kingdom were to be decommissioned, then 800 more people would die each year as a result (which does not include being seriously or slightly injured). According to several reviews, speed enforcement detection devices are promising interventions for reducing the number of road traffic injuries and deaths (Elvik and Vaa 2004; Cameron and Delaney 2006; Wilson et al. 2006), and according to *The Handbook of Road Safety Measures* (Elvik and Vaa 2004), speed cameras could reduce the number of people injured in road collisions by as much as 17%. However, according to research evaluations, the effects appear to have a greater effect in urban areas (28% reduction) than in rural areas (4% reduction) (Elvik and Vaa 2004). Research conducted to date consistently shows that speed cameras are an effective intervention for reducing road traffic injuries and deaths. One of the main issues concerning speed camera appraisal has been the actual quality of the studies. The more recent studies conducted have *methodological rigor*.

13.2.3.1 Background of Speed Cameras

The use of speed cameras is now widespread around the world. However, there is widespread variation in the nature, extent, and perceived acceptability of their use. The concept of the speed camera can be dated back to 1905. *Popular Mechanics* reported on a patent for a "Time recording camera for trapping motorists" then enabled the operator to take time-stamped images of a vehicle moving across the start and end points of a measured section of road. The time stamps enabled the speed to be calculated, and the photo enabled the identification of the driver. The Dutch company Gatsometer BV, which was founded in 1958 by rally driver Maurice Gatsonides, produced the Gatsometer. Gatsonides wished to better monitor his average speed on a racetrack and invented the device in order to improve his lap times. The company later started supplying these devices as police speed enforcement tools. The first systems introduced in the late 1960s used film cameras to take their pictures. Gatsometer introduced the first red-light camera in 1965, the first radar for use with road traffic in 1971, and the first mobile speed traffic camera in 1982.

In North America, Europe, and parts of Asia, speed cameras were introduced in the late 1980s. In the United States, an early study into speed cameras was conducted in 1989, operating in Arizona and California. Part of this study included telephone surveys, whose results concluded that majority of people were positive about the speed cameras (Freedman et al. 1990). Speed cameras were first introduced on a trial basis in southwestern British Columbia in 1988, in particular in the lower mainland, interior, and southern Vancouver Island areas. Surveys were conducted in smaller communities to determine driver perceptions of the fairness of traffic law enforcement using these devices and the perceived effectiveness of them in reducing red-light and speed infringements (Zuo and Cooper 1991).

Speed cameras were first introduced in Victoria, Australia, on a trial basis in 1985 and were aimed at detecting a large number of speeding vehicles per hour. The initial trial involved a small number of mobile cameras operating with warning signs at high collision frequency sites. The effect of this operation was minimal. No statistically significant reductions in collisions in the areas surrounding the camera sites were found. In addition, the effect on speed was limited to distances of approximately 1–2 km from the camera sites (Portans 1988). Since these cameras have come into effect, there have been a number of changes to ways in which speed cameras have been managed. The management is now outsourced to a private company, and there has been a progressive increase in operating camera time. There have been concerns that speed cameras although put in place to increase safer driving, they were seen as an opportunity to raise revenue for the government through fines.

Enforcement cameras were first introduced in the United Kingdom in 1991, when the Road Traffic Act 1991 amended the law so that courts could accept evidence of speeding from type-approved cameras accompanied only by a certificate signed on behalf of the relevant police force. This allowed speed and red traffic light cameras, collectively known as safety cameras, to be operated by police forces. The first deployment of cameras was in West London in 1992, when 21 fixed speed camera and 12 red-light camera sites were installed, and their effectiveness monitored (London Accident Analysis Unit 1997). In the early days, the take-up of automatic enforcement by police forces was slow. In 1994, there were 30 speed cameras and 54 red-light cameras, but by spring 1996, there had been continued growth with 102 cameras servicing 700 sites (475 speed and 254 red-light camera sites). By the year 2000, there were an estimated 4500 safety camera sites in use on British roads, the majority of which are fixed speed cameras, with a smaller number of red-light and mobile cameras.

There are many types of reasons for having speed cameras, which go beyond just the need to monitor speed. The primary one, for example, in London (other than speed cameras) is the bus lane camera that is used on bus lanes, which uses a sensor in the road that triggers a number plate recognition camera, compares the vehicle registration plate with a list of approved vehicles, and records images of other vehicles. Other systems use a camera mounted on the bus, for example, in London, where they monitor "red routes" on which stopping is not allowed for any purpose (other than taxis and disabled parking permit holders).

13.2.3.2 Types of Cameras

1. *Gatso camera*: These cameras are permanently positioned at the roadside and loaded regularly for enforcement.
2. *Truvelo camera*: This camera is similar to the Gatso but takes a picture of the front of the car instead of the rear. Sensors set into the road surface trigger the camera. There are white lines across the carriageway, which performs a secondary check function.
3. *Mobile enforcement/Mobile van camera*: Rather than using a fixed roadside housing, the camera technology has been installed in the backs of vans for ease of deployment.

   The cameras use a laser device to measure the speed of vehicles coming toward or traveling away from the camera and record direct onto video.

   Sites are enforced for a few hours at a time, and the mobile technology makes this a much more flexible option, allowing sites to be more easily introduced, moved, or removed.
4. *Red-light enforcement:* Where there are identifiable problems caused by *running* red lights, these cameras may be deployed. They will photograph the vehicle as it goes through a junction, while the lights are on red and act as a deterrent against this inherently risky practice. These cameras are used worldwide, in countries including Australia, Canada, the United Kingdom, Singapore, and the United States. There is continued debate about the use of red-light cameras largely focused around public safety. There are a number of road collisions that do occur when people try and *run* a red light, meaning they try to increase their speed and get through the lights, thus causing potential for a high-speed collision.

   A report in 2003 by the National Cooperative Highway Research Program examined studies from the previous 30 years in Australia, the United Kingdom, Singapore, and the United States, and concluded that red-light cameras "improve the overall safety of intersections where they are used" (McGee and Eccles 2003, 2). While the report states that evidence is not conclusive (partly due to flaws in the studies), the majority of studies show a reduction in angle collisions, a smaller increase in rear-end collisions, with some evidence of a *spillover* effect of reduced red-light running to other intersections within a jurisdiction. These findings are similar to a 2005 meta-analysis, which compared the results of 10 controlled before-and-after studies of red-light cameras in the United States, Australia, and Singapore. The analysis stated that the studies showed a reduction in collisions (up to almost 30%) in which there were injuries; however, evidence was less conclusive for a reduction in total collisions. Studies of red-light cameras worldwide show a reduction of collisions involving injury by about 25%–30%, taking into account increases in rear-end collisions, according to testimony from a meeting of the Virginia House of Delegates Militia, Police, and Public Safety Committee in 2003. These findings are supported by a review of more than 45 international studies carried out in 2010, which found that red-light cameras reduce red-light violation rates, collisions resulting from red-light running, and usually reduce right-angle collisions. To reduce the frequency and severity of collisions at intersections, many jurisdictions around the world, including Canada, have installed red-light cameras to reduce red-light running behavior. However, the effectiveness of intersection safety cameras in reducing collisions has been a topic of constant debate in the literature (Erke 2009).

5. *SPECS*: SPECS time-over-distance cameras are in use on some of the United Kingdom's roads, most commonly in areas where major road works are being undertaken. It is vital that highway workers are protected while working on the roads, and SPECS offers very well-controlled vehicle speeds that have a very positive impact on road works safety. The SPECS cameras operate as two or more sets along a fixed route. They work by recording a vehicle's number plate at each fixed camera site, using Automatic Number Plate Recognition technology. As the distance is known between these sites, the average speed can be calculated by dividing this by the time taken to travel between two points. The cameras use infrared photography, allowing them to operate both day and night.

13.2.3.3 Has Speed Dropped as a Result of Speed Cameras?

Governments, local authorities, and the police collect speed information for a number of reasons both before and after enforcement. First, speed information is collected to determine whether or not there was a problem prior to establishing a site. Second, to provide local partnerships on a site-by-site basis to determine whether the cameras were having a positive effect on the reduction of speed and collisions. Third, nationally, it is important to determine whether or not enforcement was having a positive effect on driver behavior and therefore reducing risk and collisions. It has been noted that an accepted relationship derived from research is that each 1 mph reduction in speed should result in a 5% reduction in collisions.

13.2.3.4 Case Study: UK National Speed Camera Survey and Reduction of Injuries

A statistical analysis was undertaken on the before-and-after casualty figures to estimate the effect of the introduction of speed cameras. The model allowed for underlying factors such as national trends, speed limits, and seasonality. The statistical model found that overall, "Killed and Seriously Injured" (KSIs) fell 42% at camera sites. This equated to approximately 1700 fewer KSIs per annum at camera sites. It also found that 27% of this reduction was in rural areas, and the remaining 73% reduction was in urban areas. It was also worth noting that some proportion of the reduction in KSIs was due to "regression-to-mean", but all the reductions attributable to safety cameras would remain substantial after allowing for it. The study found that fixed camera sites were more effective than mobile camera sites as reducing KSIs. The most effective combination of camera type was a fixed camera site in a rural location, which resulted in a −62% reduction in KSIs.

According to a UK Department for Transport speed camera report in 2005, the size of the "regression-to-mean" depends on a number of factors. These can include the duration of the period of observation, the minimum number of events that are required for a site to be considered, and which other criteria are used in the site selection. In the United Kingdom, the current selection criteria for sites of safety cameras include numbers of collisions, presence of speeding during off peak conditions, and speed as a casual factor in some or all of the collisions. Davis (1998) concludes that the criteria of this kind used, in addition to numbers of injuries and collisions, will tend to reduce the size of the "regression-to-mean" effect, while Gorell and Sexton (2004) outline that this use of additional data causes difficulty in identifying correctly the population of potential camera sites for use in estimating the effect.

Because cameras are rightly often sited where there have been a high number of collisions and injuries in recent years, there would most likely have been some reduction without cameras being deployed, and downward national trends would also have led to some reduction. There has been much debate about "regression-to-mean", but there is now a convergence of estimates of its typical effect at camera sites, including the result of a very extensive analysis by a strong opponent of the use of cameras. It emerges that the observed reductions in collisions and injuries at camera sites are, on average, substantially greater than that could be accounted for by "regression-to-mean" and national trends, so an average camera site is preventing the order of one injury collision every 2 years and one person being killed or seriously injured every 5 years.

13.2.3.5 Benefits, Disadvantages, Controversies, and Effectiveness

A recent study by the Royal Automobile Club in the United Kingdom in 2010 looked closely at the effectiveness of speed cameras. Speed cameras were first used for enforcement in Great Britain in 1992 as recommended by a review of road traffic law in 1988 (Allsop 2010). Their rollout was accelerated between 2001 and 2005 in a national safety camera program under the "safer speeds" theme of the road safety strategy 2000–2010. Speed camera partnerships—joint ventures between police forces, highway authorities, and magistrates' courts—were formed to do this and have since taken on a wider role as road safety partnerships. The review used a number of sources of information including a 4-year camera evaluation report published in 2005. This initial report looked at 2000 camera sites (including urban, rural, fixed, and mobile sites) where speed measurements were taken before and after camera deployment. The report found changes of a general reduction in speeding. If "regression-to-mean" is taken into account, the report found that for the year ending in March 2004, camera operations at more than 4000 sites across Great Britain prevented some 3600 personal injury collisions, saving around 1000 people from being killed or seriously injured (KSI) (Table 13.2).

The UK review's findings are consistent with the international "Cochrane Review" that looked at 35 camera sites worldwide whereby "the consistency of reported reductions in speed and crash outcomes across all studies show that speed cameras are a worthwhile intervention for reducing the number of road traffic injuries and deaths" (Wilson et al. 2010, 2).

A report by Delaney et al. (2005b) summarized the controversies of speed cameras in Table 13.3.

TABLE 13.2
Number of PIC and KSI Prevented across Great Britain in Year Ending March 2004

	Number Prevented in Year Ending March 2004	
Type of Site	PIC	KSI
Fixed urban	Between 1700 and 2200	Between 500 and 560
Fixed rural	Between 170 and 300	Between 60 and 140
Mobile urban	Between 1000 and 1400	Between 150 and 400
Mobile rural	Between 180 and 300	Between 90 and 200
All sites	Between 3050 and 4200	Between 800 and 1300

Source: Reprinted from Allsop, R., The effectiveness of speed cameras: A review of evidence, Royal Automobile Club Foundation for Motoring Limited, London, U.K., 2010, p. v. With permission.

Enforcement of speed limits by the use of speed cameras is effective but controversial. Whenever speed cameras have been used, they have been controversial. Some of the controversies have to do with attitudes about speed enforcement in general, but most relate specifically to speed cameras, specifically legal issues and privacy, surveillance, revenue versus safety, unpopularity, avoidance, and fairness.

Some opponents believe that speed cameras are unfair due to factors such as failure to identify the driver (who may or may not be the owner of the vehicle), failure to notify the offender on the spot, lack of witness to the offence, and lack of opportunity to explain the circumstances of the event on the spot to a police officer. It also sometimes argued that speed cameras are located where it is safe to speed or where speed limits are set too low in these locations. There is widespread variation in how camera schemes are operated, and these different practices may be more or less likely to provoke various controversies or affect their intensity. Clearly, worldwide, there are differences in the amount of penalty; where the money goes (to private companies, to central government, to future government road expenditures); whether cameras are placed overtly (made highly visible) or covertly (hidden), warning of speed camera presence, and the type and location of the warning signs; and how far above the speed limit a vehicle may travel before it is photographed and a penalty issued.

The next question for road safety and public health is whether the improvements of speed cameras are sustainable and repeatable elsewhere. The risk compensation theory suggests that motorists will find other ways of injuring themselves and other road users. However, the technology to measure and record vehicle identification, time, place, and speed has been with us for many years, and the automation of this technology makes enforcing speed limits more practicable. In the interest of safety, we should expect all road traffic to be regulated to safe speeds and in the near future to variably set safe speeds depending on prevailing conditions. Changes in policing over the years during which the use of speed cameras became widespread did result in fewer traffic police being visibly deployed on the roads, so it can be argued that the emphasis of police enforcement of traffic law has shifted heavily toward

TABLE 13.3
Controversies Associated with Speed Camera Use in Each of the Jurisdictions Grouped according to Goldenbeld's Dilemma Classifications

	Jurisdiction		
	Australia	**United States and Canada**	**Great Britain**
Credibility dilemma	Dual perceived role of revenue raising and road safety. Revenue from speed cameras is not reserved for use in road safety, but rather goes to consolidated revenue. Total revenue from speed cameras is excessive. Inappropriate location of speed cameras in areas where it is *safe* to speed. Overt operation of cameras is most effective in deterring speeders in unsafe locations. Covert operations aim to increase revenue.	Perceptions of speed cameras as primarily revenue-raising mechanisms. Speed cameras seen to be located on the most *lucrative* routes. Inappropriate location of speed cameras in areas where it is *safe* to speed.	Dual perceived role of revenue raising and road safety. Increasing fines after implementation leads to perceptions of a stealth tax. Overt operation of cameras is most effective in deterring speeders at unsafe locations.
Social dilemma	Belief that speeding slightly in excess of the limit is not associated with increased crash risk if otherwise driving safely. Ambivalent support for reduced enforcement tolerances.	Belief that speeding slightly in excess of the limit is not associated with increased crash risk if otherwise driving safely.	Belief that speeding slightly in excess of the limit is not associated with increased crash risk if otherwise driving safely.
Legitimacy dilemma	No opportunity afforded to explain circumstances of the event. Penalties for exceeding a speed limit by ≤10 km/h are less fair than those exceeding the limit by more than 10 km/h.	No opportunity afforded to explain circumstances of the event. Automated enforcement does not identify the driver of the vehicle. There is a delay in the notification of the offence to the driver. The process does not enable witnesses to verify the circumstances of the offences.	The level of enforcement tolerance is important in forming public opinion about the fairness of the measure. Automated enforcement is perceived as an infringement of civil liberties. Speed limits should be reviewed prior to strict enforcement to ensure enforcement is fair.

(Continued)

TABLE 13.3 (*Continued*)
Controversies Associated with Speed Camera Use in Each of the Jurisdictions Grouped according to Goldenbeld's Dilemma Classifications

	Jurisdiction		
	Australia	United States and Canada	Great Britain
Implementation dilemma	The reliability of speed cameras is bought to question when individual cameras prove faulty. Speedometers may not be sufficiently accurate to keep detected speed within enforcement tolerances.	Diversion of police resources away from more serious criminal offences. Reductions in road trauma are not seen to compensate for slower travel speeds.	Review and appropriate setting of speed limits for the conditions are required.

Source: Data from Delaney, A. et al., The history and development of speed camera use, Report No. 242, Accident Research Centre, Monash University, Melbourne, Victoria, Australia, 2005a.

the enforcement of speed limits. But police enforcement is only one part of efforts nationally and locally to reduce collisions and injuries.

Efforts like those by the motor industry to improve occupant protection, design in pedestrian protection, and provide active safety devices in cars, road safety engineering by highway authorities, and widespread efforts in road safety education, training, and publicity among road users of different kinds and in the workplace have all continued undiminished. The modest but appreciable contribution of speed cameras should be seen in the context of this whole range of activity, most of which receives much less attention in the media than do speed cameras, but which has more than halved deaths on the road since the first speed cameras were deployed in the United Kingdom.

In common with many other safety interventions, speed cameras can have unintended consequences and have given rise to some collisions and injuries that would not have occurred if the cameras had not been deployed. But the definition of camera sites is such that additional collisions and injuries of this kind in the vicinity of cameras have been taken into account in estimating the changes in numbers occurring at the camera sites and have therefore been outnumbered substantially by collisions and injuries prevented.

The effect of speed cameras on numbers of collisions and injuries on roads other than at camera sites has been analyzed only in one study, which covered the deployment of cameras on trunk roads throughout West London, using the rest of London as a control area. In this study, small increases in collisions and injuries on the nontrunk roads, increases which were small enough to have arisen easily by chance, were substantially outweighed by decreases on the trunk roads as a whole, taking camera sites and the rest of the trunk roads together. This study apart, the numbers of collisions and injuries away from camera sites prevented by or arising from the deployment of cameras remain, for advocates and opponents of cameras alike, a matter of speculation.

Considerable controversy surrounds the relationship between traffic speed, and the frequency and severity of road accidents. The laws of physics support the view that, all else being equal, higher speeds will increase both the probability that an accident will occur and the severity of its consequences.

The future of speed cameras relies on a number of different factors, namely, public attitude, urban environment and planning, and political transport action.

13.3 MANAGING DRINK-/DRUG-DRIVING

13.3.1 DRINK-DRIVING

It is widely accepted that drink-driving behaviors make significant contributions to driver risks and are associated with elevated rates of risky driving behavior, road collisions, and the associated mortality and morbidity from these collisions (Baker et al. 1992). The role that alcohol use is believed to play in collision risk is encapsulated in the legislation of many societies, which have both imposed legal restrictions on the amount of alcohol that may be present in the blood of drivers and applied heavy penalties for drink-driving behaviors.

The fundamental issue with drink-driving and road collisions is that it is a crime. Someone under the influence of alcohol (and drugs) who is involved in a road collision is committing an offence. Much of the literature focuses on drink-driving arrests rather than the study of road collisions and drink-driving. There have been limited attempts to delineate spatial location and spatial studies to drink-driving road collisions. In nearly every Western country and city, there have been many attempts at deterring drivers to *not drink and drive*. The literature focuses on fatal collisions and blood alcohol levels (as according to Rosman (2001), the linkages between blood alcohol levels and less serious collisions are less well documented).

The main approach adopted in most countries is a very high level of enforcement that is supported by intensive publicity campaigns, particularly television advertising campaigns. Also, even though the use of alcohol-related collisions as a measure of performance for anti-drink-driving enforcement and publicity campaigns has high face validity, it can have biases. First, although random breath testing programs are designed to detect and deter drink-driving, the presence of traffic police on the roads is expected to have knock-on effects into other risky and illegal driving behaviors such as speeding, joyriding, and aggressive driving. Likewise, the presence of anti-speeding traffic enforcement on the roads will have a deterrent effect on other risky and illegal driving behaviors besides speeding. The two enforcement programs are thus expected to have complementary effect on alcohol-related and speed collisions as well as knock-on effects on other types of collisions, which have thus far been ignored in the road safety literature.

There is a clear profile from worldwide data on who are drunk drivers: most offenders are likely to be male; young male manual workers (or unemployed) who drink beer in pubs have been identified as one high-risk group, but so have older professional/managerial men. One major form of data to be collected on drink-driving is roadside surveys. However, most information is deemed from road collision statistics that will report if the driver has been drinking. A study of fatally injured drivers, riders, passengers, and pedestrians detected at least one medicinal or illicit drug in 24% of the sample. Alcohol was present in 31.5% of the sample, 21.5% being over the present legal limit for driving. Whereas the incidence of alcohol in road accident fatalities had reduced from 35% that was 10 years earlier, the incidence of drugs had increased threefold. For a single drug, 11.7% of the fatal injuries tested positive and 6.3% for multiple drug presence. In males, the majority of drug use was in those aged under 40, and in women in those aged 40 and over. This reflected a difference in the type of drug consumed—a higher incidence of illicit drugs being found in males, and medicinal drugs in females. Drug use was the highest (38.5%) among fatalities reported as being unemployed, this group having a particularly high incidence of cannabis and multiple drug use.

Countermeasures include the following:

- Anti-drink-drive publicity campaigns
  One of the first anti-drink-drive campaigns was introduced by the British Government in 1967 and was an attempt to promote the introduction of breath testing. Campaigns then ended until 1975 because of the lack of funding. The UK Department of Transport officials believe that recent (1985 to present) advertising campaigns have been effective in reducing injuries. They point to a large drop in 1987, when the slant of the slogans and advertising shifted from warnings about getting caught to an emphasis on the fact that drivers who drink endanger lives—the "Drinking and Driving Wrecks Lives" slogan. Since then, there have been variations on the same theme, including in 1992 a television advert that could be broadcast only after the 9 p.m., watershed showing a girl lying on the pavement covered with blood. The campaigns are targeted primarily at young men in their late 1920s who are overrepresented in collisions, particularly at Christmas. The publicity campaign is believed to have been effective. However, it is difficult to isolate the effect of publicity from the other measures introduced over the same period, such as tougher laws and higher levels of enforcement.
- Breath tests
  In the United Kingdom, the number of breath tests has increased greatly during the 1980s but dropped again from 1999/2000. As amended by the Transport Act 1981, Section 7 of the Road Traffic Act 1972 empowers a constable in uniform who has reasonable cause to suspect that a person driving a motor vehicle on a road, who has alcohol in his body, has committed a moving traffic offence, or has been involved in a collision, to require that person to provide a specimen of breath for testing.
- Penalties
  Disqualification of 12 months. The size of the fines and the maximum length of period of disqualification depend on the seriousness of the offence, mainly the amount by which the driver is over the legal limit. The normal fine for a basic drink-drive offence is between £400 and £450.

The Road Traffic Act 1991 introduced a new offence of causing death by careless driving when under the influence of drink or drugs with a maximum of 5 years imprisonment, later increased to 10 years, and then to 14 years in 2004. The Road Safety Act 2006 contains provision for serious, including repeat, drink-drive offenders to be made to retake the driving test at the end of their period of disqualification. It also makes provision for the courts, when imposing disqualification as a penalty, to order a reduced period of disqualification if it also makes an order requiring the offender to comply with the conditions of an alcohol ignition interlock program.
- Disqualifications
- Nonlegal penalties
- High-risk offenders
- Experimental educational programs

There has been considerable research conducted on the location of alcohol distribution outlets and location of drink-related fatal collisions. Nearly all studies have found a relationship between the two. However, what we are less sure of is the study of the general patterns of the location of drink-driving fatal collisions. It makes sense that the most obvious linkage would be between bars, pubs, and alcohol outlets; however, there are a large proportion of people to whom this will not apply to. There should be renewed research into the locations of collisions of drunk drivers, examining from a multifaceted angle.

13.3.2 Drug-Driving

Driving under the influence of drugs (DUID) other than alcohol is now considered to be an increasing cause of traffic accidents worldwide. Exposure to illicit drugs impairs driving ability owing to their effects on the central nervous system, psychomotor performance, and risk-taking behavior. Studies have shown the association between the use of psychoactive substances other than alcohol and increased accident risk.

The rising prevalence of cannabis use, its increased availability and potency, lower prices, widespread social tolerance, and earlier age of onset of use have combined to increase the number of users and hence the number of people subject to cannabis use disorders. Peak initiation is at age 18, and 10 years later, 8% of users are marijuana dependent. Most cannabis use is intermittent and time limited; however, users generally stop in their mid-to-late 20s, and only a small minority continues in daily use over a period of years.

Young people also account for a disproportionate number of road traffic accidents. According to the National Center for Statistics and Analysis, the fatality rate for teenagers is four times that of drivers age 25–69, and drivers under age 25 account for a quarter of all traffic fatalities. Risk factors for having a fatal traffic accident include being a young man, having psychological characteristics such as thrill-seeking and overconfidence, driving at excessive speed, driving late at night, failing to wear a seat belt, and lacking familiarity with the vehicle. The risk factors for adolescent marijuana use are somewhat overlapping—delinquency (vandalism, shoplifting, joyriding, etc.), poor school performance, and substance use by self and peers.

The National Highway Transportation Safety Administration reported that in 25% of all motor vehicle collision (MVC) fatalities, the driver had a blood alcohol concentration of 0.01 g/dL (one-eighth the legal limit) or greater, and in 21-year-old drivers, that figure rose to 39%. Drivers with a previous *driving while impaired* conviction were responsible for 7.2% of all collisions involving alcohol.

In comparison, the percentage of road traffic accidents in which one driver tested positive for marijuana ranges from 6% to 32%. In one study, 9.7% of cannabis smokers reported having driven under the influence (during the course of a year); and that they drove an average of 8.1 times whilst intoxicated. Among those who seek treatment for cannabis problems, more than 50% report having driven while *stoned* at least once in the previous year (Sewell et al. 2010).

Sewell et al. (2010) identify three types of study that look at drug use, specifically cannabis and road collisions. The first are *cognitive studies* that measure the effects of smoking marijuana on cognitive processes that are considered to be integral to safe driving. The second are *experimental studies* on the collision risk of people under the influence of marijuana. The third are descriptive and analytic *epidemiological studies* on the relationship between cannabis use and accidents, usually performed through drug testing of injured drivers.

A roadside survey in Thailand showed that prevalence of psychoactive drug use among general drivers not involved in MVC to be 9.7%. Alcohol or psychoactive drugs were found in 4.5% of drivers in a random sampling survey in Norway. A high proportion of injured drivers have been reported to test positive in overseas studies involving psychoactive drug screening. Siliquini et al. (2007) revealed positive psychoactive substances present in 18.5% of the drivers involved in road traffic collisions in Italy. In a Swedish study, 13% of nonfatally injured drivers tested positive for pharmaceuticals that could impair driving. A study conducted in Belgium involving injured drivers showed that 12.3% screened positive for drugs, and about half of them tested positive for alcohol as well. Among injured drivers, there was a much higher prevalence of persons screened drug positive reported from the United States, ranging from 22.6% to 50.9%. In a local epidemiological study, 56% of the deceased drivers from single-vehicle collisions had alcohol and/or drugs in their bodies, 7% were positive for drugs only, and 5% were for both drugs and alcohol. However, there are no local data on the prevalence of abusive drug use in drivers of nonfatal motor vehicle injuries.

Drug abuse is a social problem in Hong Kong. Apart from heroin, psychoactive substances such as ketamine, methamphetamine, and cannabis are commonly abused. DUID has recently gained considerable attention as a potential threat to local road traffic safety. In reply to Legislative Council questions on February 24, 2010, the secretary for Transport and Housing stated that there were four traffic accidents involving drivers suspected of DUID in the past 12 months. This may be an underestimate, however, due to the limited investigation powers of the police that relate to current legislation about such driving.ABusive drugs are mostly psychoactive substances. Theoretically, they have detrimental effects on psychomotor performance and may impair driving skills. The association between psychoactive substance use and driving impairment had been investigated in various types of studies, involving laboratories, simulators, as well as on-road and field investigations.

13.4 SPATIAL IMPLICATIONS

The underlying philosophy concerning enforcement locations is also likely to impact upon the mechanisms of effect and the outcome of the program. In particular, a distinction can be drawn between a black spot approach toward speed enforcement, where enforcement occurs only at sites where there is a speed-related collision problem, and a whole of network treatment, where speed cameras are used across the entire road network and are not restricted to black spot locations. The latter approach aims to create a perception that illegal speeds can be detected at any place across the road network and thus reduce speeds and collision frequency across the network, while the former approach is concentrated on reducing speeds and collision frequency at black spot locations.

There have been many studies that have focused on the variations of enforcement over both space and time. Most recently, Yannis et al. (2007) use multilevel modeling at both national and regional scales in Greece with specific focus on drink-driving. These results suggest that there are significant spatial dependences among road collisions and enforcement. Tay (2005) focused on drink-driving enforcement and media campaigns and found a significantly higher effect on *high alcohol hours*. Chen et al. (2002) investigated the use of speed radars and found a significant effect not only around the radar locations but also along the entire *enforcement corridor*. Hauer (1982) reported what was described as a *time halo* effect of enforcement and a spatial dispersion both upstream and downstream of the enforcement sites. Jones et al. (2008) looked at rural mobile speed cameras in rural United Kingdom and found the cameras had a positive influence in reducing the number of road collisions (with taking into account the "regression-to-mean" effect).

13.5 CONCLUSION

It is important to study the health impacts on policy interventions; however, it can be argued that why speed cameras are required to prove themselves in ways that other law enforcement methods are not. A further issue is that the cameras might produce slower speeds, and hence more road use by pedestrians and cyclists. This could lead to more injuries but lower injury rates. This shows the problems with focusing on reducing injuries/deaths without challenging car dependency. An increase in pedestrian injuries could paradoxically be part of an improvement in population health if it occurred due to a substantial increase in walking. But traditionally, responses to the danger posed by cars is to remove other road users—forgetting all the other health benefits of a modal shift including air pollution, equity, human rights, resource use, noise, and climate change. There is an interesting shift in thought with regard to speed enforcement. In the United Kingdom, the speed limit for the motorways is 70 mph. Due to the nature of motorways (separation of traffic through barriers, etc.), a large proportion of drivers exceed this limit safely by adapting to the situation (less traffic, etc.). In urban scenarios, for example, a speed limit of 30 mph may be totally inappropriate for the conditions. There are many areas where it would be foolish to travel anywhere near 30 mph. The idea being that it might be better to educate the driver in the first place, that they should be more acutely aware of the road environment, pedestrians, and/or anything else that is put at risk as a result of their presence. The main point of the argument is that due to the increased automation of traffic enforcement, the need for a driver to react and observe is diminished; ultimately, they are stripped of their accountability in the road environment. As more and more automated traffic controls are introduced, the need for a driver to think (and hone his driving skills) is reduced. Many roundabouts are now controlled by traffic lights—so the need to learn to merge into a stream is reduced—the knock-on effect being people no longer match the speed of the traffic on a fast road slip road but expect the traffic already there to move over and allow them to enter at a speed differential of maybe 20 mph slower.

REFERENCES

Allsop, R. 2010. The effectiveness of speed cameras: A review of evidence. Royal Automobile Club Foundation for Motoring Limited, London, U.K.

ATSB. 2003. ATSB annual review 2003. Australian Transport Safety Bureau, Canberra, Australian Capital Territory, Australia.

Baker, S. P., B. O'Neill, M. J. Ginsburg, and G. Li. 1992. *The Injury Fact Book*. Oxford, U.K.: Oxford University Press.

Barker, J., S. Farmer, and D. Nicholls. 1998. Injury accidents on rural single-carriageway roads, 1994–95: An analysis of STATS19 data. Transport Research Laboratory, Crowthorne, U.K.

Cameron, M. H. and A. Delaney. 2006. Development of strategies for best practice in speed enforcement in Western Australia. Accident Research Centre, Monash University, Melbourne, Victoria, Australia.

Chen, G., W. Meckle, and J. Wilson. 2002. Speed and safety effect of photo radar enforcement on a highway corridor in British Columbia. *Accident Analysis & Prevention* 34 (2): 129–138.

Chin, H. C. 1999. Investigation into the effectiveness of the speed camera. *Transport* 135 (2): 93–101.

Clarke, R. V. G. 1980. Situational crime prevention: Theory and practice. *British Journal of Criminology* 20: 136–147.

Clarke, R. V. G. 1992. *Situational Crime Prevention: Successful Case Studies*. Albany, NY: Harrow and Heston.

Davis, G. A. 1998. Method for estimating effect of traffic volume and speed on pedestrian safety for residential streets. *Transportation Research Record* 1636: 110–115.

Dean, J. S. 1929. Murder most foul…: A study of the road deaths problem. First published 1947 for The Public Affairs News Service by George Allen & Unwin Ltd, London, U.K.

Delaney, A., H. Ward, and M. Cameron. 2005a. The history and development of speed camera use. Report No. 242. Accident Research Centre, Monash University, Melbourne, Victoria, Australia.

Delaney, A., H. Ward, M. Cameron, and A. F. Williams. 2005b. Controversies and speed cameras: Lessons learnt internationally. *Journal of Public Health Policy* 26 (4): 404–415.

De Paola, M., V. Scoppa, and M. Falcone. 2010. The deterrent effects of penalty point system in driving licenses: A regression discontinuity approach. Working Paper No. 04-2010, University of Calabria, Arcavacata, Italy.

Department for the Environment, Transport and the Regions (DETR). 2001. Instructions for the completion of road accident reports. DETR, London, U.K.

Department for Transport. 2004. *Handbook of Rules and Guidance for the National Safety Camera Programme for England and Wales for 2005/06*. London, U.K.: DfT.

Elvik, R. and T. Vaa, eds. 2004. *The Handbook of Road Safety Measures*. Amsterdam, the Netherlands: Elsevier.

Erke, A. 2009. Red light for red-light cameras?: A meta-analysis of the effects of red-light cameras on crashes. *Accident Analysis & Prevention* 41 (5): 897–905.

Freedman, M., A. F. Williams, and A. K. Lund. 1990. Public opinion regarding photo radar. *Transportation Research Record* 1270: 59–65.

Garber, S. and J. D. Graham. 1990. The effects of the new 65 mile-per-hour speed limit on rural highway fatalities: A state-by-state analysis. *Accident Analysis & Prevention* 22 (2): 137–149.

Gorell, R. and B. Sexton. 2004. Performance of safety cameras in London: Final report. TRL Report PPR027. Transport Research Laboratory, Crowthorne, U.K.

Graham, S. 1996. Will higher speed limits kill? *Traffic Safety* 96 (3): 6–10.

Hauer, E. 1982. Traffic conflicts and exposure. *Accident Analysis & Prevention* 14 (5): 359–364.

Holland, C. A. and M. T. Conner. 1996. Exceeding the speed limit: An evaluation of the effectiveness of a police intervention. *Accident Analysis & Prevention* 28 (5): 587–597.

Jones, A. P., V. Sauerzapf, and R. Haynes. 2008. The effects of mobile speed camera introduction on road traffic crashes and casualties in a rural county of England. *Journal of Safety Research* 39 (1): 101–110.

Kloeden, C. N., A. J. McLean, V. M. Moore, and G. Ponte. 1997. Travelling speed and the risk of crash involvement: Volume 1—Findings. NHMRC Road Accident Research Unit, The University of Adelaide.

Lave, C. and P. Elias. 1994. Did the 65 mph speed limit save lives? *Accident Analysis & Prevention* 26 (1): 49–62.

Leggett, L. 1997. Using police enforcement to prevent road crashes: The randomised scheduled management system. In R. Homel, ed. *Policing for Prevention: Reducing Crime, Public Intoxication and Injury. Crime Prevention Studies*, Vol. 7. Monsey, NewYork: Criminal Justice Press Clarke, 1992, pp. 175–197.

London Accident Analysis Unit. 1997. West London speed camera demonstration project: Analysis of accident and casualty data 36 months 'after' implementation and comparison with the 36 months 'before' data. London Accident Analysis Unit, London Research Centre, London, U.K.

McGee, H. W. and K. A. Eccles. 2003. Impact of red light camera enforcement on crash experience. NCHRP Synthesis of Highway Practice 310. Transportation Research Board, Washington, DC.

Morrison, D. S., M. Petticrew, and H. Thomson. 2003. What are the most effective ways of improving population health through transport interventions? Evidence from systematic reviews. *Journal of Epidemiology and Community Health* 57 (5): 327–333.

NHTSA. 2004. Traffic safety facts, 2004 data: Speeding. National Highway Transportation Safety Administration, U.S. Department of Transportation, Washington, DC.

NHTSA. 2009. An analysis of speeding-related crashes: Definitions and the effects of road environments. National Highway Transportation Safety Administration, U.S. Department of Transportation, Washington, DC.

Nilsson, G. 1982. The effects of speed limits on traffic accidents in Sweden. In *Proceedings of the International Symposium on the Effects of Speed Limits on Traffic Accidents and Transport Energy Use*, Dublin, Ireland, October 6–8, 1982, pp. 1–8. Paris, France: Organisation for Economic Co-operation and Development.

Peden, M., R. Scurfield, D. Sleet, D. Mohan, A. Adnan, E. Jarawan, and C. Mathers, eds. 2004. World report on the road traffic injury prevention. World Health Organization, Geneva, Switzerland.

Peters, E. 1998. Highways are safe at any speed. *Wall Street Journal*, November 24, A22.

Pilkington, P. 2002. Increasing visibility of speed cameras might increase deaths and injuries on roads. *British Medical Journal* 324: 1153.

Portans, I. 1988. The potential value of speed cameras. Report No. SR/88/2. Road Traffic Authority, Melbourne, Victoria, Australia.

Quimby, A., G. Maycock, C. Palmer, and S. Buttress. 1999. The factors that influence a driver's choice of speed: A questionnaire study. Transport Research Laboratory, Crowthorne, U.K.

Research Triangle Institute. 1970. Speed and accidents, Vol. I. National Highway Safety Bureau, U.S. Department of Transportation, Washington, DC.

RoadPeace. 2012. http://www.roadpeace.org/why/our_vision/ (accessed February 17, 2012).

Rosman, D. L. 2001. The Western Australian Road Injury Database (1987–1996): Ten years of linked police, hospital and death records of road crashes and injuries. *Accident Analysis & Prevention* 33 (1): 81–88.

Rumar, K. 1985. The role of perceptual and cognitive filters in observed behavior. In Evans, L. and Schwing, R.C. eds. Human Behavior and Traffic Safety. Plenum Press, N.Y., U.S., pp. 151–165.

Sewell, R. A., P. D. Skosnik, I. Garcia-Sosa, M. Ranganathan, and D. C. D'Souza. 2010. Behavioral, cognitive and psychophysiological effects of cannabinoids: Relevance to psychosis and schizophrenia. *Revista Brasileira de Psiquiatria* 32: 515–530.

Siliquini, R., S. Chiadò Piat, M. M. Gianino, and G. Renga. 2007. Drivers involved in road traffic accidents in Piedmont Region: Psychoactive substances consumption. *Journal of Preventive Medicine and Hygiene* 48 (4): 123–128.

Spek, A. C. E., P. A. Wieringa, and W. H. Janssen. 2006. Intersection approach speed and accident probability. *Transportation Research Part F: Traffic Psychology and Behaviour* 9 (2): 155–171.

Stuster, J., Z. Coffman, and D. Warren. 1998. Synthesis of safety research related to speed and speed management. No. FHWA-RD-98-154. Federal Highway Administration, McLean, VA.

Tay, R. 2005. Drink driving enforcement and publicity campaigns: Are the policy recommendations sensitive to model specification? *Accident Analysis & Prevention* 37 (2): 259–266.

Taylor, M. C. 2001. Managing vehicle speeds for safety: Why? How? *Traffic Engineering & Control* 42 (7): 226–229.

Taylor, M. C., D. A. Lynam, and A. Baruya. 2000. The effects of drivers' speed on the frequency of road accidents. TRL Report 421. Transport Research Laboratory, Crowthorne, U.K.

Tingvall, C. and N. Haworth. 1999. Vision zero—An ethical approach to safety and mobility. Paper presented at the *6th Institute of Transportation Engineers International Conference, Road Safety and Traffic Enforcement: Beyond 2000*, Melbourne, Victoria, Australia, September 6–7, 1999.

TRB. 1998. Managing speed: Review of current practice for setting and enforcing speed limits. Special Report 254. Transportation Research Board, Washington, DC.

VicRoads. 2013. Traffic engineering manual, Volume 1: Chapter 7—Speed zoning guidelines, 5th edn., pp. 7–8, VicRoads, Kew, Victoria, Australia.

Wagenaar, W. A. and J. T. Reason. 1990. Types and tokens in road accident causation. *Ergonomics* 33 (10–11): 1365–1375.
Wegman, F. and L. Aarts, eds. 2006. Advancing sustainable safety: National road safety outlook for 2005–2020. SWOV Institute for Road Safety Research, Leidschendam, the Netherlands.
Liberal Democratic Party. 2014. Traffic laws. http://www.ldp.org.au/index.php/policies/1165-traffic-laws (accessed January 14, 2014).
Wilson, C., C. Willis, J. K. Hendrikz, and N. Bellamy. 2006. Speed enforcement detection devices for preventing road traffic injuries. *Cochrane Database of Systematic Reviews* 2006 (2): Article no. CD004607.
Wilson, C., C. Willis, J. K. Hendrikz, R. Le Brocque, and N. Bellamy. 2010. Speed cameras for the prevention of road traffic injuries and deaths. *Cochrane Database of Systematic Reviews* 2010 (10): Article no. CD004607.
Zuo, Y. and P. J. Cooper. 1991. Public reaction to police use of automatic cameras to reduce traffic control infractions and driving speeds in British Columbia. Paper presented at the *Canadian Multidisciplinary Road Safety Conference*, Vancouver, British Columbia, Canada, June 17–19, 1991.
Yannis, G., E. Papadimitriou, and C. Antoniou. 2007. Multilevel modelling for the regional effect of enforcement on road accidents. *Accident Analysis & Prevention* 39 (4): 818–825.

14 Engineering

14.1 INTRODUCTION

Engineering safety on the road has been a major concern of road safety professionals for decades. Road safety engineering focuses on the interaction between the driver and the road environment. No engineering safety measure is complete without assessing the driver element. It is challenging to predict how drivers will react in different circumstances, as every driver or participant in the road environment (pedestrian, cyclist, etc.) will react in different ways. Engineering measures seek to make people react *safely*, whether this is reducing speed, slowing down at an intersection, being aware of signage, and so on. In this chapter, we attempt to outline the linkages between road safety engineering, geography, and spatial analysis. Many studies of road safety engineering measures are limited by methodological flaws, such as failure to account for "regression-to-mean" associated with the treatment of high-collision locations and reliance on simple before-and-after measurements without suitable controls.

Collision reduction in the United Kingdom is achieved through the application of cost-effective measures on existing roads: the investigative procedure is detailed in RoSPA's *Road Safety Engineering Manual*, which covers techniques for the identification of hazardous road locations (HRLs), diagnosis of problems, selection of treatment, and evaluation. Collision prevention and reduction is achieved through the application of safety principles in the provision, improvement, and maintenance of roads: some of these procedures are outlined in Institute of Highways and Transportation, Road Safety Audit Guideline (2008), which summarizes safety principles for geometric design, road surfaces, road markings, road furniture and signs, and traffic management.

Road safety engineering measures are complex and extensive. They are often categorized by different methods. In this book, we focus on single, mass, route, and area-wide actions. This geographical approach is common among road safety professionals. These road safety engineering measures cover the following areas:

- *Road design and road equipment*: Cycle lanes, motorways, bypasses, urban arterial roads, channelization of junctions, roundabouts, redesigning junctions, staggering junctions, grade separated junctions, black spot treatment, improving road alignment, guardrails and collision rails, road lighting, tunnel safety.
- *Road maintenance*: Resurfacing of roads, treatment of uneven roads, improving road surface friction, bright road surfaces, landslide protection measures, winter maintenance of roads, traffic signs.
- *Traffic control*: Area-wide traffic calming, speed limits, pedestrian controls, stop signs, pedestrian crossings, bus lanes.

For all the measures, the effect varies enormously depending on the design and site conditions. For example, according to Elvik et al. (2009), roundabouts reduce the number of injury collisions but increase the number of property-damage-only collisions. Certain construction measures, including walking and cycling tracks and the design of arterial roads, do not appear to reduce injury collisions, instead potentially creating even more traffic. In some cases, there is evidence that road engineering measures have reduced the number of collisions at specific sites but they have increased elsewhere. This type of collision displacement is called collision migration, which is common when treating black spot areas.

Two key factors inhibit the effectiveness of road safety management. First, many different organizations, public and private, national, regional, and local, are involved in improving the interaction and reducing the collision risk between motor vehicles and road users on public highways, and coordination is often a major issue. The second factor, and perhaps even more importantly, is that road safety is not the first priority of any of the statutory agencies involved. Key priorities include road maintenance and network development for the roads authorities, registration of motor vehicles and drivers for the road transport departments, and crime prevention and prosecution for the police and Justice ministry. As discussed in the chapter, road safety engineering is very expensive compared to educational and speed-reducing measures (by enforcement). Therefore, one of the main priorities of engineering measures is evaluating and modifying rather than creating new modifications in the road environment.

Ogden (1996) determines that a safe road can be defined as one that is designed and managed so that it

- Warns the driver of any substandard or unusual features.
- Informs the driver of conditions to be encountered.
- Guides the driver through unusual sections.
- Controls the driver's passage through conflict points and road links.
- Forgives a driver's errant or inappropriate behavior.

The key to the selection of countermeasures at a particular site, route, area, or for mass application is to concentrate on the particular collision types that would have been identified previously (see previous chapters and also Table 14.1).

Criteria for countermeasure selection include the following:

- *Technical feasibility*: Can the countermeasure provide an answer to the collision problems that have been diagnosed, and does it have technical basis for success?
- *Economic efficiency*: Is the countermeasure likely to be cost-effective, and will it produce benefits to exceed costs?
- *Affordability*: Can it be accommodated within the program budget; if not, can it be deferred or should a cheaper interim solution be adopted?
- *Acceptability*: Does the countermeasure clearly target the identified problem? And will it be readily understandable by the community?

TABLE 14.1
Collision Situation and Engineering Remedies

General Accident Situation	Remedial Measures
Skidding	Restoring surface texture
	Resurfacing
	Improve drainage
Collisions with roadside objects	Better delineation
	Guardrails or fencing
	Frangible posts
	Remove objects
Pedestrian/vehicle conflicts	Pedestrian/vehicle segregation
	Pedestrian crossing facilities
	Pedestrian fences
Loss of control	Bigger or better road signs
	Road markings
	Speed controls
	Safety fencing
	Super elevation
Nighttime collisions	Reflective signs
	Delineation
	Road markings
	Street lighting
Poor visibility	Trim or remove vegetation
	Improved sightlines
	Realignment
Poor driving behavior	Road markings
	Enforcement
	Median barriers
	Overtaking lanes

Source: Reprinted from Asian Development Bank, Road safety guidelines for the Asian and Pacific region: Guidelines for decision makers on road safety policy, ADB, Manila, Philippines, 2003, pp. 4.5-5. With permission.

- *Practicable*: Is there likely to be a problem of noncompliance? And can the measure work without unreasonable enforcement effort?
- *Political and institutional acceptability*: Is the countermeasure likely to attract political support? And will it be supported by the organization responsible for its installation and ongoing management?
- *Legal*: Is the countermeasure a legal device? And will users be breaking any law by using it the way intended?
- *Compatibility*: Is the countermeasure compatible and consistent with other strategies, either in the same locality or that have been applied in similar situations elsewhere?

14.2 LOCATION-SPECIFIC TREATMENTS

14.2.1 Single Site

The most effective technique for tackling clustered collisions is to identify the factor in common and apply the appropriate remedial solution. This is commonly known as "hot spot" treatment or hazardous site analysis. When dealing with single site locations, it is very important that road safety engineers deal with the human factors. Most of the time with single site locations, it is the perceptual and visual problems experienced by the road users that lead the engineer to improve measures.

14.2.2 Mass Action

Perceptual problems are widespread, even though they do not necessarily lead to cluster of accidents at any one site. The second example illustrates how the mass action approach can be used to overcome one such problem (Figure 14.1). This concerns a study of crossroads of similar layout—where straight minor roads crossed major roads. Although not one of the group of sites could be classed as a black spot, an analysis of the collisions in total indicated a dominant factor: drivers unintentionally overran from the minor road, again associated with the very straight unbroken alignment of the minor road ahead. In a controlled trial, offset traffic islands with upstanding Give Way signs were installed in the line of sight of the minor road driver at 13 such junctions. The *before* accident frequency at individual sites, averaging 2.8 injury collisions/year, was halved after treatment. The cost of treatment at all sites totaled £25,000—the economic saving from collision reduction over 4 years was 10 times this figure.

14.2.3 Route Action

Route action may be simply an aggregation of single site treatments (usually in urban areas) or an application along a length, such as road marking to deter overtaking, or improving skidding resistance. The example is a trial on a 3 km length of road, a through route in an urban area. For several years, about 50 injury collisions had occurred annually, nearly one-half of these being associated with a right-turn maneuver and one-quarter involving pedestrians. The treatment was to improve the control of traffic (by installing roundabouts) at key junctions providing access to the adjacent residential neighborhoods, imposing right-turn bans at other junctions and adding three new light-controlled crossings.

14.2.4 Area-Wide Action

Area-wide action plans usually consist of a larger-scale approach, potentially managing small areas of a country or multiple towns/cities. While *area-wide action plans* are not commonly mentioned in the road safety literature and research, the large majority of engineering projects and research actually uses them as a unit of measure.

Engineering

(a)

(b)

FIGURE 14.1 Perspective view of *straight-through* crossroads: (a) original layout and (b) after remedy. (Reproduced from Sabey, B., *Injury Prev.*, 1(3), 185, 1995. With permission from BMJ Publishing Group Ltd.)

Sabey (1995) discusses a major five-town trial undertaken by the Transport Research Laboratory in the 1980s that demonstrated how overall savings in road collisions can be achieved over a whole area by strategic application of low-cost measures, even when collisions are scattered about the network. The principle is illustrated for one of the five towns, Sheffield. The trial area of approximately 9 km^2, with a population of around 50,000, covered a network of roads bounded by clearly defined traffic routes, but within that boundary, local roads all had equal status. The management of safety in such an urban area outside the center hinges on first defining a hierarchy of roads according to their function.

14.3 ENGINEERING MEASURES

14.3.1 Physical Engineering Measures

14.3.1.1 Low Cost versus High Cost

The economic justification for installing a safety system is usually based on its economic return. This is generally calculated as an estimated first-year rate of return, which is an estimate of the monetary benefits to be gained in collision savings in the first year, set against the cost of the scheme. While many schemes will save only a small number of collisions per year, this can still produce a good rate of return (Tables 14.2–14.4).

Simple low-cost engineering can save thousands of lives. In an effort to reduce collisions and ease traffic congestion on U.S. highways, traffic engineers and planners have traditionally pursued a wide range of actions. In some cases, the most cost-effective solution requires a significant investment in public funds. In other cases, the most cost-effective solution can be achieved through implementation of lower-cost solutions. "Low cost" is a relative term. Agencies implementing large projects with large budgets may perceive a *low-cost* project differently from an agency with a limited budget. For purposes of this discussion, "low cost" is defined as a project or strategy that generally requires an investment below £30,000. Many of the strategies discussed range from several hundred dollars to several thousand dollars in

TABLE 14.2
Collision Reduction Schemes in Oxfordshire, United Kingdom, 2007

Treatment	Reduction in Collisions (%)	Number of Sites
Urban—pelican crossing	25	39
Urban—traffic signals	50	12
Urban—mini roundabout	40	34
Urban—road humps	50	49
Urban—speed cameras	25	46
Rural—right-hand lanes	60	10
Rural—signing treatments	30	103
Rural—antiskid junction treatment	30	11
Rural—visibility improvement at junction	20	18
Rural—visibility improvement on bend	40	13
Rural—bend signing	30	140
Rural—antiskid bend treatment	50	13
Rural—30 mph village speed limits	25	180
Rural—speed cameras	15	16

Source: Reprinted from Royal Society for the Prevention of Accidents (RoSPA), Road safety engineering: Cost effective local safety schemes, September 2012, http://www.rospa.com/roadsafety/adviceandinformation/highway/road-safety-engineering.aspx, (accessed October 29, 2014). With permission.

TABLE 14.3
Potential Reductions (%) in Various Injury Collision Types

Treatment	Head-On Collisions	Run-Off-Road Collisions	Intersection Collisions	Relative Cost
Road signs and delineation	25–40	25–40	25–40	$
Rumble strips	10–25	10–25		$–$$
Central median hatching	10–25			$
Speed reduction (per 10 km/h)	15–40	15–40	15–40	$
Dedicated lanes for turning traffic			25–40	$–$$
Removal of roadside objects		25–40		$$
Roadside barriers		25–40		$$
Shoulder sealing	25–40	25–40		$$
Intersection—roundabout			60+	$$–$$$
Straighten curvy roads	25–40	25–40		$$$
Overtaking lanes	10–25	10–25		$$$
Divided roads and/or median barriers	40–60	40–60		$$$
Intersection—grade separation			40–60	$

Source: Data from New Zealand Road Assessment Programme (KiwiRAP), How safe are our roads?: Star rating New Zealand's State Highways, KiwiRAP, Auckland, New Zealand, 2010, p. 12.

Key:
$: Less than $50,000 per km, or low cost.
$$: $50,000–$500,000 per km, or medium cost.
$$$: Greater than $500,000 per km, or high cost.

magnitude. The research conducted for the study, however, indicates that "low cost" does not mean "low benefit."

Low-cost traffic engineering improvement techniques are typically HRL applications or are limited to shorter sections of roadway that do not cover an entire length of an arterial corridor. Some of these strategies include pavement markings, static and dynamic signing, roadway lighting, raised medians, curb cuts, roadway geometric changes, or lane controls. These strategies provide the guidance, warning, and control needed for drivers to ensure safe and informed operation through traffic bottlenecks or congested areas.

Low-cost treatments implemented at HRLs included

- Creating a left-turn lane within the confines of an existing roadway.
- Adding left-turn phases to existing signals.
- Replacing "Yield" signs with "Stop" signs at intersections.
- Replacing two-way stops with multi-way stops.
- Installing traffic signals.
- Using bigger and/or better signs.
- Installing short segments of center line and stop bars at "Stop" locations.
- Installing double-indicating "Stop" signs (adding a left-side sign).
- Painting the message "Stop Ahead" and "Stop" on pavement.

TABLE 14.4
List of Selected Road Engineering Safety Countermeasures

Low-Cost Engineering Road Safety Countermeasures	Medium-Cost Engineering Road Safety Countermeasures	High-Cost Engineering Road Safety Countermeasures
Splitter islands for yield- or stop-controlled intersections	Gateways	End of queue detection
Advanced green for pedestrians	Channelization of opposing traffic flows	New Jersey Jug Handle Intersection
Overhead stop sign	Puffins	Speed cameras
Yellow bar marking	Safety edge	Vehicle-activated warning signs
Colored bike lanes through intersections	2+1 roadway design without cable barriers	Dynamic rerouting with automatic traffic jam warning
Drowsy driving sign	Freeway median cable barrier system	2+1 roadway designs with cable barriers separating opposite traffic flows
Bus boarders and one-lane bus stops	30 km/h zone	Variable speed limits
Audio tactile line marking	Waving road surfaces	
	Three-dimensional road marking	
	Infrared animal detection systems	

Source: Data from AECOM Canada Ltd., CIMA+, and Lund University, International road engineering safety countermeasures and their applications in the Canadian context, Transport Canada, Mississauga, Ontario, Canada, 2009, pp. 10–11.

- Removing signals from late night/early morning programmed flashing operation.
- Adding back plates to existing signal installations.
- Adding a signal head to an existing display.
- Replacing 8 in. signal heads with 12 in. signal heads.
- Adding "Signal Ahead" signs.
- Installing red "T" displays (two red signal heads mounted horizontally over an amber and green).
- Installing an all-red interval.
- Replacing protected/permissive left-turn phases with full-protected left-turn phases.

In the United Kingdom, the relative cost of a *high-cost* engineering scheme is between £50,000 and £400,000. This is used only once potential low-cost solutions are exhausted, as it can involve a protracted process of consultation and implementation. Though lower than for low-cost schemes, they still achieve impressive rates of return of around 250%. Five schemes were implemented in 2000/2001, totaling £1.148 million and saving an estimated 29 injuries (the 2000/2001 program is the

Engineering 249

most recent with monitoring for 3 years pre-and-post implementation). There is detailed monitoring of schemes back to 1993.

14.3.1.2 Roundabouts

Most European countries apply roundabouts at junctions, and their numbers are increasing rapidly. Recent research (Elvik 2003; Brabander et al. 2005) suggests that roundabouts have been effective at reducing road collisions. Brabander et al. (2005) report that since 1986, over 2000 roundabouts have been built in the Netherlands, mostly in urban areas, and more are being planned. Sweden had 150 roundabouts in the beginning of 1980s and currently has 2000. Roundabouts are aimed at lowering junction speeds and removing right-angle and head-on collisions. Roundabouts also have a greater capacity than normal give-way or signalized junctions. A driver approaching a roundabout is forced to lower his entry speed, which reduces collision severity. The roundabouts in Europe are characterized by a pure circular design, a narrow carriageway, radially oriented entry roads, and right-of-way of the traffic on the roundabout. One of the major modifications that takes place is replacing a junction with a roundabout. Elvik (2003) reported that converting intersections to roundabouts greatly reduced the number of road collisions. Roundabouts also reduce the severity of collisions, for example, Elvik (2003) concluded that in a study of 28 research papers, roundabouts reduced fatal collisions by between 50% and 70%. When converting an ordinary junction to a roundabout, injury collisions will decrease by 32% for a three-leg junction and 41% for a four-leg junction. Corresponding figures are 11% and 17% when converting a signalized junction to a roundabout. The benefit–cost ratio when converting a typical three- or four-leg junction to a roundabout is around 211.

14.3.2 MANAGEMENT MEASURES

We have seen notable shifts in road safety management over the last 50 years. This has largely been due to better engineering, computer, and spatial knowledge. The World Health Organization (Peden et al. 2004) outlines four significant phases to road safety management that have been progressively more ambitious. The following section outlines briefly the four stages:

Phase 1: Focus on driver interventions
In the 1950s and 1960s, safety management was generally characterized by dispersed, uncoordinated, and insufficiently resourced institutional units performing isolated single functions (Koornstra et al. 2002). Road safety policies placed considerable emphasis on the driver by establishing legislative rules and penalties and expecting subsequent changes in behavior, supported by information and publicity. It was argued that since human error contributed mostly to collision causation, it could be addressed most effectively by educating and training the road user to behave better. Placing the onus of blame on the road traffic victim acted as a major impediment to the appropriate authorities fully embracing their responsibilities for a safer road traffic system (Rumar 1999).

Phase 2: Focus on system-wide interventions

In the 1970s and 1980s, these earlier approaches gave way to strategies that recognized the need for a systems approach to intervention. Dr. William Haddon, an American epidemiologist, developed a systematic framework for road safety based on the disease model that encompassed infrastructure, vehicles, and users in the pre-collision, in-collision, and post-collision stages (Haddon 1968). Central to the original framework was the emphasis on effectively managing the exchange of kinetic energy in a collision that leads to injury, to ensure that the thresholds of human tolerances to injury were not exceeded. The focus of policy broadened from an emphasis on the driver in the pre-collision phase to also include in-collision protection (both for roadsides and for vehicles) and post-collision care. This broadened it to a system-wide approach to intervention and the complex interaction of factors that influence injury outcomes. It underpinned a major shift in road safety practice that took several decades to evolve. However, the focus remained at the level of systematic intervention and did not directly address the institutional management functions producing these interventions or the results that were desired from them.

Phase 3: Focus on system-wide interventions, targeted results, and institutional leaderships

By the early 1990s, good practice countries were using action-focused plans with numerical outcome targets to be achieved with broad packages of system-wide measures based on monitoring and evaluation. Ongoing monitoring established that growing motorization need not inevitably lead to increases in death rates but could be reversed by continuous and planned investment in improving the quality of the traffic system. The United Kingdom, for example, halved its death rate (per 100,000 head of population) between 1972 and 1999 despite a doubling in motorized vehicles. Key institutional management functions were also becoming more effective. Institutional leadership roles were identified, intergovernmental coordination processes were established, and funding and resource allocation mechanisms and processes were becoming better aligned with the results required. Developments in Australasian jurisdictions (e.g., Victoria and New Zealand) further enhanced institutional management functions concerning results, focus, multi-sectoral coordination, delivery partnerships, and funding mechanisms (Trinca et al. 1988; Bliss 2004; Peden et al. 2004; Wegman and Aarts 2006). Accountability arrangements were enhanced by the use of target hierarchies linking institutional outputs with intermediate and final outcomes to coordinate and integrate multi-sectoral activities. This phase laid the foundation for today's best practice and reflects the state of development found in many higher-performing countries today.

Phase 4: Focus on system-wide interventions, long-term elimination of deaths and serious injuries, and shared responsibility

By the late 1990s, two of the best performing countries had determined that improving upon the ambitious targets that had already been set would require rethinking of interventions and institutional arrangements. The Dutch Sustainable Safety (Wegman and Elsenaar 1997; Wegman et al.

2008) and Swedish Vision Zero (Tingvall 1995; Committee of Inquiry into Road Traffic Responsibility 2000) strategies redefined the level of ambition and set a goal to make the road system intrinsically safe. The implications of this level of ambition are currently being worked through in the countries concerned and elsewhere. These strategies recognize that speed management is central and have refocused attention on road and vehicle design and related protective features. The "blame the victim" culture is superseded by "blaming the traffic system," which throws the spotlight on operator accountability. These examples of *Safe System* approaches have influenced strategies in Norway, Finland, Denmark, Switzerland, and Australia.

Today, the growing view is that road safety is a system-wide and shared multisectoral responsibility, which is becoming increasingly ambitious in terms of its results focus. Sustaining the level of ambition now evident in high-income countries requires a road safety management system based on effective institutional management functions that can deliver evidence-based interventions to achieve desired results. Achievement of the ultimate goal of eliminating death and serious injury will require continued application of good practice developed in the third phase of targeted programs, coupled with innovative solutions that are yet to be determined, based on well-established safety principles.

Safety management should start with a safety impact assessment before a decision is made to site a new road. Safety audit at the design and construction stage is needed to ensure all aspects of detailed design that might affect safety are addressed. Once the road is built, highway authorities have a responsibility to ensure its safe operation. This is best done through a combination of accident investigation and on-road inspection to enable cost-effective remedial programs to be developed; many tools exist to support these activities. The skid resistance of a road surface is an important road safety factor: both micro-texture and macro-texture of the surface play a part.

Safety is produced just like other goods and services, and the production process is viewed as a management system with three levels: *institutional management functions*, which produce *interventions*, which in turn produce *results* (Figure 14.2). The New Zealand framework was adopted by the European Transport Safety Council (Wegman 2001), which highlighted its results management framework, and it was further elaborated by the SUNflower Project (Koornstra et al. 2002), which located the institutional implementation arrangements in the broader context of country's *structure and culture* (Koornstra et al. 2002, 4). The first World Bank guideline concerning the implementation of the *World Report* recommendations (Bliss 2004) used the framework to introduce prototype safety management capacity review tools. This updated guideline refines these tools and further defines the organizational manifestation of the SUNflower Project's *structure and culture* in terms of seven institutional management functions.

Institutional management functions: The seven identified institutional management functions are the foundation on which road safety management systems are built. They are essential for the production of interventions that,

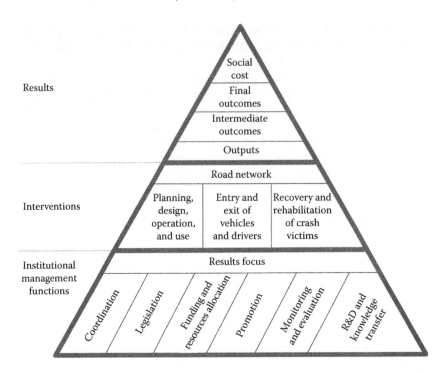

FIGURE 14.2 Road safety management model. (Reprinted from Bliss, T. and Breen, J., Country guidelines for the conduct of road safety management capacity reviews and the specification of lead agency reforms, investment strategies and safe system projects: Implementing the recommendations of the world report on road traffic injury prevention, Transport Note No. TN-1, World Bank Global Road Safety Facility, Washington, DC, 2009, p. 10. With permission.)

in turn, achieve road safety results, and for this reason, they must receive the highest priority in road safety planning and policy initiatives. The institutional management functions relate to all government, civil society, and business entities that produce interventions and ultimately results.

Interventions: Broadly, these comprise system-wide strategies and programs of interventions to address safety targets. Interventions cover the planning, design, and operation of the road network, the entry and exit of vehicles, and users into the road network, and the recovery and rehabilitation of collision victims. They seek to manage exposure to the risk of collisions, prevent collisions, and reduce collision injury severity and the consequences of collision injury. They comprise safety designs, standards, rules, as well as a combination of activity to secure compliance with these such as information, publicity, enforcement, and incentive.

Results: In good practice management systems, road safety results are expressed in the form of long-term goals and interim quantitative targets. Targets specify the desired safety performance endorsed by governments at all levels, stakeholders, and the community. To be credible, interim targets must be achievable with cost-effective interventions. Targets are usually set in terms of final outcomes. They can also include intermediate outcomes consistent with their achievement and institutional output measures required to achieve the intermediate results.

14.3.2.1 Generic Characteristics of the Road Safety Management System

It places an emphasis on the production of road safety and recognizes that safety is produced just like other goods and services. The production process is viewed as a management system with three levels: institutional management functions, which produce interventions, which in turn produce results. Much of the day-to-day road safety discussion is concerned with interventions alone, and use of the management system opens up the discussion to the important and often neglected issues of institutional ownership and accountability for results.

It is neutral to country structures and cultures that will shape the way institutions function and the goals to be set and achieved. Any country can use this framework and adapt their road safety initiatives to it. It accommodates evolutionary development. This is illustrated by the evolving focus on results that has been evident in high-income countries through its ultimate expression in the *Safe System* approach. In any particular period of development, the system can be used to review road safety management capacity and prepare related strategies and programs. It applies to any given land use/transportation system and takes as given the current and projected exposure to risk arising from that system. However, it can also manage the land use/transport trade-offs by considering these as options in the desired focus on results and addressing them with interventions concerning the planning, design, operation, and use of the road network, and the entry and exit of vehicles and road users to this network. It takes the road network as its frame of reference and locates the deaths and injuries that are avoidable. The three broad categories of intervention are defined in terms of the road network and have strong spatial dimensions. This distinguishes the system from earlier frameworks that emphasized safer roads, safer vehicles, and safer people, without locating them specifically in the network contexts where deaths and serious injuries occur.

14.3.2.2 Reduction and Prevention

Roads should be designed to cater for a defined function, separating roads for through traffic, roads for distribution of traffic within an area, and local access roads. By adopting a consistent and clearly differentiated design for each function group, driver's subjective assessment of risk can be better than actual risk; however, this is rarely the case. This in turn *should* encourage road user behavior consistent with the safety standard of the road. The same general functional management principles should be applied in both urban and rural networks. Road infrastructure should be designed taking account of the same injury tolerance criteria as those developed for vehicle occupant protection and pedestrian impacts, so that roads and vehicles together provide an effective safety system. Collision rates vary with road alignment, road width, roadside, and median treatment, and depending on junction type and design. Appropriate design choices are needed for roads serving each function to minimize the number of collisions likely to occur and to mitigate injury severity, particularly on higher-speed roads. Ironically, motorways/highways are some of the safest roads although they have higher speeds. The separation of traffic using collision barriers in the center of the road itself reduces head-on collisions significantly.

The design of roads should be adapted to the limitations of human capacity. Among pedestrians, the young and the elderly are most at risk. Risk to cyclists varies substantially between countries, mainly reflecting the infrastructure provided for them and the motorized traffic levels they interact with. Risk for motorized two-wheelers is particularly high, and solutions are needed to minimize the severity of injuries resulting from their impact with roadside furniture. Road designers should recognize the diminished physical and cognitive capabilities of elderly road users.

Safety is likely to be the main objective to all traffic management schemes. However, Ogden (1996) calls into question what is meant by this objective? He argues that in most applications, the objective of seeking to have safer roads leads directly to a need for data on collision occurrences and motoring programs to ensure that collisions are reduced. This is the philosophy underlying most of road safety engineering.

However, in dealing with local streets and mainly those in urban areas, these finite definitions become a little blurred. In this instance of modifying urban streets, we are dealing with people's living space, of which mobility and access are important elements of someone's living space. However, there is not a clear distinction between the transport-related needs that the living space fills and other needs. People tend to take a very *holistic* view of their local environment and more likely to find it acceptable if it feels *safe* and secure than if it does not. Therefore, as Ogden (1996) points out, the safety issues become part of a wider amenity issue. The distinction between safety and security (Wallwork 1993) is an important one and is critical in local areas.

Therefore, it is obvious that traffic management schemes that aim to reduce speeds on local roads and may reduce traffic volumes are likely to be supported by the local community because they feel more secure. Many local communities are becoming more involved in their road safety of traffic management of their area. This sense of community and being able to make the area safer and secure is an important empowering democratic process, which we are only just seeing come to light as more and more people are becoming concerned with their local area. The objective, however, of all these communities is *safety*, and most people will likely say they feel "safer." This is important according to Ogden (1996), because many well-designed traffic management schemes have achieved safety benefits that have been measured in the usual way—this is only one indication of success and may not be the indicator that is the most important for achieving community acceptance of the scheme.

One area of potential interest that has been overlooked is the power of the local community in terms of road safety engineering. In the United Kingdom, there are many small villages, local roads, and areas that are subject to their own unique risks, which are understood by the local community more than anyone else. There would be significant interest into those communities who have influenced and changed/reduced road collisions using their own type of "Not In My Back Yard" solutions.

Most schemes involving local areas aim to enhance to the totality of the local environment; they may need to go beyond measures that simply aim to reduce vehicle speeds and traffic volumes. In other words, traffic management and road safety schemes have multiple objectives. The problem is that the safety/security objectives could very often be achieved easily but in a way that contributes nothing to other objectives and indeed in some cases even detracts from them. For example, by very crude road speed bumps or street closures, which may make the locality look ugly or downgraded. To deal with the combination of objectives is challenging, largely due to the funding sources being separate.

Most traffic engineering schemes, especially those concerning *traffic calming* on local streets, involve retrofitting devices to existing streets to rectify problems caused by previous poor decisions or modifying street networks that were not designed for contemporary living and mobility patterns (Table 14.5). The most important questions road safety engineering schemes face now is this: what can be done to ensure these problems are not built into new networks? There needs to be a significant element of policy planning that tries to ensure for future changes. Although it can be impossible to predict the future of our mobility, rising and changing populations mean that we have a general forecast. We know for a fact that in most Western countries, the population over 65 is growing and will continue to grow. As it stands now, our provision for this aging population in road safety is nonexistent. Research has shown that these are among the most vulnerable in society in terms of road collisions; however, there is limited traffic management and road safety management that is being done to address these changes.

TABLE 14.5
Darwin Matrix for Traffic Calming

	Type of Measure	
Scope of Measure	**Physical/Environmental (Technique)**	**Social/Cultural (Ethos)**
L: Local (street or neighborhood)	LE: Local area traffic management Speed control devices Most reported speed and accident physical countermeasures	LC: Neighborhood speed watch Community action Attitudinal change
I: Intermediate (zone, precinct, corridor, regional)	IE: Environmentally adapted through roads Shared zones, lower speed zones Pedestrian shopping precincts Corridors	IC: Voluntary behavior change Mode choice, speed
M: Macro (city-wide)	ME: Transportation systems management (TSM) Total system measures (fares policy, city-wide road pricing)	MC: Travel demand management Urban form and structure

Source: Reprinted from Brindle, R.E., *Aust. Road Res.*, 21(2), 37, 1991. With permission.

Wallwork (1993, 9) said that "traffic calming is a negative reaction to a problem cause by bad planning/zoning and/or street design. We need to be proactive in our approach and learn more from the past and others." The OECD summarized the key road network planning principles for new residential areas as the following:

- Strict differentiation of streets according to their traffic function leads to safer residential areas.
- Distribution of traffic into a residential area with multiple access from a ring road is safer than central distribution.
- Full segregation of vehicle, pedestrian, and bicycle movements is accompanied by very low collision rates.
- Cul-de-sacs are safer than loop streets, which are safer than ordinary through streets.

When thinking about road collision engineering solutions in the case of this chapter, there are three major areas for consideration, each requiring differing levels of approach in terms of the spatial analysis. First is road design and road equipment; second, road maintenance; and third, traffic control. Let us deal with each one in turn.

Elvik et al. (2009) in *The Handbook of Road Safety Measures* outline a number of different measures based on road design and road equipment, including but not limited to motorways, cycle lanes, bypasses, roundabouts, black spot treatment, road lightening, and staggered junctions. Elvik et al. (2009) in their book examine the meta-analysis of a number of measures and highlight that many studies have not looked at the before-and-after studies of different measures that have been implemented; instead, many of the studies (ranging from 1997 to 2009) compare collisions on road sections or at junctions with different properties. For a number of measures, the effect will obviously vary substantially, depending on the design of the measure and the site conditions.

Road maintenance can include road surfaces, treatment of uneven road surfaces, landslide protection measures, winter maintenance, and flooding measures. Most of the road maintenance measures are carried out on existing roads and do not usually involve long-term changes of the road itself. Depending on the locality, seasonal road maintenance of roads is crucial for the prevention of road collisions. Frequently, road maintenance is limited to fixing potholes and cleaning drainage facilities, without replacing missing traffic signs, guardrails, road markings, and other safety features essential to create a safe road network. Elvik et al. (2009) highlighted that the amount of research in this area of road engineering and road safety is highly variable. It often gets *overlooked* in terms of importance at a research level, largely because it is seen as an obvious treatment for roads. There are many studies that have evaluated the road safety effects of road resurfacing and improving friction of road surfaces and winter maintenance of roads. However, according to Elvik et al. (2009), relatively few studies have evaluated other measures. Interestingly, the protection against landslides and road collisions has not been quantified. Although this is largely due to the majority of collisions with regard to landslides occurring in less urbanized countries with limited data and information, the majority of the research focuses on winter maintenance, resurfacing, and reducing friction.

The resurfacing of roads usually involves re-asphalting and, according to research, only leads to small decreases in road collisions. Some studies have actually indicated that there has been an increase in road collisions in the period immediately after asphalting, largely due to the ability to increase speed on a smoother surface, more positive effects have been found in the long term however. There is an unclear relationship between road collisions and uneven surfaces, however there are indications that multi-vehicle collisions increase on uneven surfaces. Also, increased rut depth is related to increased road collisions. Improving the road surface friction reduces the number of road collisions, with greatest effects being on wet roads and sharp bends, where friction is generally very low. According to the research, friction is more important for collision rates than unevenness. Overall winter maintenance of roads improves road safety. Both salting roads and raising the standard for snow clearance are important to reducing road collision rates. Included in this would also be winter maintenance of pavements, footpaths, cycle paths, and other public areas, which does not necessarily appear to reduce the number of road collisions but can contribute to the safety of the road environment as a whole. The management of winter maintenance is often by local councils insofar it is managed locally rather than nationally. It will also depend on the climate in the urban area. For example, cities such as Hong Kong and Singapore do not need to manage roads for snow and ice; however, these cities are more likely to be affected by flooding due to the monsoon nature of their climates. In countries such as the United Kingdom and the United States, winter maintenance is crucial to the prevention of road collisions in adverse weather. In recent years, it has become a political issue, with at times limited resources to provide adequate salting and management of roads during snow and ice periods. Many councils in the United Kingdom have what they term a "Winter Service Plan," which involves different management procedures depending on the type of road. Priority is given to major routes and will involve treating the road with salt in response to forecasted weather conditions. Often councils will use an ice detection and prediction weather tool in conjunction with thermal mapping linked to weather stations, to assist in the decision-making process of deciding which roads should be treated. This in turn is broken into high-risk months (usually being November, December, and January) and low-risk months (months either side). One of the key political issues being faced is the cost–benefit analysis of winter road maintenance. To salt and grit roads is very expensive, and the prioritization of which roads to treat is crucial.

The accident risk is also a function of the public awareness of the problem. Studies have shown that drivers are not good at adjusting the speed of their vehicle to the prevailing road conditions, even if the hazard is clearly visible (Öberg et al. 1991; Wallman 1997). Wintertime road conditions also vary between areas and change with time. Johansson (1997) showed that the traffic collision rate during hazardous situations is higher in southern Sweden, where slipperiness on roads is rarer, than in Northern Sweden, where people are used to driving in winter road conditions. It is important for winter road maintenance personnel to know which type of road's slipperiness has the highest collision risk in their area.

A number of studies have shown that drivers of motor vehicles do reduce their speed enough in slippery driving conditions to maintain the braking distance. Overall, the most important measures during the winter months are snow clearance, sanding, and salting. In places such as the Nordic countries (Sweden, Norway, and Finland), where snowfall and ice are common in winter months, most roads and highways are divided into maintenance classes on the basis of traffic volume.

One of the major research considerations is that the size of the effect of a winter maintenance measure depends on the length of time being considered. The effect is greatest immediately after a measure has been implemented, but it will be *watered down* if a longer period is considered. The effect throughout the whole winter season depends on how often precipitation or weather conditions that require maintenance occur, and how quickly the measure is implemented.

Winter maintenance of roads has a huge effect on mobility, and it is good mobility that is the main objective of winter maintenance measures. It is also worth noting briefly that winter maintenance measures can have a number of effects on the environment. Salting roads greatly increases the salt content of groundwater and in the soil near to the road. Damage to vegetation in particular types of trees has also been found due to the increased salt content. Many people are critical of salting roads largely due to environmental and cost–benefit concerns.

In the United Kingdom, over 2 million tons of salt are spread onto the roads, and it costs over £150 million each year; however, research has indicated that without salting, delays would cause over £2 billion per year. Due to the fact that winter maintenance measures are solely reliant on the weather, it makes determining expenditure very difficult.

The final type of road safety engineering that Elvik et al. (2009) outline is traffic control. Traffic control has already been approached somewhat in this chapter as it includes measures such as traffic calming, and also it includes speed; however, this is the subject of another chapter in this book. Here, we focus on measures at intersections and traffic control of cyclists and pedestrians. The main characteristic of most forms of traffic control is that they are local, that is, they apply to a given crossroads, a city quarter, or another clearly defined part of the road network. It has been suggested in the research that traffic control measures do not solve problems but merely move them elsewhere. However, in the majority of cases, the effects have been studied only at the locations where a specific type of traffic control has been introduced. Traffic control measures are intended to change road user behavior, and generally speaking, the greater the change in behavior, the greater the expected changes of road collisions.

Traffic control measures have varying effects on the number of road collisions. Measures that have been found to reduce the number of collisions are area-wide traffic calming, environmental streets, pedestrian streets, urban play streets, access control, stop signs at junctions with signalized pedestrian crossings; however, as Elvik et al. (2009) point out, some of the results can be affected by "regression-to-mean." It seems as well that traffic control measures primarily intended to increase mobility or improve traffic flow do not necessarily reduce the number of road collisions. Such measures include increased speed limits and reversible traffic lanes.

Engineering

The road network in older parts of town and cities was often constructed for less traffic that it carries today. Older areas were not planned according to the principles for separation and differentiation of the road network (Forskargruppen Scaft 1972). Increased through traffic in residential areas increases collision rates and reduces security and children's opportunities for play and outdoor activities. Traditionally, black spot treatment has been an important safety measure in towns and cities. However, this type of strategy cannot always solve traffic safety problems in areas with an undifferentiated road network. On typical residential roads, collisions are as a rule more randomly spread across the road network than on main roads.

Area-wide traffic calming is the systematic use of the principles of separation and differentiation of the road network in developed areas. By means of traffic control measures, area-wide traffic calming is intended to remove through traffic from residential districts and direct it onto a main road network upgraded to carry increased traffic without an increase in the collision rate.

Measures of area-wide traffic calming include the following by Elvik et al. (2009):

- A ban on through traffic in residential streets using traffic signs or physical closure
- Speed-reducing devices in residential streets (either using signposts or physical measures outlined in this chapter)
- One-way traffic in residential streets to reduce traffic flow
- Improving main roads
- Changing parking regulations in residential streets and access roads

For inhabitants and others who use roads with heavy traffic, traffic is often experienced as a problem, especially when speeds are high. Heavy traffic and high speeds lead to high collision rates, creating noise, pollution, and a feeling of insecurity. The road becomes a barrier, and opportunities for social contact are reduced. In order to reduce the conflict between a road's transport function and the need for safety and a livable environment in towns, the roads can be redesigned to reduce speed and at the same time the traffic environment can be more pleasant. Converting a main road into an environmental street is intended to improve the environment in towns by reducing road collisions and the feeling of insecurity and the environmental problems caused by traffic.

An environmental street is a road where through traffic is permitted but where the road is built in such a way that it leads to low speed and a high degree of alertness and consideration with regard to local traffic. Elements might include the following:

- Tracks for walking or cycling
- Speed humps and raised pedestrian crossings
- Widening of pavement at intersections
- Bus bays
- Marked parking places
- Lighting
- Planting and furnishing of pavements and traffic islands

Pedestrian streets are another common method in urban areas to reduce road collisions. Essentially, this will be completely separating pedestrians and vehicles. In many towns and cities in the United Kingdom, this scheme has been adopted, so pedestrians are able to move around the inner city without relating to the risk of vehicles and buses. Pedestrian streets are often rebuilt by removing curb stones and signposting, and the effect of pedestrian streets on the reduction of road collisions has been significant. However, often research has been carried out in areas where there have been a high number of pedestrian collisions to begin with, so the effects at smaller places might be less where there have been fewer pedestrian collisions.

Urban play streets are by definition intended to encourage recreation and outdoor play in areas where vehicle traffic is limited. Urban play areas permit mixed traffic at walking speed and may be one of a number of measures used in area-wide traffic calming for a specific area. They are often planted with trees, shrubs, and equipment such as tables and chairs. Urban play streets are primarily an environmental measure even though they can improve road safety.

One of the major themes that lie at the forefront of any road safety engineering scheme is that of its sustainability. This book has acknowledged that road traffic collisions are a complex and multifaceted event. We continually see road safety initiatives dealing with the issues such as speeding, drink-driving, seat belts, and other human factors associated with road collisions. What we do not see significantly is the continued modification and development of the physical road environment. The professional responsibility of road safety practitioners is to create a safer road environment at the macroscale (the safety performance of the road network) and the micro perspective (the level of safety at road sections). This chapter has shown that road safety engineering should be applied to all stages of the road or transport development, from the planning of new developments, in the design of new roads, in safety improvements for existing roads, in remedial treatments of hazardous locations, and in routine maintenance programs. Collision investigation activities and related studies are well developed. In New Zealand, for example, there is a well-established program of collision reduction studies in which local road-controlling authorities are involved. The collision locations treated under this program are separately monitored, and the effects are evaluated (LTSA 2004). We have also seen the application of the road safety audit philosophy to existing road sections in attempting to assess the inherent safety of the road environment. This approach, together with the measurement of the actual safety performance of road sections (in terms of collision and casualty rates), has led to the development of road assessment programs, of which EuroRAP and more recently AusRAP are examples (AAA 2005). Much is known about applying safety principles (e.g., conflict reduction, hazard management, road user information management) in road design and remedial treatment and about the effectiveness of proven countermeasures. The issue is not what to do, but rather how to ensure that this is put effectively into practice on a continuous—and sustainable—basis.

We may define *sustainable road safety engineering* as "a process whereby the development of the road environment keeps pace with the demands of the transport task, such that crashes and injuries on the road network are continually reduced, ultimately to acceptable levels of risk" (Croft 2005, 4).

Engineering

The *road environment* refers to the roadway itself, the roadside environment, the management of traffic on it, and the adjacent land use. The *demands of the transport task* refers to the requirements for movement of people and goods, the traffic mix (road users, vehicle types), traffic growth, freight trends, and intermodal (e.g., road/rail) aspects. Analyzing *collisions* requires an understanding of the location and incidence of events, details of road user or vehicle types involved, factors that contribute, and an assessment of the outcomes and costs. The *road network* refers to the diversity of road types from a functional perspective—major arterial routes at one end, local access streets at the other—and the different requirements in terms of capacity, design, and safety. We also need to recognize the different factors (e.g., land use, terrain, geology) that modify or influence how we meet those requirements. The *development* of the road environment refers not only to the planning, design, and construction of new roads and facilities. It also refers to how the existing network may be adapted to accommodate changing transport demands, how we manage the traffic it carries, and how we maintain and manage the asset it represents. *Keeping pace* with the demands of the transport task from a safety perspective implies keeping abreast of latest developments—as indicated by research and analysis, and experience in comparable jurisdictions. This includes identifying and developing opportunities for technology transfer in a variety of areas—for example, real-time variable signing, data communications, and enforcement techniques. An underlying imperative is for the network and its management to be responsive to change and the needs identified from monitoring safety performance. The notion of *keeping pace* may be regarded as reactive. One should also consider *setting the pace*—this implies the need to develop transport policy settings that directly influence the demands on the transport system and the way it operates. These would be reflected in demand management strategies ranging from encouragement of travel behavior change to direct pricing initiatives. *Continual reduction* of collisions and their consequences—within certain limits and/or in pursuit of set target levels (frequencies, rates, other outcomes)—implies an ability to properly measure performance and analyze details of performance, so that progress and improvements can be clearly demonstrated. This requires a strategic approach to setting desired and achievable outcomes. In terms of the *risk levels* presented by the road environment, we must consider the propensity for collisions to occur and the extent of their consequences (injuries and costs)—and define what is meant by acceptable. This can be interpreted in terms of managing the energy exchange aspects of the road environment, such that levels of injury that can occur are tolerable by the human body—a direct interpretation of the Vision Zero approach.

14.3.3 Vulnerable Road Users

It is important to understand the engineering needs of vulnerable road users and promote cycling and walking, for example, road safety engineers would consider the following:

- Footways, cycle lanes, cycle tracks, bridleways
- Controlled or uncontrolled crossing facilities

- Grade-separated crossings
- Crossing points for slip roads on major roads
- Facilities for cyclists where vehicles merge at high speed
- Roundabout designs to benefit cyclists and pedestrians
- Segregation

14.3.3.1 Bicyclists

Bicycle use varies considerably from country to country and city to city. In our highly urbanized cities, cycling is seen as a risky method of transport. In London, cycling has become a popular commuting alterative due to increased transport and fuel costs and the advantages of health and exercise. However, in the highly urbanized Hong Kong, cycling is less popular largely due to the road network and little incentive to provide cycle lanes and cycle safety options. Cycling (especially to work) is more common in Australia and New Zealand. Arguably, their road network is newer with more planning toward incorporating cyclists into the road environment. Cyclists are generally overrepresented in casualty collisions, and the most at risk are the young. Strategies for improving cycle safety include using protection (helmets, using reflective clothing), training, and also modification of the road and traffic environment. These include traffic calming, provision for cyclists, explicit considerations for cyclists at intersections, and street lighting.

Austroads (1993) have indicated that there are four basic requirements of cyclists in relation to the physical facilities they use:

1. A space to ride, particularly adequate lateral clearances
2. A smooth surface
3. The ability to maintain speed (minimizing the need to stop or slow down)
4. Route connectivity and continuity

Ogden (1996) outlines a useful taxonomy of cycling facilities:

1. On-street mixed traffic cycling
2. On-street cycle-only lanes
3. On-street shared bus-cycle lanes
4. Cycle use of road shoulders
5. Cycle paths

The key issue with cycle engineering is to ensure not only their dedicated transport facilities for cyclists, but also how cyclists can interact safely with other traffic modes. Various methods of altering or reallocating the roadway right-of-way to facilitate bicycling and create bikeways have been added to many of the manuals used by transport planners and engineers. We can divide the engineering mechanisms into segregated and nonsegregated cycle facilities. Segregated cycling facilities include side path or shared use path (with pedestrians and is separated from traffic by a barrier of land or green space). A cycle track is a separated cycle facility that can incorporate bicycle-only signal phases at intersections.

14.3.3.2 Pedestrians

Pedestrians have been largely ignored or given minimal consideration in the design of much of the U.S. roadway system (Retting et al. 2003). When the built environment assigns low priority to pedestrians, it can be difficult for vehicles and pedestrians to share the road safely. When we talk about the issue of *assigning*, what is meant can often mean that over time the role of pedestrians is diminished due to changing modifications in the road environment that prioritize the flow of traffic. Modifications to the built environment can reduce the risk and severity of vehicle–pedestrian collisions. According to Retting et al. (2003), engineering modifications generally can be classified into three broad categories: separation of pedestrians from vehicles by time or space, measures that increase the visibility and conspicuity of pedestrians, and reductions in vehicle speeds.

Separation countermeasures reduce the exposure of pedestrians to potential harm, both on the roadside and when they are crossing streets. Because in many pedestrian collisions, the driver reportedly does not see the pedestrian before the accident, measures are needed to increase the visibility and conspicuity of pedestrians. Higher vehicle speeds are strongly associated with a greater likelihood of collisions involving pedestrians as well as more serious pedestrian injuries.

We undertook a thorough review of traffic engineering countermeasures documented in the scientific literature as effective in reducing the risk of collisions involving pedestrians. The primary search engine used was the National Academy of Sciences' Transportation Research Information Services (TRIS) database. TRIS is the world's largest and most comprehensive bibliographic resource on transportation information. Keywords were *pedestrians* along with *injuries, safety, reduction, countermeasures,* and *crosswalks*. In terms of study types, we included before-and-after, case–control, and cross-sectional studies of the effects of speed reduction, separation, or visibility enhancement measures on the occurrence of pedestrian–vehicle collisions or conflicts.

Many studies of traffic engineering measures are limited by methodological flaws, such as failure to account for "regression-to-mean" associated with treatment of high-collision locations and reliance on simple before-and-after measurements without suitable controls. To the extent possible, we included in our review studies based on adequate scientific criteria, such as use of comparison sites to control for confounding factors. In the case of several promising countermeasures, only limited evaluations with somewhat less reliable methodologies were available.

A common weakness in many collision-based before-and-after evaluations of traffic engineering countermeasures is the failure to account for "regression-to-mean," which can result in overestimation of the effects of an intervention, when treatment sites are selected because they have involved high numbers of collisions. Selection of comparison sites with similar characteristics can partially, but not fully, address "regression-to-mean." We included in our review several studies with methodological weaknesses. In these cases, we make note of their limitations.

Some researchers conducting observational road safety studies evaluate pedestrian–motor vehicle conflicts in lieu of collision data to evaluate roadway countermeasures, in part because collisions are rare events and because conflict studies provide information about potential collision causes. Conflicts generally are defined as "near-miss" situations in which a vehicle had to abruptly brake or swerve to avoid striking a pedestrian, or a pedestrian had to take sudden evasive action to avoid being struck. The validity of using conflicts to estimate collisions was examined by Hauer and Garder (1986). Hauer and Garder formulated and tested statistical methods to measure the validity of traffic conflicts on the basis of empirical evidence. According to Garder, it can be shown that a 1-day conflict count provides a more accurate estimate of the expected number of collisions than a 1-year collision history, if the expected number of collisions is less than 5 per year. In conflict studies and other short-term before-and-after evaluations of road user behavior, "regression-to-mean" associated with the treatment of high-collision locations is not a factor.

One study reported that installation of traffic signals substantially reduced conflicts occurring at high-speed intersections, where previously no signals were present and pedestrians had difficulty crossing (Garder 1989). At intersections with traffic signals, exclusive traffic signal phasings—which stop all vehicle traffic for part or all of the pedestrian crossing signals—have been shown to significantly reduce conflicts. A comparative analysis of intersections with and without exclusive pedestrian signal phasings reported that the risk of pedestrian–vehicle collisions at intersections with exclusive timing was approximately half that at intersections with standard pedestrian signals (Zegeer et al. 1982).

14.4 BEFORE-AND-AFTER STUDIES

There is clearly a need for monitoring road safety engineering measures. Monitoring may be simply defined as the systematic collection of data about performance of road safety treatments after their implementation. Post-implementation monitoring is essential to ascertain the effects (both positive and negative) of a treatment. It is also important to monitor a scheme to assess whether or not it might have led to an increase in road collisions. It can also be considered to be a professional responsibility to share the results of experience with peers, so that knowledge and skills can be mutually developed.

Ward and Allsop (1982) suggested that road safety schemes potentially affect the following parameters, and therefore some or all of them need to be monitored:

- The number and type of road collisions
- The severity of road collisions
- The distribution of road collisions
- Traffic flows and travel times
- Turning movements and delays at intersections
- Access times and distances within residential areas
- Routes taken by motorists, cyclists, and pedestrians
- Operation of buses

One challenge of monitoring collisions alone is that because collisions are relatively rare events, it may take a long time for a statistically reliable sample to accrue.

Engineering

This can be overcome by using proxy measures or indirect measures such as insurance company claims.

According to Ogden (1996), the essence of monitoring is to measure for each of the performance indicators what is actually happening in the real world and then, in evaluation phase, attempt to compare that with what we expect would have happened if the treatment had not been introduced. There are several experimental design challenges in doing this:

- There may be changes in the road environment such as a change in speed limit, change in traffic flow, and change in land use. All these are possible over a 3- to 5-year time period and virtually certain over an area or route.
- Road collisions are rare and randomly occurring events. There will be fluctuations year by year that might have nothing to do with the treatment being analyzed. Data for short time periods are highly unreliable. These random year-by-year fluctuations, which not necessarily biasing the result of a monitoring exercise, introduce variability that must be accounted for in the statistical analysis. A particular problem is that of "regression-to-mean."
- It is necessary to monitor all significant factors that would possibly affect the outcome; otherwise, the outcome could be wrongly attributed to the treatment. If the variation in the treatment (e.g., the speed limit) varies systematically with another variable (e.g., design standard), it may not be possible to isolate the effects of one from the other. However, if only one is measured, it is likely that all of the change will be attributed to it.
- If the two variables that are systematically related are in fact both measured, then it will not be possible to reliably isolate their independent effects. This is particularly a problem if multiple linear regression techniques are used, since these require that the various independent variables are not correlated with one another.
- Statistical correlation does not necessarily imply logical correlation. For example, Haight and Olsen (1981) quoted in a case where the law giving pedestrians the right-of-way over vehicles was considerably strengthened in 1977, and the number of pedestrian deaths dropped from 365 in 1977 to 283 in 1983. However, the new law was not enforced and thus had no effect on behavior, so the improvement in the pedestrian situation must have been due to some other factors. This underlies the important of ensuring a linkage between the treatment being monitored and the change in the performance measure.
- Seasonal factors must be taken into account. Some factors may vary diurnally (natural light, street lighting), and others will vary seasonally (rain, hours of daylight, traffic flow). The selection of factors such as control sites and before-and-after periods must take these variations into account. It would be incorrect to compare summer collision record with a winter collision record.
- Collision reporting levels may also change over time, and there may be inconsistencies in the data that would need to be considered. For example,

definitions attached to specific pieces of data may change over time, or the requirement to report collisions may have changed.
- There may be a long-term trend in collision occurrence, and thus changes over time in the number or rate of collisions at a site may merely reflect global trends.

The simplest method is to compare the collision record at the site before and after the implementation of the change. This according to critics is the least satisfactory method because of the lack of control of extraneous factors. For example, during the decade of the 1980s, several countries experienced a very substantial reduction in total casualty collisions. If a treatment installed in the middle of the decade was evaluated using, for example, 3- to 5-year before-and-after periods, it would quite possibly have shown significant reduction in the "after" period compared with the "before" period. However, in reality, this may have mere reflected nationwide trends and had very little to do with the conditions at the site. Nevertheless, this method involves the following:

- Determining in advance the relevant objectives and corresponding evaluation criteria (Table 14.6).
- Monitoring the site or area to obtain numerical values of these criteria before and after the treatment.

TABLE 14.6
Statistical Tests or Procedures for Different Designs and Criteria

Evaluation Design	Criterion	Tests or Procedures
Before-and-after	Frequencies	χ^2 for Poisson
		Paired t-test
	Rates	Paired t-test
	Proportions	Z-test for proportions
	Variances	F-test
	Distribution shifts	Ridit
		Kolmogorov–Smirnov
Before-and-after with randomized controls, comparison groups, or with correction for "regression-to-mean"	Frequencies	χ^2 for Poisson frequency
		Paired t-test for before/after within group
		t-test for group vs. group
		Analysis of covariance
		Median test (categorical data)
		Mann–Whitney (categorical data)
	Proportions	Z-test for proportions
	Rates	Paired t-test for before/after within group
		t-test for group vs. group
		Analysis of covariance
	Variances	F-test
	Distribution shifts	F-test
		Kolmogorov–Smirnov

- Comparing the before and after results.
- Considering where there are any plausible explanations for the changes and correcting for them if possible.

This process highlights the importance of determining in advance what the evaluation criteria are to be. While unexpected results might appear and the data should be examined carefully, the prime criterion is whether the treatment has had the desired effect or not. To this end, it is necessary to distinguish collision by type, and possibly time of day, or weather.

A major drawback of before-and-after studies approach is that it takes no account of trends of changes across the network as a whole. This can be overcome through the use of control sites. There are two variations of this method: the first using control groups that are randomly determined, and the second selecting comparison groups. The first method involves a controlled experiment whereby several candidate sites for a particular treatment are identified in advance. They are then randomly split into two groups; all the sites in the first group are treated and no sites in the second group are treated. Their purpose therefore is attempting to make the control and treatment groups equal on all factors except the execution of the treatment. This method has a significant power as an investigative tool. However, it has limited validity for most applications faced by a road safety engineer, because there will rarely be the opportunity to conduct a controlled experiment of this nature.

Therefore, the second methodology is of much more relevance. The process involves the following:

- Determining in advance the relevant objectives (e.g., collision types intended to be effected) and corresponding evaluation criteria.
- Identifying a control site or set of control sites, where no remedial works have been or are intended to be introduced.
- Monitoring both the treated sites and the control sites to obtain numerical values of these criteria before the treatment and after the treatment.
- Comparing the before and after results at both the treated and control sites.
- Considering whether there are other plausible explanations for the changes and correcting them if possible.

Selection of control sites is very important, and ideally they would be randomly selected. However, this is rarely possible, unless a large number of control sites can be identified, and a random selection made from these. The control sites should satisfy the following criteria:

- Be similar to the treated sites in general characteristics.
- Be geographically close.
- Have the same or similar traffic flows.
- Not be affected by the treatment at the test site.
- Not be treated in any way themselves for the period of the before-and-after study.
- Have collision records or other data that are consistent in collection criteria and coding.

Typical control sites include an adjacent section of rural highway or nearby network of urban streets.

14.5 CONCLUSION

This chapter has sought to approach and discuss road safety engineering analysis. In some ways, the discussion itself would warrant a whole book. The wealth of information and approaches is immense. It is important to acknowledge that road safety engineering, from the point of view of this book, focuses on how road safety professionals choose sites for intervention or what the intervention might be. Many themes surrounding road safety engineering discuss the importance of data; we felt that this topic has been covered extensively in the book, as well as the dissemination of hot spots. Road safety engineering is a complex theme in *road safety* itself, comprising of many different participants and policies. Elvik et al. (2009) outline some of the factors that make road safety problems difficult to solve, the first of which is speeding. Many drivers do not regard this as a road safety problem, however speeding will remain very much a social dilemma regardless of how much we spend on traffic calming measures. Geographical information systems and spatial analysis remain at the heart of road safety engineering, and many of the technical spatial analyses already discussed in this book lead up to road safety engineering solutions. Many government and local government organizations will have a version of a geographic collision analysis software with which they will be able to analyze the spatial data and plan countermeasures effectively.

REFERENCES

AAA. 2005. How safe are our roads? Rating Australia's national highways for risk. Australian Automobile Association, Canberra, Australian Capital Territory, Australia.

AECOM Canada Ltd., CIMA+, and Lund University. 2009. International road engineering safety countermeasures and their applications in the Canadian context. Transport Canada, Mississauga, Ontario, Canada.

Asian Development Bank. 2003. Road safety guidelines for the Asian and Pacific region: Guidelines for decision makers on road safety policy. ADB, Manila, Philippines.

Austroads. 1993. Guide to traffic engineering practice. Part 14, bicycles. Austroads, Sydney, Australia.

Bliss, T. 2004. Implementing the recommendations of the world report on road traffic injury prevention. Transport Note No. TN-1. World Bank Global Road Safety Facility, Washington, DC.

Brabander, B. D., E. Nuyts, and L. Vereeck. 2005. Road safety effects of roundabouts in Flanders. Report No. RA-2005-63. Onderzoekslijn Handhaving en Beleid, Steunpunt Verkeersveiligheid, Belgium.

Committee of Inquiry into Road Traffic Responsibility. 2000. Ett gemensamt ansvar för trafiksäkerheten (Shared responsibility for road safety). Report SOU 2000:43, 19–27 (summary in English). Ministry of Industry, Employment and Communications, Stockholm, Sweden.

Croft, P. 2005. Sustainable road safety engineering. In *Proceedings of European Transport Conference*, Strasbourg, France. Henley-in-Arden, U.K.: Association for European Transport.

Elvik, R. 2003. Effects on road safety of converting intersections to roundabouts: Review of evidence from non-US studies. *Transportation Research Record* 1847: 1–10.
Elvik, R., A. Høye, T. Vaa, and M. Sørensen. 2009. *The Handbook of Road Safety Measures*, 2nd edn. Bingley, U.K.: Emerald Group Publishing.
Forskargruppen Scaft. 1972. Principer för trafiksanering med hensyn till trafiksäkerhet. Meddelande 55. Chalmers Tekniska Högskola, Institutionen för stadsbyggnad, Göteborg, Sweden.
Garder, P. 1989. Pedestrian safety at traffic signals: A study carried out with the help of a traffic conflicts technique. *Accident Analysis & Prevention* 21 (5): 435–444.
Haddon, W. 1968. The precrash, crash, and postcrash parts of the highway safety program. SAE Technical Paper 680237. doi: 10.4271/680237.
Haight, F. A. and R. A. Olsen. 1981. Pedestrian safety in the United States: Some recent trends. *Accident Analysis & Prevention* 13 (1): 43–55.
Hauer, E. and P. Garder. 1986. Research into the validity of the traffic conflicts technique. *Accident Analysis & Prevention* 18 (6): 471–481.
Institute of Highways and Transportation. 2008. Road safety audit guideline, 3rd edn. IHT, London, U.K.
Johansson, Ö. 1997. Olycksutfall och skadeutfall vintertid: En jämförelse av olyckor under 3 vintrar med helt olika klimatbetingelser. *Vägverket Publikation* 1997: 50.
Koornstra, M., D. Lynam, G. Nilsson, P. Noordzij, H.-E. Pettersson, F. Wegman, and P. Wouters. 2002. *SUNflower: A Comparative Study of the Development of Road Safety in Sweden, the United Kingdom, and the Netherlands*. Leidschendam, the Netherlands: SWOV Institute for Road Safety Research.
LTSA. 2004. Overall results of crash reduction study safety improvements. http://www.landtransport.govt.nz/roads/crash-reduction-programme.html (accessed October 2004).
New Zealand Road Assessment Programme (KiwiRAP). 2010. How safe are our roads?: Star rating New Zealand's State Highways. KiwiRAP, Auckland, New Zealand.
Öberg, G., K. Gustafson, and L. Axelson. 1991. More effective de-icing with less salt: Final report on the MINSALT-project. *VTI Rapport* 369: 1–58.
Ogden, K. W. 1996. *Safer Roads: A Guide to Road Safety Engineering*. Burlington, VT: Ashgate Publishing Company.
Peden, M., R. Scurfield, D. Sleet, D. Mohan, A. Adnan, E. Jarawan, and C. Mathers, eds. 2004. World report on the road traffic injury prevention. World Health Organization, Geneva, Switzerland.
Retting, R. A., S. A. Ferguson, and A. T. McCartt. 2003. A review of evidence-based traffic engineering measures designed to reduce pedestrian–motor vehicle crashes. *American Journal of Public Health* 93 (9): 1456–1463.
Royal Society for the Prevention of Accidents (RoSPA). 2012. Road safety engineering: Cost effective local safety schemes, September 2012. http://www.rospa.com/roadsafety/adviceandinformation/highway/road-safety-engineering.aspx (accessed October 29, 2014).
Rumar, K. 1999. Transport safety visions, targets and strategies: Beyond 2000. European Transport Safety Lecture, European Transport Safety Council (ETSC), Brussels, Belgium.
Sabey, B. 1995. Engineering safety on the road. *Injury Prevention* 1 (3): 182–186.
Tingvall, C. 1995. The zero vision. In *Proceedings of the First International Conference on Transportation, Traffic Safety and Health: The New Mobility*, Gothenburg, Sweden, eds. H. von Holst, Å Nygren, and R. Thord, pp. 35–57. Berlin, Germany: Springer-Verlag.
Trinca, G. W., I. R. Johnston, B. J. Campbell, F. A. Haight, P. R. Knight, G. M. Mackay, J. McLean, and E. Petrucelli. 1988. Reducing traffic injury—A global challenge. Royal Australasian College of Surgeons, Melbourne, Australia.

Wallman, C. G. 1997. Driver behaviour on winter roads—A driving simulation study. *VTI Rapport* 419A. Swedish National Road and Transport Research Institute, Linköping, Sweden.

Wallwork, M. J. 1993. Traffic calming: A guide to street sharing. In *The Traffic Safety Toolbox*, pp. 234–245. Washington, DC: Institute of Transportation Engineers.

Ward, H. and R. Allsop. 1982. Area-wide approach to urban road safety—Evaluation of schemes by monitoring of traffic and accidents. *Traffic Engineering & Control* 23 (9): 424–428.

Wegman, F. 2001. A road safety information system: From concept to implementation. Contribution to the Road Safety Training Course of the World Bank, Washington, DC, May 1, 2001. Report number D-2001-14. SWOV Institute for Road Safety Research, Leidschendam, the Netherlands.

Wegman, F. and L. Aarts, eds. 2006. *Advancing Sustainable Safety: National Road Safety Outlook for 2005–2020*. Leidschendam, the Netherlands: SWOV Institute for Road Safety Research, Leidschendam, the Netherlands.

Wegman, F., L. Aarts, and C. Bax. 2008. Advancing sustainable safety: National road safety outlook for the Netherlands for 2005–2020. *Safety Science* 46 (2): 323–343.

Wegman, F. and P. Elsenaar. 1997. *Sustainable Solutions to Improve Road Safety in the Netherlands*. Leidschendam, the Netherlands: SWOV Institute for Road Safety Research.

Zegeer, C. V., K. S. Opiela, and M. J. Cynecki. 1982. Effect of pedestrian signals and signal timing on pedestrian accidents. *Transportation Research Record* 847: 62–72.

15 Education

15.1 INTRODUCTION

Road safety education is one of the cornerstones for reducing road collisions. The growing exhaustion of engineering and new enforcement methods means road safety professionals are in an ever-increasing demand to improve education through direct (education) and indirect (publicity) methods. Most of the educational measures are aimed at school children, and advertising is aimed at society as a whole or as specific groups (such as motorcyclists). In this chapter, we will focus on discussing different elements around the use of education in road safety. We discuss different groups of society, different methods, and outcomes, and interpret some of the findings from research reports. Generally speaking in terms of spatial analysis and road safety education and publicity, there has been limited research. Only in recent years we have seen an attempt at using demographics to target specific high-risk groups in society (this will be discussed later in the chapter). In terms of the "spatiality" of education and publicity, more innovative work needs to be achieved. This chapter will therefore focus on the analysis that has been achieved and what the current trends, challenges, and statistics show us.

Road safety education is fundamentally focused on children. This results from an obvious linkage between being new to the road environment and getting children to understand their risks in and around the road as pedestrians, cyclists, and later as drivers. In general, it is argued that road safety education is often given a low-status priority largely due to other health and safety issues being deemed more important. Issues such as health and crime are seen as more important than road safety in our ever-complex society. There is also a potential issue of being *overburdened* with public health messages and information. Children at school age (through their parents) are under increasing pressure in terms of healthy eating, exercise, technology, and so on. Adult consumers are bombarded on a daily basis to "do this" or "not do this." The other issue in road safety education is the variability in delivery. In the United Kingdom, road safety education is provided by road safety officers to different degrees and successes. Integrated road safety education is highlighted as best practice in contrast to occasional talks or other less integrated approaches. The need for differentiated publicity to raise risk awareness, particularly among young teenagers, is also highlighted. There is a call for road safety education to be accepted as a lifelong process, with implications for all road users, which we will discuss later in the chapter. Children's perspectives and behaviors in traffic should be taken into account, both in driver training and also by vehicle designers.

A wide range of driver education and training reviews have taken place globally. It is worth taking note of terminology. While it would readily be possible to distinguish training (which is concerned with skill acquisition) from education (which is concerned with knowledge acquisition) in the driving field, there is little evidence that people note the difference. A wide range of interventions have been employed, including on- and off-road training and in-class education, both pre- and post-license. There does appear to be a prevailing assumption that both child education and driver education are effective.

The fact that young male drivers are the group who most readily pass the driving test might be cause for concern, when it is realized that it is this very same group that has the highest fatality rate. In theory, improved vehicle control skills might indeed prompt safe behavior, for example, by improving hazard perception skills; or they might prompt unsafe behavior, for example, by encouraging faster speeds. It is an empirical question as to which result occurs.

The sheer number of available interventions is perhaps testament to the broad awareness of the problem. However, there is much less indication that the interventions are evidence based. It has been noted on a number of occasions that there is broad and uncritical support for work on driver training, which appears to proceed on the basis that if it does not do any good, then at least it does no harm (Christie 2001; Stradling et al. 2006). Hauer (2007, 330) concluded that the prevailing culture is to think that "… road safety can be delivered on the basis of opinion, folklore, tradition, intuition and personal experience." Relatively few programs are based on either theory or evidence, and relatively little is evaluated (Smith and Shannon 2003). When road safety education interventions have been evaluated, a range of reviews have failed to indicate the success of safety education (Brown et al. 1987; Mayhew et al. 1998; Vernick et al. 1999; Christie 2001; Mayhew and Simpson 2002; Ker et al. 2003). A number of authors have noted that not only is there little evidence to support driver education, but "even more discouraging, a few studies even showed a safety disbenefit—that is, an increase, rather than a decrease in crash involvement" (Mayhew and Simpson 2002, ii3). Ker et al. (2003, 9) sum up their systematic review of post-license training, concluding that there is "no evidence that post-licence driver education programmes are effective in preventing road traffic injuries or crashes."

The Department for Transport, through *Children's Road Traffic Safety* (Christie et al. 2004), attempted to identify good practice and innovation from other countries that could improve road safety education for children in the United Kingdom. This report found that the United Kingdom has developed good practice in a number of areas, but that specific areas need strengthening. Its call for national support for curriculum-linked, Ofsted (Office for Standards in Education, Children's Services and Skills)-inspected school travel plans prefigures the current requirements for school travel planning and the requirements of the National Healthy Schools Standard. More widely, the report emphasized the need for greater consistency in environmental modification and suggested the more rigorous enforcement of low speed-limit zones around schools. It also stressed the need for legislation to focus attention on driver responsibility for pedestrian collisions. The report suggests greater consideration of best practice in other countries in terms of secondary safety behavior. Finally, the need for greater awareness of the causal relationship between economic disadvantage and increased road safety risk is highlighted within the report.

15.2 CHILDREN AND YOUTH

A wide range of literature has been dedicated to studying road safety education in children. The principal aim of this education is to lead to measureable changes in the child's behavior. There is discussion among academics and road safety professionals regarding knowledge versus practical methods in order to educate children. Practical measures are often time consuming and costly. In reducing road injuries, most countries including the United Kingdom have a national scheme, and often this is integrated into the school curriculum. Sometimes additional local programs are available that deal with specific issues in that area. Generally, the schemes vary quite considerably in terms of methods they use, especially when comparing internationally, but are often unified in program content. Generally speaking, the main aims are to provide children with a broad conceptual knowledge of the road and traffic environment and educate simple attitudes toward road safety. The idea being that children will develop a framework in which to build on for different road environment situations; therefore, the approach is often a top-down one, going from teaching about general situations to more specific ones.

In the case of child pedestrian collisions, educational measures have been introduced as a means of teaching children how to cope with traffic, and substantial resources have been devoted to their development and provision. However, although collision rates have undoubtedly decreased over the last 30 years, it is remarkably difficult to know how much of this may be attributed to the effects of education and how much to other factors (Thomson et al. 1996). The situation has been exacerbated by the fact that few educational measures introduced at either local or national level have been evaluated with regard to their effectiveness in improving children's traffic behavior (Singh 1982; Thomson 1991), although in recent years, this has started to change. Moreover, Hillman et al. (1990) have cogently argued that a major factor underlying the decrease in child injuries between 1971 and 1991 was a dramatic reduction in *exposure* to traffic, caused by greatly increased parental anxiety toward their children's safety. In general, there seems to be a widespread feeling at the present time that educational measures have not achieved as much as had been hoped and that there may even be quite strict limits on what can be achieved through education. This would shift the emphasis away from education altogether toward engineering or urban planning measures aimed at creating an intrinsically safer pedestrian environment, where the need for education would be reduced or even eliminated. One of the major barriers documented by a recent report by the UK Department for Transport (DfT 2008) in terms of child road safety education is engaging with the community, especially those from ethnic minorities. The report suggests that while education in primary schools is successful to a large degree, it is neglected in secondary schools. Coupled with this is a tension in encouraging more walking and cycling while reducing road injuries, since the former results in a greater exposure to risk. There has been criticism in the United Kingdom by road safety professionals that road safety education is weighted heavily toward children and youth.

What are the main objectives of road safety education for children? Every educational program needs clearly defined objectives. The goal for road safety education for children is to reduce child road collisions. Most countries have a national scheme, and often this might be integrated into the school curriculum. Such schemes vary quite widely with respect to the methods they employ but are much more unified when it comes to program content. In the vast majority of cases, the aim is to provide children with broadly conceptual knowledge about the traffic environment and instill suitable attitudes toward safety. The approach is thus a *top-down* one, in which learning is assumed to proceed from the general to the specific. In fact, this view contradicts virtually every theory of learning and development, all of which agree that learning typically progresses in the converse direction: that is, from specific concrete situations to the gradual elaboration of abstract conceptual knowledge.

According to Thomson (2006), there has been little or no effort to teach children a more integrated approach to road safety. There is so much advancement in terms of school-based education; often, road safety gets left behind in terms of innovation and advancements. Numerous methods of teaching this material have been attempted, but, in practice, the vast bulk of road safety education takes place in the classroom. The material is almost exclusively verbal: that is, children learn by being told what to do rather than by actually doing anything. Thomson (2006) outlines the major techniques for teaching road safety in the classroom, which include using books or other printed material, film techniques, behavioral techniques, training visual timing skills, the pretend road method, perceptions of safe and dangerous road locations, and training to deal with specific situations. Road safety, as with health, is accountable to a number of different parties in a child's life. First and foremost are the parents, then the school, and then the society as a whole. Whose responsibility should teaching road safety be, the parents or the teachers? Most people would argue it is both. However, the question to be asked is whether there is enough collaboration between the two to educate children about road safety? The other major educational issue for young people is that road safety education often ends in primary school and albeit being educated the parents. The youth (categorized here by children aged between 13 and 17) face very different challenges and exposure risks than children. Young adults or *youth* have more freedom that comes with learning to drive and being allowed more *spatial* freedom. With this comes a different interaction in the road environment, one which is not approached fully (in terms of a clear defined set of objectives) by schools and parents alike.

Internationally, the majority of road safety education takes place in the classroom. Often, the way in which the material is delivered is vocal with children taking a very passive role in learning about road safety. Generally speaking, there is little practical training for children. The increasing numbers of children who are driven to school by their parents, or who are not allowed to play outside due to the dangerous traffic environment, mean that children do not actively learn road environment skills. It is a vicious circle with parents not confident letting their child interact with the road environment, and when the children do, their behavior and knowledge is limited. The following diagram highlights an Australian approach to road safety educational programs, the successful outcome of which is hard to quantify, which will be discussed later in the chapter.

15.2.1 School Education: Cycle Safety

Children are much more likely to either walk or cycle as it gives them freedom and independence. The patterns of children and cycling vary enormously in terms of geography, socio-economics, and infrastructure. Moreover, it varies internationally; in this section, we will focus on the United Kingdom, Hong Kong, Australia, and New Zealand in order to understand the nature of cycling education for children. With rising levels of obesity and inactivity in children in the Western world, cycling and walking to school have been actively encouraged by governments. While this is commendable, the importance of cycling safety and education needs to be addressed. Children are often unaware of the risks and dangers they can face within the road environment, and this poses a significant obstacle when educating children. One of the key ways in which governments have tried to reduce this risk is to introduce the mandatory use of cycle helmets. There is a wide source of literature that focuses on the advantages of children wearing helmets; however, not all academics and policy makers agree with the use. The following countries have mandatory helmet laws: Australia, Canada, Czech Republic, Finland, Iceland, New Zealand, Sweden, and the United States. In a recent paper by Christie et al. (2011), which looks at the nature of cycling and its disadvantage in the United Kingdom, there is a low proportion from disadvantaged areas who cycle either to school or for leisure. Christie et al. (2011) outlined two potential factors that caused this lack of cycling, and these were poor-quality environments and parental fear for their child's safety. It was determined that actually more than a one-third of the children questioned would like to travel by bicycle to school; however, parental fears made this not viable. With this in mind, it leads us to wonder whether cycle safety for children at school should also incorporate the parents as well. A large reason for not cycling is the speed and road structure. Within residential areas, the speed and the volume of traffic need to be reduced so that they become less hostile for cyclists. This needs to be addressed through engineering and enforcement. There is evidence that the rates of child injuries are reduced by area-wide traffic calming (Mackie et al. 1990; Cloke et al. 1999; Wheeler and Taylor 2000; Jones et al. 2005) and that 20 mph zones may reduce child road casualty rates for both child pedestrians and child pedal cyclists (Webster and Mackie 1996; Grayling et al. 2002; Webster and Layfield 2003; Layfield et al. 2005; Tilly et al. 2005; Webster et al. 2005; Grundy et al. 2008). There is also evidence from the United States that "safe routes to school" programs based on engineering measures may reduce child casualty rates as pedestrians or cyclists (Blomberg et al. 2008; Gutierrez et al. 2008). It is not our intent to outline in detail *who* cycles, but more the level of information (or education) received by children and youth who cycle. We know that serious injuries for child cyclists peak between 10 and 13 (Chambers 2007). Best practice acknowledges that cyclists on roads must be able to understand and apply road rules. While it might be reasonable to say some children can start cycling on quiet roads at ages 10 or 11, evidence suggests that the majority of children will not be ready to ride in usual urban traffic until they are much older, about the same time we begin to consider them old enough to drive a car.

There is a continued need to address the education of young cyclists for their own safety and for the parent's confidence in the child's ability to cope in a busy road environment. The United Kingdom operates a program called "Bikeability," which is a cycling skills course (formerly the "Cycling Proficiency Test"), which is a targeted program for children at primary school to undertake a cycling test. The United Kingdom is one of a handful of countries that use this targeted approach for child cyclists. Whereas New Zealand and Australia have mandatory cycle helmet use, they do not advocate any cycling tests. Cycling tests are optional in New Zealand and are not as advocated as they are in the United Kingdom. In Australia, each state has different cycling courses that, again, are not mandatory. The drawback with all these courses is that they all cost extra money. In disadvantaged areas, this will prove to be a negative aspect. Hong Kong has a similar situation, insofar as there are criteria for child cyclists, but it is not openly encouraged. The environment of Hong Kong (especially Hong Kong Island and Kowloon) means that dense traffic and infrastructure are not suited for child cyclists. There is no *mandatory* cycle test for children; the only stipulation is that children under 11 must be accompanied by an adult. Most of the cycling by children takes place in the less dense New Territories; however, helmets are not a legal requirement. Although the main aim of this section is to discuss children and cycling, there is a very real danger and educational issue for adult cyclists.

15.2.2 Probationary License

Probationary license or learner's permit is a restricted license for a person who is learning to drive but has not yet taken the necessary exams to obtain a full driver's license. Globally, the age at which a person is able to apply for this license varies considerably. For example, the age in Hong Kong is 18, compared to 15 years in New Zealand. Often, there is strict criterion while driving on a probationary license, the most common having a supervisor driver in the car at all times. Young drivers have a higher risk of road collisions than older, more experienced drivers (Mayhew et al. 1998). For example, Williams (1998) reported that in 1995 in the United States, 16- to 19-year-old drivers were involved in 17 million collisions per million miles of travel, compared to drivers in their early 20s and those 40–44, who were involved in four collisions per million miles, respectively. It is generally accepted therefore that experience and age are significant factors that contribute to the overrepresentation, because risk declines as age increases.

In a study by Mayhew et al. (2003), they calculated the monthly collision rates for drivers with learner's permits and those on full driver's licenses, and the collision rates for learners as opposed to novices are very low. Thus, it is often concluded that leaner drivers are relatively safe drivers as they are often driving under constant supervision, compared to those types of license permits or novice drivers (essentially without anyone supervising). There is a fairly strong research agenda that looks at the probationary license and collision involvement, in particular that of age and gender.

Often countries now have a graduated driver license system, which corresponds to different levels of responsibility while learning to drive. This could begin with supervisory accompaniment, following by unsupervised driving but not at night or carry passengers. New Zealand introduced a Graduate Drivers License in 1997.

The question that is paramount to this subfield of road collision analysis is whether improvements to road collision statistics in youth can be attributed to the graduated license permit or as a result of global trends in improved motor vehicle safety and prevention. Research by Kingham et al. (2008) draws upon comparisons between both New Zealand and Great Britain for analyzing the graduated driver license scheme. New Zealand has the lowest minimum driving age (introduced as 15 in 1924) outside of the United States. Statistics showed an immediate reduction in road collisions after the introduction of the scheme in 1987 (29% decrease compared to 1986) (Langley et al. 1996). However, the cause for concern rested with that New Zealand's fatalities in the 15- to 24-year-old age group remained starkly higher than that of the United States, Canada, and United Kingdom.

Overall analysis of the probationary license system depends on the country. Not all countries have adopted this scheme: in Europe, for example, many countries have lowered the age limit for supervised practice before the licensing age (Page 1995; Gregersen et al. 2000). Nearly all of the evaluation studies have used road collision involvement or risk as the effect variable. In Swedish and Finnish studies, road collisions after licensing have been used while Graduated License System (GLS) evaluations have mostly analyzed the total number of road collisions including those occurring in the different stages of the system. Few studies have focused on distinguishing road collisions during practice from those after licensing. There are consistent limitations concerning the analysis of the GLS on road collisions, as it is often very difficult to separate out the effects of the GLS on road collision statistics from other influencing factors such as the rise in motor vehicle safety or the general decrease in road collisions (specifically in Western countries). Kingham et al. (2008) concluded that it was impossible to determine whether the GLS had reduced road collision mortality among youth. Alongside the statistical analysis of road collision statistics for youth is the analysis of contributory risk factors. These data mechanisms through literature surveys, cohort studies, focus groups, telephone surveys, questionnaires, and supplementary collision data show overall, globally, the youth have very high risks with regard to road collisions.

From a statistical standpoint, the large majority of the studies focus on comparative studies and simple regression models (Karaca-Mandic and Ridgeway 2010). A study by Begg and Stephenson (2003) used a cohort of young people who had been involved from birth in a longitudinal study of health and development. The cohort asked questionnaires about their attitudes to the newly introduced GLS, with the majority supporting the policy (aged 15). The general idea behind GLS is to reduce exposure to collision risk in terms of night driving and carrying passengers.

15.3 ELDERLY

Evidence has shown us that the elderly have an increased risk of being fatality injured in road traffic collisions (Loo and Tsui 2009). If they are involved in a collision, the risk of severe injury is considerably higher because of their physical vulnerability. In addition, functional limitations also increase the risk of a collision. In the coming decades, the number of elderly road users will increase considerably. Despite their higher risk, it remains important for the elderly to participate in traffic for as long as possible. In that case, the car is less dangerous than the bicycle or walking. Measures to improve the road safety of this group must particularly be looked for in adaptations of the infrastructure, in education and information, and in adapting the vehicle with increasing attention for the application of intelligent transport system.

One of the major issues the Western world faces is a growing population of elderly people. Worldwide, the number of elderly (over 65) is expected to double by 2050 (OECD 2011) with an even higher increase of the number of people over 80 years in the same time period. Morgan and King (1995) determined that collisions are 3.5 times more likely to be fatal for elderly drivers. This is largely due to not only the increased frailty but also issues such as reaction speeds and confidence on the road. The types of collision that the elderly are likely to be in as drivers vary considerably. They are more likely to travel shorter distances and at slower speeds, which can be actually more dangerous than traveling at higher speeds with potentially more confidence.

With a progressive decline in functions, adaptations to the road and vehicle surroundings cannot always prevent individuals becoming unfit to drive a vehicle. Therefore, a procedure that leads to a timely withdrawal from traffic is necessary. The problem is determining the threshold: when is someone still fit to drive and under which preconditions (vehicle adaptations, aids, training, limited driving license)? It is important to remember as well we are talking about not only the elderly as drivers but also generally elderly interacting with the road environment.

Older drivers need information on the physical and cognitive changes that accompany aging and on the implications of ceasing to drive. In particular, it is important to inform older drivers of the following:

- The potential for declining sensory and cognitive abilities, difficulties that may arise in traffic as a result of these declining abilities, and how to modify driving strategies to avoid these difficulties. Recognition by the individual driver is the essential first step in effective remedial action. At the same time, information must be available to provide reassurance that with care and planning, drivers can continue to drive safely well into old age.
- Vehicle equipment that is available to make driving easier.
- Increased vulnerability and the importance of using protection devices.
- Influence of age-related illnesses and prescribed medication on driving abilities.
- Information about the procedure to be followed to extend the driving license.
- Possible decision to no longer drive a car: making this debatable, and discussing the roles that relatives and family doctor can play.
- How and where to seek and access mobility alternatives to the car.

In practice, what this must mean is that program designers must be aware of the potential risks of getting it wrong. While most road safety activities aimed at elderly drivers are to give feedback and help drivers to use the road safely, there is a chance that an activity may identify a driver with a condition that could mean they cannot. Programs need to be sensitive enough to help the driver, their family and friends, and relevant health professionals to have a large part in the decision to give up driving, with the results of any assessment used to help guide that discussion.

Often, the most effective means for preventing and reducing road collisions in the elderly is using cognitive tests in terms of fitness to drive. Education generally is a sensitive matter: people who are over 65 have often experienced a wide range of road and driving environments. Self-regulation is often the key factor in reducing risk, and one study found that it was women who reported lower confidence levels when driving. Research has shown that older drivers approve of education programs for older drivers, which means that there is the potential for high take-up of such schemes. One evaluated driver education scheme, the 55 Alive/Mature Driver Education Programme in Canada, was designed to provide information on the rules of the road, hazard recognition, and changes that affect driving. The program also encourages drivers to reduce exposure to risky environments and to plan for the time when they will have to give up driving. The evaluation had three phases—first to look at the self-selection bias of drivers enrolling on the study, second to examine collision rates, and third to run focus group sessions with men who had attended the program.

It found that there was a self-selection bias on the program, with drivers who had been involved in an at-fault collision being more likely to attend the course. Perhaps surprisingly, the study found that attendance on the program was associated with higher collision risk for those over the age of 75 years, although there was no change in the collisions among younger mature drivers. It was unclear from collision records which of the strategies did influence collision involvement, or whether it was a further factor such as reduced exposure to collision risk. The focus groups invited drivers to talk about their driving habits and attitudes. The main finding was that the men who had not been involved in a collision following the program (whether or not they had been in one before the course) used many more of the strategies presented in the program to self-regulate their driving.

A report by Evgenikos et al. (2009) on road safety and the elderly in Europe looked at basic road safety parameters with regard to the elderly population. While there has been an overall reduction in road collisions involving the elderly, 30% between 1997 and 2006, they are still the biggest risk group in society and one that often goes neglected in road safety campaigns, education, and advertising. Generally, the report found that the elderly were not actually of major risk to other road users and that the most high-risk age group was between 75 and 84.

15.3.1 PUBLICITY AND CAMPAIGNS

Road safety campaigns can be defined as purposeful attempts to inform, persuade, and motivate a population (or subgroup of a population) to change its attitudes and/or behaviors to improve road safety using organized communications involving specific media channels within a given time period. It can have many and multiple purposes, such as informing the public of new or little known traffic rules, increasing problem awareness, or convincing people to refrain from hazardous behaviors and adopting safe ones instead. It is not our aim here to delve into the psychologies of road safety campaigning but to evaluate the purpose and outcomes of campaigning and to understand the analysis behind the advertising. Road safety campaigns globally are focused on a small group of specific themes. These include speed, alcohol, seat belts, visibility, tail-gating, mobile phones, drugs, children, cyclists, and pedestrians.

Road safety campaigns vary enormously from country to country and who they focus on. While there is a growing body of evidence that traffic law enforcement programs, such as random breath testing (RBT) and speed cameras, are effective in reducing illegal high-risk behaviors (e.g., Cameron et al. 1992), mass media advertising plays an important role in addressing these behaviors. First, mass media advertising can be used to maximize the deterrent effects achieved by enforcement programs by increasing the driving public's perceived risk of apprehension (Elliott 2011). Second, mass media advertising can work independently to educate and persuade road users to adopt safer behavior(s) and related lifestyles. Consequently, ensuring that advertising approaches are achieving their persuasive goals is paramount.

Of the approaches utilized in road safety publicity campaigns, shock tactics, which aim to evoke strong fear responses in individuals, feature prominently (Tay 1999; Tay and Watson 2002). These shock-based "fear appeals," or more accurately, fear-arousing threat appeals present individuals with the negative outcomes that they may experience as a result of engaging in the depicted unsafe and/or illegal behaviors. Nonetheless, there needs to be more debate on whether or not road collisions ought to be used as the basis of success or failure. The issue concerning how road safety advertising seemingly influences behavior also needs to be addressed, as does the issue of how frequently road users need to be exposed to a particular road safety communication for it to have an effect. Judging by the number of road safety campaigns that make use of fear appeals, there is a firm belief in the ability to "scare people straight." Lewis et al. (2007) ask the question of whether we should move *beyond* these fear-evoking appeals in road safety. Many Western countries including Australia, New Zealand, and the United Kingdom use these tactics of depicting collision scenes resulting from unsafe or illegal driving behavior.

For any mass media campaign to meet its planners' objectives, it must have an influence on behavior to have an effect, even if that effect was not the intended one. For example, a campaign about drink-drive enforcement aimed at drink-drivers may inadvertently encourage greater levels of RBT to be carried out because the police now see RBT enforcement as accepted by the community and in turn influencing drivers to moderate their drink-driving. Whether or not there is a reduction in alcohol-related collisions should not be the criterion of campaign success. The campaign was designed to influence drink-driving behavior either directly or indirectly. Did the campaign overall have the desired effects on the advocated behaviors even if the change cannot be detected in the collision database? Beyond knowing the campaign had a measurable effect (on behavior), it would be invaluable to determine *how* or *why* it has had the intended effect. A review of how to measure success can be found in Morgan and Poorta's (2008) review of successful public service media campaigns in the United Kingdom.

There is quite a substantial body of opinion, based on evaluations of individual campaigns, that mass media campaigns will not usually reduce collisions. Strecher et al. (2006, 35) argue that "One-size-fits-all mass media interventions that run independently of other strategies have demonstrated little or no behavioural improvement." On the other hand, advocates of advertising point to significant changes in some attitudes over the past 30 years, for example, less tolerance of smoking and drink-driving. Indeed, research suggests that mass media campaigns are generally more successful in fulfilling an agenda-setting role (i.e., changing social norms) by increasing awareness of an issue or problem rather than altering behavior. To determine the effectiveness of recent mass media campaigns, a scientific outcome-based evaluation is desirable. However, a rigorous evaluation is difficult and costly to achieve and may not necessarily provide definitive answers. In the absence of such an evaluation, a more constructive approach is to review the literature to determine what conditions are necessary for mass media campaigns to successfully change road safety–related outcomes. In the past, a number of studies have used meta-analytic techniques or have reviewed the literature to determine key elements associated with effective road safety mass media campaigns.

With regard to drink-driving, Wakefield et al. (2010) estimated that the average associated decline in vehicle collisions has been estimated to be at least 7% and of alcohol-impaired driving to be 13%. Results of designated driver programs have been less conclusive. The most notable road safety campaigns (in the United States) have promoted seat belt use. The Click It or Ticket program in North Carolina, USA, was associated with an increase in seat belt use from 63% to 80% and lowered rates of highway deaths, and it became a model for other state and national programs.

Most of the literature and research on road safety campaigns and advertising is evaluation. Without proper evaluation, there is no understanding of whether the campaigns are working. Of course, one of the major issues is how you go about evaluating campaigns. Is it linked to the reduction of road collisions and fatalities? If so, who is to say there were no other factors that came into play? One of the main ways by which researchers approach this is often through meta-analysis. Hoekstra and Wegman (2011) are one of many research papers that focus on this meta-analysis of road safety campaigns. Hoekstra and Wegman outlined that a meta-analysis (Elvik et al. 2009) showed that the effects of mass media campaigns alone are small, especially when compared to the effects of campaigns that were combined with other measures. Without enforcement and/or education, a mass media campaign has virtually no effect in terms of reducing the number of road accidents, while adding either of both those measures ensures a reduction of over 10% (see Table 15.1). Interestingly enough, it is the local, personally directed campaigns that show by far the biggest effect on road collisions.

Because reports on the evaluation results of road safety campaigns are few and far between, there is still little insight available into the effectiveness of campaigns in general, let alone which ingredients have proven to be successful and which have not. This in turn makes it hard to determine if and how the practice and effectiveness of road safety campaigns might be improved, thereby depriving the organizations behind road safety campaigns of the opportunity to learn from their successes and their mistakes and make a bigger difference. Evaluations of road safety campaigns may, for example, shed some light on the more controversial of current practices (such as the use of fear appeals) and help determine if and when these practices are really effective.

TABLE 15.1
Effects of Road Safety Campaigns on Road Collisions

	Best Estimate (%)	95% Confidence Level
General effect	−9	(−13; −5)
Mass media alone	+1	(−9; +12)
Mass media + enforcement	−13	(−19; −6)
Mass media + enforcement + education	−14	(−22; −5)
Local individual campaigns	−39	(−56; −17)

Source: Data from Elvik, R. et al., *The Handbook of Road Safety Measures*, 2nd edn., Emerald Group Publishing, Bingley, U.K., 2009.

Campaigns tend to be focused in Western countries, using the mass media and lasting for more than 200 days (Phillips et al. 2011). The majority of the campaigns are focused on reducing speed and drink-driving and are accompanied by some level of enforcement:

- Drug taking and driving
- Drink-driving
- Fatigue
- Mobile phones
- Seat belts
- Campaign objectives (general)

Road safety publicity can be used to achieve various aims and objectives. In general, the aims of such publicity are to change the road user's behavior, attitude, or knowledge in order to increase road safety. However, usually, "road safety campaigns can succeed if advertising is only one of the elements in the total campaign and usually not the key element" (Elliott 1989, 8). According to Elliott, mass media campaigns can achieve the following:

- Increase awareness of a problem or a behavior.
- Raise the level of information about a topic or issue.
- Help form beliefs, especially where they are not firmly held.
- Make a topic more salient and sensitize the audience to other forms of communication.
- Stimulate interpersonal influences via conversations with others (e.g., police, teachers, or parents).
- Generate information seeking by individuals.
- Reinforce existing beliefs and behaviors.

A report by the World Bank in 2008 looked at road safety in China. While we notice a stark difference in advertising between Hong Kong and crossing the border in China, there is probably more of a need for road safety advertising and education in China. There has been evidence of campaigns focused on children, drivers, and rural residents, but the sheer increase in vehicles in China makes it almost impossible to evaluate and manage successfully. China has been plunged into a fast-paced vehicle-buying frenzy, and with that comes inexperience on the road, and largely not for young drivers. The whole education and campaigning for road safety therefore becomes one of uniqueness, quite unlike what most Western countries have experienced. In 2006, China had 150 million drivers (World Bank 2008), and we suspect that this figure has approached at least 200 million by 2012.

Historically, in many countries, road safety publicity campaigns have not been approached in a scientific manner. There are various types of data that can be utilized:

- *Collision and casualty data*
- *Observation*
- *Attitude testing*
- *Knowledge testing*

When planning a road safety campaign, there are various elements that need to be addressed. The first is target behavior, which would be apparent from the collision data. Second is the target audience. The target audiences are not necessarily the people who are behaving in an inappropriate way. It is more effective to target other people who influence the road user in question, the *significant others* might be family, friends, or colleagues. Third is the audience motivation. Consideration needs to be given to what will actually motivate an audience to change their attitude or behavior. Fourth is the message content. Literature has advised that the message should be clear, unambiguous, and directional. Fifth is media selection, which will depend on the target audience. It is necessary to consider where the targeted road users are likely to see a message.

In order to decide on the most appropriate method(s) of evaluation, it is necessary to first know the objectives of the campaign. In most cases, the overall objective will be to reduce collisions or injuries. It is necessary to use appropriate means of evaluating publicity campaigns. If collision prevention/reduction is to be used as a measure, then the time interval must be great enough to pick up any effects. While use of collision or casualty statistics may be appropriate, especially in the case of long-term (5- or 10-year) campaigns, in the shorter term, it is not appropriate to use collision data alone. The use of collision rates as a measure can be awkward for all kinds of reasons such as underreporting, time scale, and influence of other factors. Instead, there are other measures that can be used. Wherever possible, multiple measures should be used:

- Popular liking for a message
- Popular opinion of message effectiveness
- Expert opinion of message effectiveness
- The numbers and types of road users reached
- Recall of the message used
- Change in traffic knowledge
- Change in attitudes

- Change in behavior as reported by the individual
- Change in observed behavior
- Change in violation rates
- Change in collision rates

Thus far, the discussion of road safety campaigns has centered on the status quo: what has been done, how effective has it been, and what steps can be taken to improve upon some of the current practices, both in the method and in the evaluation of road safety campaign; in the next section, we discuss the use of social marketing for road safety campaigns and the advantages and disadvantages of these methods.

15.3.2 Using Geodemographics to Target Road Users

A recent premise of road safety education is of "social marketing." There have been many unfavorable reviews concerning the use of social marketing for road safety. Elliott (2011) documented the failure of social marketing over three decades, concluding that while marketing had some useful tools to offer (a marketing analysis), its theory of persuasion "Make what the customer wants and will buy" is fundamentally different from the persuasion task faced by road safety practitioners, where the task is about getting people to *start* or *stop* specific behaviors. In road safety as in health, crime, and other public sector activities, the main objective of the communication is to advocate people to change by starting or stopping something because it is good for them and society. The aim is to change their own behavior. Correspondingly, in the private sector social marketing, a brand, product, or service merely requests a modification of their existing behavior, an element of "chose our brand not theirs" mentality. Edited by Lannon (2008), it includes campaigns on exercise, taxation, domestic violence, stroke, burglary, unwanted pregnancy, unbelted rear passengers, car theft, household fire, child literacy, drink-drivers killing pedestrians, cancer, binge drinking, mobile phones while driving, organ donation, blood donation, anti-social noise, pedal cycle accidents, smoking, illegal mini cabs, drink-driving, chip pan fires, and TV licenses. Elliott (2011) outlines that road safety advertising has three main functions:

1. It has to be noticed, gaining some degree of attention, and the best way of achieving this is to generate an emotional response.
2. Ensure it is remembered, and this is tied in with how it is seen.
3. Influence road user behavior either directly or indirectly.

A recent project has been undertaken by the Thames Valley Safer Roads Partnership, which uses a web-based interface and road accident data interlinked with geodemographic data. This tool allows practitioners to select data interactively and use it with Experian's Mosaic software. It can be user customized and will offer *customer insight* (Road Safety Analysis 2010). This is the first attempt to successfully link geodemographics and road accident data for road safety practitioners. It shows the promise of such tools to help analyze and communicate information to the public with regard to reducing road traffic injury. Market Analysis and Segmentation Tools (MAST) is web based and is a pioneering method of using corporate public sector methods for public service. Road Safety Analysis Limited (RSA) is a not-for-profit company, set up specifically to run MAST Online. The Department for Transport and the Thames Valley Safer Roads Partnership, as original sponsors of the MAST Project, agreed to transfer MAST Online and other project assets to RSA as from April 2010. The company directors are the original project team, who will ensure that opportunities created by the MAST Project will continue to benefit road safety in the future. Each director will work for the company on a part-time basis. Any surplus will be reinvested in the development of MAST Online or other similar products that serve the interests of road safety.

One of the key parts of the strategy MAST employs is the idea of *customer insight*. Road safety is facing a crucial time of transition. The traditional focus on new engineering schemes to improve the road network is yielding fewer and fewer viable casualty reduction solutions. In future, the real gains will be made by changing road user behavior. In other policy areas such as public health, huge progress has been made through investment in social marketing—but road safety has lagged some way behind in developing this approach. The UK government's recent consultation on road safety strategy sets out the clear intention to make much better customer insight data available to road safety practitioners. MAST fulfils that pledge by giving local, regional, and national programs the same kind of high-quality data that the commercial sector relies upon to communicate effectively with its target audience. A lot of the reasons behind the idea of *customer insight* are due to the increasing financial and budgetary constraints placed on local councils and local government. They constantly need to ensure that resources are used effectively and efficiently. Knowing how road users can stay safe is just one part of their aim; professionals need to translate the knowledge of collision and behavior information to their *customers*. The interesting and key point of using the geodemographics (which we have discussed earlier in this book) is determining the type of media that are used the most by those in different groups. The idea is to find what the most appropriate and effective method of engagement is, across society and for those involved in certain types of road collisions. Interestingly, MAST took the strategy of determining that the role of the TV was essentially a national approach, with the majority of the United Kingdom watching TV, and from there, you make general education inferences about drink-driving, driving while texting, and so on. The next media format for promoting education about road safety was the Internet and online methods, and then finally, the role of the *real-world* community (including sports clubs and local organizations). Examples of community-led educational initiatives were based around local football clubs, where they had road safety competitions, football-focused road safety themes, local TV commercials and posters, and launched events.

MAST undertook in 2008 the "Safer Motorcycle Rider Campaign." In the Thames Valley of the United Kingdom, there are over 1000 motorcycle collisions every year, and the existing campaign in the regional area was creatively tired, and there was no direction or targeting. There was also a significant lack of understanding with regard to who the *audience* was, and a new approach was needed. The educational campaign was overhauled based on casualty analysis and Mosaic insight, with created "at-risk types" that we explained in Chapter 4. They created two personas, the first being "Dangerous Dads" and the second being "Young Worker Wreckage." In the first year of this campaign, MAST set up radio advertising for the segmented personas, outdoor advertising to correlate to collision routes, washroom advertising (aligned with other traits in the target audience), and a website that was updated based on the understanding of the target audience. So it focused on hitting the target audience, most of which were men (80%), and most of them aged between 34 and 45 years. The outcome of this was that motorcycle collisions reduced by 26% in the area where the media advertising and education had taken place.

15.4 LOST GENERATION

There is a wide gap in road safety education and awareness that has been neglected. The term "lost generation" used here refers to the majority of the population who are at risk from road collisions (either as a pedestrian or driver or cyclist), and these people are 18–40 and are the high-risk group. However, due to road safety education being fundamentally focused on children and adolescents, it is generally thought that this group in society does not need education because they have already received it. It is clear that when focused on other public health education programs (health, crime, drugs, and alcohol), the education does not stop at children, so why should road safety? Historically, support for education, training, and information campaigns in the road safety field has been largely unquestioned, even in the professional community. The public, together with some professionals, politicians, and media commentators, believes that road safety education, training, and information reduce collisions and save lives. It is of note that those who have questioned the use of education, training, and advertising as collision and injury reduction measures have been criticized as defeatist, misguided, or worse (Insurance Institute for Highway safety [IIHS] 2001). The general public, and many professionals, has been happy to expect and assume that if you tell or show people the correct/safe thing to do and/or point out what is unsafe, they will heed the message and change their behavior accordingly. An additional assumption would appear to be that even if education, training, and information do not help, then they at least do not hurt either. These assumptions have little evidence to support them and perhaps should have died out decades ago (IIHS 2001; Elliott 2011).

Driver education and training programs often improve driver knowledge and skill, but this does not always lead to a change in on-road behavior or reduced collision risk among trainees. While skill and knowledge are important, particularly for novice drivers, they have little influence on the driving environment or conditions under which driving behavior occurs post training. Conventional driver training is also likely to neither undo firmly established past learning laid down over weeks, months, and years of practice and experience, nor durably alter motivation or modify underlying personal values. On-road driving experience is the medium via which higher-order cognitive skills related to driving (e.g., hazard perception) are developed and maintained.

15.4.1 EDUCATION

The spatial analysis methods surrounding road safety education are complex. In short, there has not been much progress into the spatial patterns of road safety education in society. Understandably, road safety professionals and researchers are predominantly focused on determining the site of road collisions and profiling the spatiality of these areas rather than focus on a passive notion of road safety education. At this point though, one might ask, what is the spatial analysis of road safety education? There are many facets to this answer. The key point to think about is that road safety education will vary spatially. People will be exposed to different road safety education depending on their age, location, employment, background, and so on. How does a person's road safety education affect the likelihood of being involved in a road collision? These are the sorts of questions that we want to ask. One of the other spatial issues concerning road safety education is determining the types of road collisions occurring to certain people and profiling their education specifically for them. An example might be a high proportion of road collisions occurring due to drug taking in a certain area or school or community. In this situation, it would be appropriate to target this area for drug taking and road user education specifically. The following are some questions we might want to ask:

- Should we prioritize specific road safety education to specific areas/groups of society?
- How can we analyze the role of road safety education with regard to road collisions?
- What different types of road safety education are there?
- How do people respond to road safety education?
- How do children respond to road safety education? What proportion goes on to be involved in road collisions based on the level of education (road safety) they have had?
- What are you trying to modify with education, attitude, and behavior?
- Is it better to provide *shock* treatment (like with drug education, having former drug addicts give presentations)? What about road collision victims?
- Whose responsibility is it to provide this education? Parents, schools, local councils?
- How do you measure the effectiveness of road safety education?

Road safety education generally has received a lot of attention, both politically and at a research level. There are many reports and studies that have outlined the best ways to educate children in the safe use of the road environment. However, there is an issue that road safety education should not stop with children. Of course, children are the most important people to educate in society with regard to the safe use of roads; however, the road environment is constantly changing, and there are many groups in society who would benefit from continued education. As with lifestyle issues such as obesity and smoking, education continues throughout society. It does not stop with children.

15.4.2 Strategic Targeting

It is difficult to approach the subject of strategic targeting, when most of the literature and research focus on evaluation and historical links between education/advertising and injury rates. When combining the research and tactical reports by government, it is clear that there are some major obstacles to overcome in terms of successful road safety education. Outlined as follows are some of the more strategic barriers to road safety education:

- *Community engagement*: This has been seen as one of the major issues by road safety professionals.
- *Evaluation*: Evaluation of the impact of road safety education and campaigning is difficult. Largely because it is difficult to differentiate from other factors that might have played a role in reducing (or increasing) road injuries.
- *Analysis*: It is difficult to get information about the effect that education plays in preventing road collisions to a fine degree.
- *Delivery*: We are facing an increase in elderly and a potential greater divide in terms of socio-economics among just a few. With this must come more innovation in how we approach the delivery of education. The use of social media, elderly support, different methods push the boundaries of advertising.
- *Shared responsibility*: This includes not only professionals in road safety, government, health, transport, and education but also society as a whole.
- *Scale*: Scale is often neglected in road safety education and advertising. There is a tendency to focus on the national approach; however, arguably, a more community or smaller-scale approach might benefit in certain circumstances.

15.5 ISSUES OF ETHNICITY

Inequalities between ethnic groups exist, but there is little evidence to suggest that targeted interventions can reduce the inequalities (Kendrick et al. 2007). To target specific ethnic groups can also be a politically sensitive issue. A large proportion of the literature and research is dedicated to young ethnic minorities and education. However, there is a large proportion of the risk population who are adults and the elderly. This pattern corresponds not only to the United Kingdom but also to other developed cities and countries. First-generation migrants into any country face major differences in the road environment and cultural differences in the road network. Prevention of being involved in a road collision for this group of society rests largely on education. One of the main challenges, with regard to the overall education of ethnic minorities and road safety, is the identification exactly which communities are at high risk and the levels of understanding of how and why there is an overrepresentation of some groups in road injuries. There are so many different cultures and countries, so to identify exact people and groups of people can be a daunting task. A large issue for road safety professionals and researchers is the lack of evidence to support the relationship between ethnic minorities and injury risk. It is only just coming to light the links between certain health issues and ethnic minorities, and road safety has traditionally fallen behind in this social targeting.

In a recent study in London on ethnic minorities and road collisions, there was an active agreement of road safety professionals to move away from translating educational materials into other languages, because it was not seen as cost-effective with so many local languages and often unnecessary, because the main beneficiaries of educational materials (children) had good English language skills. Although specific materials might be translated for newly arrived communities or to publicize consultation events in general, translation of promotional materials was not seen as productive method of addressing the diverse needs of local ethnic minority communities. In London, it has been seen as important to have detailed local knowledge to *tailor* for interventions; good and sustainable links with the local communities were also considered important. For long-settled communities, these linkages were not a problem. The challenge was working with more newly arrived communities, or those without the organizational resources and knowledge needed to liaise with statutory authorities. Road safety for young people in general is not a huge priority. The London study outlined that it was unlikely that knowledge was a key factor in explaining the differences in risk across ethnic groups, given that all young people in the study knew safe places to cross the road and cycle helmets protect you, and there was no suggestion that this knowledge was differentially distributed across London's ethnic groups. However, there was little direct relationship between knowledge and behavior, and how this knowledge might be put into practice might be different. This leads us to "socializing and safety" and how young people optimize their travel options and convenience for, when traveling around London for socializing.

15.6 CONCLUSION

This chapter has attempted to unpick the role of road safety education and the methods of education such as advertising, campaigning, and targeting different groups in society. McKenna (2009) says that it can also be concluded that education plays an indirect role, as it appears to have been helpful in legitimizing changes in drink-driving legislation, but to hold true this finding would need to be demonstrated consistently. If education does in fact play a direct role in changing behaviors and reducing risk, the profession needs to find ways of showing this while explaining existing levels of scheme ineffectiveness. It is clear that education has an important role in road safety, but there has been little to understand what this role is and to have a clearer idea of the objectives road safety education has. Society is changing quickly, and road safety education needs to change with it in terms of advancement of analysis and evaluation to the delivery. Road safety education programs frequently compete against wider social norms, which are difficult to influence. Where smoking is concerned, a change in attitude was achieved by educational intervention alongside a ban on smoking advertisements and smoking in public settings, illustrating the importance of looking at education alongside other policy levers.

REFERENCES

Begg, D. and S. Stephenson. 2003. Graduated driver licensing: The New Zealand experience. *Journal of Safety Research* 34 (1): 99–105.

Blomberg, R. D., A. M. Cleven, F. D. Thomas III, and R. C. Peck. 2008. Evaluation of the safety benefits of Legacy Safe Routes to School Programs. Report no. DOT-HS-811-013. Dunlap and Associates, Incorporated, and National Highway Traffic Safety Administration, Washington, DC.

Brown, I. D., J. A. Groeger, and B. Biehl. 1987. Is driver training contributing enough towards road safety? In *Road Users and Traffic Safety*, eds. J. A. Rothengatter and R. A. de Bruin. Assen, the Netherlands: Van Gorcum.

Cameron, M. H., A. Cavallo, and A. Gilbert. 1992. Crash-based evaluation of the speed camera program in Victoria 1990–1991. Phase 1: General effects. Phase 2: Effects of program mechanisms. Report No. 42. Accident Research Centre, Monash University, Melbourne, Victoria, Australia.

Chambers, J. 2007. Time to revisit New Zealand cycle safety. *Safekids News* 39: 1, December 2007. Safekids New Zealand.

Christie, N., R. Kimberlee, E. Towner, S. Rodgers, H. Ward, J. Sleney, and R. Lyons. 2011. Children aged 9–14 living in disadvantaged areas in England: Opportunities and barriers for cycling. *Journal of Transport Geography* 19 (4): 943–949.

Christie, N., E. Towner, S. Cairns, and H. Ward. 2004. Children's road traffic safety: An international survey of policy and practice. Road Safety Research Report No. 47. Department for Transport, London, U.K.

Christie, R. 2001. The effectiveness of driver training as a road safety measure: A review of the literature. No. 01/03. Royal Automobile Club of Victoria (RACV) Ltd., Nobel Park, Victoria, Australia.

Cloke, J., D. Webster, P. Boulter, G. Harris, R. Stait, P. Abbott, and L. Chinn. 1999. Traffic calming: Environmental assessment of the Leigh Park Area Safety Scheme in Havant. TRL Report 397. Transport Research Laboratory, Crowthorne, U.K.

DfT. 2008. Widening the reach of road safety—Emerging practice in road safety in disadvantaged communities: Practitioners' guide. Department for Transport, London, U.K.

Elliott, B. 1989. Effective road safety campaigns: A practical handbook. CR80. Federal Office of Road Safety, Canberra, Australian Capital Territory, Australia.

Elliott, B. 2011. Beyond reviews of road safety mass media campaigns: Looking elsewhere for new insights. *Journal of the Australasian College of Road Safety* 22 (4): 11–17.

Elvik, R., A. Høye, T. Vaa, and M. Sørensen. 2009. *The Handbook of Road Safety Measures*, 2nd edn. Bingley, U.K.: Emerald Group Publishing.

Evgenikos, P., G. Yannis, T. Leitner, S. Hoeglinger, J. Broughton, and B. Lawton. 2009. Road safety and the elderly in Europe. Paper presented at the *Fourth IRTAD Conference: Road Safety Data; Collection and Analysis for Target Setting and Monitoring Performances and Progress*, Seoul, South Korea, September 16–17, 2009.

Grayling, T., K. Hallam, D. Graham, R. Anderson, and S. Glaister. 2002. *Streets Ahead: Safe and Liveable Streets for Children*. London, U.K.: Institute for Public Policy Research.

Gregersen, N. P., H.-Y. Berg, I. Engströma, S. Noléna, A. Nyberg, and P.-A. Rimmö. 2000. Sixteen years age limit for learner drivers in Sweden—An evaluation of safety effects. *Accident Analysis & Prevention* 32 (1): 25–35.

Grundy, C., R. Steinbach, P. Edwards, P. Wilkinson, and J. Green. 2008. 20 mph zones and road safety in London: A report to the London Road Safety Unit. LSHTM, London, U.K.

Gutierrez, N., M. Orenstein, J. Cooper, T. Rice, and D. R. Ragland. 2008. Pedestrian and bicyclist safety effects of the California Safe Routes to School program. Paper presented at the *Transportation Research Board 87th Annual Meeting*, Washington, DC.

Hauer, E. 2007. A case for evidence-based road safety delivery. In *Improving Traffic Safety Culture in the United States: The Journey Forward*, pp. 329–343. Washington, DC: American Automobile Association Foundation for Traffic Safety.

Hillman, M., J. Adams, and J. Whitelegg. 1990. *One False Move…: A Study of Children's Independent Mobility*. London, U.K.: Policy Studies Institute.

Hoekstra, T. and F. Wegman. 2011. Improving the effectiveness of road safety campaigns: Current and new practices. *IATSS Research* 34 (2): 80–86.

Insurance Institute for Highway Safety. 2001. Education alone won't make drivers safer. *Status Report* 36 (5): 1–8.

Jones, S. J., R. A. Lyons, A. John, and S. R. Palmer. 2005. Traffic calming policy can reduce inequalities in child pedestrian injuries: Database study. *Injury Prevention* 11 (3): 152–156.

Karaca-Mandic, P. and G. Ridgeway. 2010. Behavioral impact of graduated driver licensing on teenage driving risk and exposure. *Journal of Health Economics* 29 (1): 48–61.

Kendrick, D., J. Barlow, A. Hampshire, L. Polnay, and S. Stewart-Brown. 2007. Parenting interventions for the prevention of unintentional injuries in childhood. *Cochrane Database of Systematic Reviews* (4): Article no. CD006020.

Ker, K., I. G. Roberts, T. Collier, F. R. Beyer, F. Bunn, and C. Frost. 2003. Post-licence driver education for the prevention of road traffic crashes. *Cochrane Database of Systematic Reviews* (3): Article no. CD003734.

Kingham, S., J. Pearce, D. Dorling, and M. Faulk. 2008. The impact of the graduated driver licence scheme on road traffic accident youth mortality in New Zealand. *Journal of Transport Geography* 16 (2): 134–141.

Langley, J. D., A. C. Wagenaar, and D. J. Begg. 1996. An evaluation of the New Zealand graduated driver licensing system. *Accident Analysis & Prevention* 28 (2): 139–146.

Lannon, J., ed. 2008. *How Public Service Advertising Works*. Oxfordshire, U.K.: World Advertising Research Center.

Layfield, R., D. Webster, and S. Buttress. 2005. Pilot home zone schemes: Evaluation of Magor Village, Monmouthshire. Report No. 633. Transport Research Laboratory, Crowthorne, U.K.

Lewis, I. M., B. Watson, K. M. White, and R. Tay. 2007. Promoting public health messages: Should we move beyond fear-evoking appeals in road safety? *Qualitative Health Research* 17 (1): 61–74.

Loo, B. P. Y. and K. L. Tsui. 2009. Pedestrian injuries in an ageing society: Insights from hospital trauma registry. *Journal of Trauma-Injury Infection & Critical Care* 66: 1196–1201.

Mackie, A. M., H. A. Ward, and R. T. Walker. 1990. Urban safety project. 3: Overall evaluation of area wide schemes. Report No. RR263. Transport and Road Research Laboratory, Crowthorne, U.K.

Mayhew, D. R. and H. M. Simpson. 2002. The safety value of driver education and training. *Injury Prevention* 8 (Suppl. 2): ii3–ii8.

Mayhew, D. R., H. M. Simpson, and A. Pak. 2003. Changes in collision rates among novice drivers during the first months of driving. *Accident Analysis & Prevention* 35 (5): 683–691.

Mayhew, D. R., H. M. Simpson, A. F. Williams, and S. A. Ferguson. 1998. Effectiveness and role of driver education and training in a graduated licensing system. *Journal of Public Health Policy* 19 (1): 51–67.

McKenna, F. P. 2009. Do attitudes and intentions change across a speed awareness workshop? *Behavioural Research in Road Safety* 2007 (17): 265. Department for Transport, London, U.K.

Morgan, R. and D. King. 1995. The older driver—A review. *Postgraduate Medical Journal* 71 (839): 525–528.

Morgan, R. and J. Poorta. 2008. Measuring success in government advertising. In *How Public Service Advertising Works*, ed. J. Lannon, pp. 163–198. Oxfordshire, U.K.: World Advertising Research Center.

OECD. 2011. Help wanted? Providing and paying for long-term care. Paris, France: Organisation for Economic Co-operation and Development.

Page, Y. 1995. Jeunes conducteurs, apprentissage anticipé de la conduit et accidents de la route. *Les Cahiers de l'Observatoire* 2: 15–55.

Phillips, R. O., P. Ulleberg, and T. Vaa. 2011. Meta-analysis of the effect of road safety campaigns on accidents. *Accident Analysis & Prevention* 43 (3): 1204–1218.

Road Safety Analysis. 2010. http://www.roadsafetyanalysis.org/ (accessed March 31, 2010).

Singh, A. 1982. Pedestrian education. In *Pedestrian Accidents*, eds. A. J. Chapman, F. M. Wade, and H. C. Foot. Chichester, U.K.: John Wiley & Sons.

Smith, C. A. and H. S. Shannon. 2003. How much science is there in injury prevention and control? *Injury Prevention* 9 (1): 89–90.

Stradling, S., N. Kinnear, and H. Mann. 2006. Evaluation of Brake Young Driver Education scheme. Transport Research Institute, Napier University, Edinburgh, U.K.

Strecher, V. J., J. A. Bauermeister, J. Shope, C. Chang, M. Newport-Berra, A. Giroux, and E. Guay. 2006. Interventions to promote safe driving behaviour: Lessons learned from other health-related behaviours. In *Behavioural Research in Road Safety 2006: Sixteenth Seminar*, pp. 28–38. London, U.K.: Department for Transport.

Tay, R. 1999. Effectiveness of the anti-drink driving advertising campaign in New Zealand. *Road and Transport Research* 8 (4): 3–15.

Tay, R. and B. Watson. 2002. Changing drivers' intentions and behaviours using fear-based driver fatigue advertisements. *Health Marketing Quarterly* 19 (4): 55–68.

Thomson, J. A. 1991. *The Facts About Child Pedestrian Accidents*. London, U.K.: Cassell.

Thomson, J. A. 2006. Issues in safety education interventions. *Injury Prevention* 12 (3): 138–139.

Thomson, J. A., A. Tolmie, H. C. Foot, and B. McLaren. 1996. Child development and the aims of road safety education—A review and analysis. Road Safety Research Report No. 1. Department of Transport, London, U.K.

Tilly, A., D. Webster, and S. Buttress. 2005. Pilot home zone schemes: Evaluation of Northmoor, Manchester. Report No. 625. Transport Research Laboratory, Crowthorne, U.K.

Vernick, J. S., G. Li, S. Ogaitis, E. J. MacKenzie, S. P. Baker, and A. C. Gielen. 1999. Effects of high-school driver education on motor vehicle crashes, violations, and licensure. *American Journal of Preventive Medicine* 16 (1): 40–46.

Wakefield, M. A., B. Loken, and R. C. Hornik. 2010. Use of mass media campaigns to change health behaviour. *The Lancet* 376 (9748): 1261–1271.

Webster, D. C. and R. E. Layfield. 2003. Review of 20 mph zones in London boroughs. Report No. PPR243. Transport Research Laboratory, Crowthorne, U.K.

Webster, D. C. and A. M. Mackie. 1996. Review of traffic calming schemes in 20 mph zones. TRL Report 215. Transport Research Laboratory, Crowthorne, U.K.

Webster, D., A. Tilly, and S. Buttress. 2005. Pilot home zone schemes: Evaluation of Cavell Way, Sittingbourne. Report No. 626. Transport Research Laboratory, Crowthorne, U.K.

Wheeler, A. H. and M. C. Taylor. 2000. Changes in accident frequency following the introduction of traffic calming in villages. Report No. TRL452. Transport Research Laboratory, Crowthorne, U.K.

Williams, A. F. 1998. Risky driving behaviour among adolescents. In *New Perspectives on Adolescent Risk Behaviour*, ed. R. Jessors. Cambridge, U.K.: Cambridge University Press.

World Bank. 2008. China: Road traffic safety, the achievements, the challenges, and the way ahead. Washington, DC: World Bank.

16 Road Safety Strategy

16.1 INTRODUCTION

Despite all measures to improve road safety performances, many road safety administrations will face problems of stagnated improvements after successfully addressing some of the enforcement, engineering, education, and emergency and response aspects. When individual ad hoc and focused measures have been introduced to tackle the most obvious road safety problems (e.g., the introduction of seat belt, helmet wearing, and drink-driving legislations), a holistic framework is needed to make continuous improvement of road safety sustainable in the long term. Such a holistic framework is embodied in a road safety strategy of the road safety administration. Specific road safety strategies will vary depending on the actual geographic context (including territorial size, terrain, climate, level of urbanization, level of development, and culture) of individual road safety administrations. Nonetheless, a road safety strategy will have nine major components. They are (1) vision, (2) objectives, (3) targets, (4) action plan, (5) evaluation and monitoring, (6) research and development, (7) quantitative modeling, (8) institutional framework, and (9) funding (Loo et al. 2005). In this chapter, we shall examine and discuss the relevancy of each of these components in a road safety strategy. Examples from different parts of the world will be provided to substantiate the arguments. Particular attention will be paid to highlight the spatial dimension or geographical variability, such as the rural–urban divide, in the formulation of road safety strategies.

16.2 TRADITIONAL APPROACHES

In injury prevention, the Haddon matrix (1970) is one of the most well known among both academics and policy makers. The matrix serves as a tool to analyze public health problems by factors (columns) that identify the interacting factors contributing to the injury process and phases (rows) following the pre-event, event, and post-event time frames. Factors are, in turn, listed by the host, agent/vehicle, and the environment (may be further divided into physical and social environments). With the Haddon matrix, the causes and effects of traffic collisions can be more easily identified, listed, and analyzed. Moreover, public health interventions relevant to each of the cells can be devised. Sometimes, they are called pre-event strategy, event strategy, and post-event strategy. Runyan (1998) has introduced a third dimension of decision criteria to make the Haddon matrix more useful to public health policy makers. Nonetheless, the matrix remains a broad conceptual framework, which is not directly amenable to an integrated approach to mobilize people in the society to improve road safety.

Another major contribution was made by Frank Haight, who served as the founding editor of *Accident Analysis & Prevention* for over 35 years from 1969 to 2004.

In his seminal work (Haight 1985), he highlighted eight important changes in the ways that our understanding of road safety has improved. They are as follows:

1. Recognizing that there can be no *cure*.
2. Abandoning the language of cause and blame.
3. Understanding that we need to go beyond collisions to address their consequences.
4. Admitting the importance of exposure.
5. Taking into account statistical pitfalls and trends.
6. Rejecting arguments based on what stands to reason.
7. Attempting project evaluation.
8. Admitting that cost, not carnage, is the issue.

Based on the earlier understanding, a program with six components was proposed. They are the following:

1. The reorganization of road safety under an independent public health agency.
2. Planning for the long term.
3. The disengagement of road safety from public concern and public relations.
4. Commitment to full truthful disclosure to the public and political leaders.
5. The reorganization of professional education and public information.
6. The formulation of a coherent modern research program.

Though these comments form a good base and provide guidance in developing an overall strategy for road safety, the need to make drastic changes (e.g., in reorganizing government institutions and establishing an independent public health agency for road safety) has made the program not fully implementable in many administrations.

Throughout the 1990s, efforts to search for better and more integrated road safety strategies have been made but mostly focus on the specific contexts of individual countries/administrations. Among them, a new approach that focused on *intrinsic* road safety has been proposed by the Dutch administration (van Uden and Heijkamp 1995). This approach emphasized the need to address the root causes of road collisions and, hence, collision prevention at the facility planning and development stage. Greater attention is given for an integral approach to traffic safety, through influencing all kinds of decision makers outside the realm of road safety. Nonetheless, the discussion is rather fragmented, and there is a lack of agreement on the essential elements of an overall road safety strategy. A recent report of the Global Road Safety Partnership (Aeron-Thomas et al. 2002) has focused on four major aspects (organization, plans, funding, and private sector participation) of road safety management in nine case studies, including both developed and developing countries. For historical and circumstantial reasons, different countries adopt different approaches and have different emphases. Over the years, good practices and experiences have been accumulated to benefit countries or regions that are contemplating their own road safety strategies. Key factors for the success or failure of road safety initiatives have been

Road Safety Strategy

identified (Wegman et al. 1991; Halden and Harland 1997). Nonetheless, there is a lack of a comprehensive framework for the formulation, implementation, and evaluation of road safety strategies.

16.3 NINE COMPONENTS OF THE ROAD SAFETY STRATEGY

In order to develop a systematic analytic framework for the development of road safety strategies and to review the practices in selected administrations, a nine-component comparative framework for developing, comparing, and evaluating road safety strategies was proposed by Loo et al. (2005). The nine components are vision, objectives, targets, action plan, evaluation and monitoring, research and development, quantitative modeling, institutional framework, and funding. While the first four components are essential for the formulation of a road safety strategy, the remaining components are critical factors affecting its successful implementation.

16.3.1 Vision

A road safety vision describes an innovative future traffic system or a desired direction of safety development. A good vision should be understandable, desirable, feasible, guiding, motivating, and flexible (OECD 2002). With a vision, the road safety strategy can be seen as a collection of plans that aim to fulfill the vision. In other words, the design of all the other components of the road safety strategy is governed/led by this long-term vision rather than the practical considerations of "what can be done?". Having a vision ensures that road safety gains a prominent position in the government agenda, raises public interest, and creates public support for road safety improvements. With the vision as a long-term goal, short-term objectives, targets, and action plans can be set accordingly.

16.3.2 Objectives

What are the results (qualitative and quantitative) that the vision aims for? These should be clearly stated as objectives. These objectives, in turn, guide policies and mobilize relevant stakeholders to take action and work together. In comparison with road safety programs without objectives, those with clear objectives have the advantages of being more realistic and having a wider scope, which resulted in a more concerted institutional efforts and more focused resource allocation (OECD 2002).

16.3.3 Targets

Targets are quantitative objectives of the road safety strategy. Road safety targets can be set directly on key indicators, such as the number of road traffic fatalities, number of road traffic injuries, traffic collision frequency, or other collision rates either on absolute (numbers) or relative (percentage) terms. Alternatively, targets may be set for secondary safety indicators, such as the seat belt usage rate, helmet-wearing rate, red-light violation, or drink-driving offences. The OECD Scientific Expert

Group (1994) reviewed the practices, purposes, and effects of setting quantitative targets in road safety work and showed how the setting of quantitative road safety targets can lead to the formulation of more realistic traffic safety programs and the better use of public funds and other resources. Moreover, Allsop et al. (2011), Elvik (1993, 2001), and Wong et al. (2006) found that there is a positive statistical association between the setting of quantitative road safety targets and the percentage reduction of road accident fatalities in the administrations concerned.

16.3.4 Action Plan

In order to achieve the objectives, an action plan that lists the collection of measures or actions to be taken is necessary. Action plan is usually formulated by the respective road safety administrations after engaging with the key stakeholders to ensure that actions are applicable and cost-effective. These actions should tackle the road safety problems of the administration effectively. Rumar (2002) identified three levels of road safety problems that are common to most countries. First-order problems are the more widely recognized problems, such as speeding and drink-driving. Second-order problems are less obvious, but they have negative implications on road safety. They include poor road designs and aspects of traffic management that may give rise to road hazards. Third-order problems are the most difficult to identify and address. They include a general lack of knowledge and/or interest about road safety among the general public and government officials, and the lack of coordination in road safety efforts. It is important that the action plan will consider road safety problems at all three levels.

16.3.5 Evaluation and Monitoring

No road safety strategy is complete without evaluation and monitoring. Evaluation and monitoring may best be conducted independently of the major agency(ies) responsible for designing and implementing the action plan. Moreover, it should not just be conducted at the end of action plan or target date(s) that road safety targets have been set. It should be a periodic and continuous process to identify ineffective measures in the action plan and to replace them with other new measures. In addition, it may include a mechanism of ensuring that existing and future road or traffic projects are satisfactory in terms of safety performance. Road safety audit is one of the means of achieving the earlier.

16.3.6 Research and Development

Research and development is taken widely to include not only research and development involving technology and hardware designs (such as on collision worthiness of vehicles and pavement materials) but also the wider aspects of humanities and social sciences (such as policy analysis and behavior modeling) and medicine (such as post-trauma recovery). Research and development should inform every step in the formulation, implementation, and evaluation of the road safety strategy, so that the pros and cons of different options are known, and wise decisions can be made.

16.3.7 Quantitative Modeling

Quantitative modeling is helpful in giving an objective and scientific picture of the current scale and trend of the road safety problems, as well as the relationships among various factors that have bearings on road safety. One example is collision cost estimation and projection. Traffic collisions inflict substantial cost to the society in terms of both direct loss (including property damage and medical expenses) and indirect loss (including lost productivity or quality of life). Having a better idea of the price of the traffic injury burden helps to justify the efforts devoted to improving road safety.

16.3.8 Institutional Framework

At the national and subnational levels, governments are usually the proponents of road safety strategies. Within governments, political commitment and interdepartmental coordination are essential. However, the successful implementation of a road safety strategy will depend on the efforts of everyone in the society beyond the public sector. Hence, an institutional framework that provides various platforms for all stakeholders (notably, the public sector, private sector, and the general public) to communicate and to agree on the road safety strategy is important. Through these platforms, key questions like "What road safety targets should be set?" and "What new legislations are to be made?" can be debated. Getting major stakeholders' inputs and supports is often crucial for the success of many road safety initiatives. For instance, an initiative to build safer vehicles cannot possibly be achieved without the active participation of automobile manufacturers. In addition, the institutional framework is important in keeping the momentum of the road safety strategy beyond the initial launch, but to keep it alive within the community thereafter.

16.3.9 Funding

Resources (whether private or public) in a society are scarce. Every step in the formulation, implementation, and evaluation of the road safety strategy requires resources. In particular, sufficient government funding should be dedicated to the road safety strategy to ensure its success. The nature of the road safety strategy means that funding cannot be ad hoc and once-off but regular and persistent. Once again, having a realistic estimation of collision cost in the society would justify the amount of resources required to deliver the road safety strategy for the betterment of the society. In relation, cost–benefit analysis of specific road safety measures/initiatives may be necessary to ascertain their cost-effectiveness and to justify the allocation of fund.

16.4 IMPORTANCE OF BENCHMARKING AND INCORPORATING GEOGRAPHICAL VARIABILITY

While road safety strategies at the national and subnational levels are often developed based on the specific local contexts, it is important for road safety administrations to understand how they are doing when compared to others, especially in the hope of doing better. Hence, the dual importance of benchmarking and incorporating geographical variability needs to be underlined. In the benchmarking of road safety strategies, the earlier comparative framework is again very useful. For each of the nine components, the evaluation yardsticks are (1) the levels of details: whether the contents, together with any underlying philosophy/rationale, of each road safety strategy component are clearly defined, well explained, and made accessible to the general public; (2) the scope: whether the road safety practices under each component are comprehensive and well balanced in addressing related issues and problems; and (3) the degree of sophistication: whether the planning, formulation, and implementation of that road safety component reflect careful thinking, thorough design, and in-depth analysis. The evaluation of case studies can form a benchmarking platform for the planning, formulation, and implementation of good practices for road safety strategies.

16.4.1 International Best Practices

Loo et al. (2005) compared the road safety strategies of six selected administrations: Australia, California, Great Britain (GB), Japan, New Zealand, and Sweden. In these case studies, a score ranging from 1 to 4 is given to each component of a road safety strategy of an administration. As the six case studies are either at the intermediate or

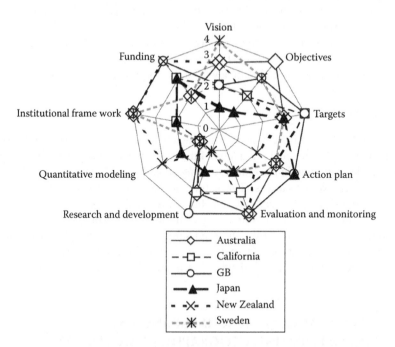

FIGURE 16.1 Evaluation of the road safety strategies for six administrations. (Reprinted from Loo, B.P.Y. et al., *Transp. Rev.*, 25(5), 633, 2005. With permission from Taylor & Francis Group Ltd., http://www.tandfonline.com/.)

advanced stages of road safety development (UN-ESCAP 2002), a score of 1 is interpreted as "average," 2 as "fair," 3 as "good," and 4 as "excellent." As good practices in a component of a road safety strategy may be adopted in more than one area, the score of 4 is not limited to one administration only.

Figure 16.1 shows the results of the analysis in a radar chart. None of the six administrations excels in all nine road safety strategy components. Each of the six administrations provides valuable lessons for other administrations. Based on the analytical framework of the nine components of a road safety strategy, the administrations with the best practices can also be listed under the respective road safety strategy components of (1) vision—Sweden; (2) objectives—Australia; (3) targets—California and GB; (4) action plan—GB and Japan; (5) evaluation and monitoring—Australia, GB, and New Zealand; (6) research and development—GB; (7) quantitative modeling—New Zealand; (8) institutional framework—Australia, GB, New Zealand, and Sweden; and (9) funding—GB and New Zealand. The good practices that have been identified in these case studies should be useful to other administrations in the planning, formulation, and implementation of their road safety strategies.

Based on the earlier analysis, Australia is good at setting objectives, evaluation, and monitoring, and developing an institutional framework (Figure 16.1). First, Australia's road safety strategy has comprehensive and clear strategic objectives. The eight major areas covered are road user behavior, the safety of roads, vehicles, human error, equity among road users, medical and retrieval services, alternatives to motor vehicle use, and research about policy and programs. Second, the objectives and targets of road safety strategy are assessed. Each action plan is reviewed at the end of its 2-year period, and a further action plan is then developed. Third, there is a well-established institutional framework for coordinating the road safety initiatives of the Commonwealth, state, territory, and local governments, as well as other organizations that may influence road safety outcomes (NTRSC 2001; ATC 2002).

California, USA, is good at setting targets (Figure 16.1). Apart from the overall road accident fatality goal, there are four other levels of performance targets. They are program goals (in improving the police traffic services, reducing alcohol and other drugs usage, improving child restraints, speed control, and pedestrian and bicycle safety), administrative goals (such as for emergency medical services, traffic engineering and operations, and traffic records), legislative goals, and public affairs goals (OTS 1998; CHP 2002).

GB is good at setting targets, action plans, evaluation and monitoring, research and development, developing an institutional framework, and funding (Figure 16.1). First of all, GB is pragmatic in setting targets. Its overall objective can be easily understood from the title of its road safety strategy paper "Tomorrow's roads: Safer for everyone." Second, its action plan carries 10 main themes: safety for children, drivers, infrastructure, speeds, vehicles, motorcycling, pedestrians, cyclists and horse-riders, better enforcement, and promoting safer road use. Third, the evaluation was conducted in a timely manner, and the review is made accessible to the general public. The Department for Transport (DfT) published the first 3-year review in 2004, which reports and evaluates the progress toward the targets and the effectiveness of measures undertaken under the major 10 themes

(DfT 2004). Fourth, the action plan is well supported by research and development programs. Research has been commissioned in the three major areas of analyzing and understanding accident causation, developing and evaluating road safety measures, and monitoring the effects of road safety policy. Fifth, there is a clear line of responsibility running from local to central road safety bodies. While DfT is responsible for developing and coordinating the implementation of the national strategy, partnership among the central government and its agencies, local authorities, police forces, voluntary groups and road user associations, motor manufacturers, and individual road users is stressed. Finally, funding comes from all sectors, including public and private sources. In comparison to the other five administrations, GB has the most well-balanced road safety strategy (DfT 2001, 2002a,b; DTLR 2002).

Japan is good at the road safety component of formulating and implementing action plans (Figure 16.1). Since the Traffic Safety Policies Law was enacted in 1971, the administration has implemented a series of 5-year road safety plans. Its Seventh Fundamental Traffic Safety Program covers the period from 2001 to 2005. The major measures that are outlined include traffic safety for the elderly, the use of seat belts and child seats, road facilities, traffic safety education, vehicle safety, traffic guidance and control, rescue and first-aid systems, measures for accident victims, traffic accident investigation analysis, and traffic safety activities with public participation (DGPPC 2001).

New Zealand's road safety strategy is good at conducting evaluation, quantitative modeling, institutional framework, and funding (Figure 16.1). In terms of evaluation and monitoring, progress toward achieving targets is tracked through a series of quarterly reports, which record the current performance of interventions and suggest remedial actions if necessary. The National Road Safety Committee also conducts an annual progress review as part of its planning season (NRSC 2000). The proposed strategy is reviewed every 3 years. Each review reexamines the entire strategy, including the assumptions on which its targets are based and the reasoning underlying the estimation of the targets. In terms of quantitative modeling, predictive road safety models are used. Computerized simulation models are built to help identify the proposed strategy's targets and predict the cost of achieving them. In terms of the institutional framework, New Zealand has many dedicated organizations, from the working to policy setting level, which are responsible for road safety works. On funding, the funding sources and apportioning of each program in the road safety strategy are clear and consistent.

In Sweden, the parliament passed the road safety vision "Vision Zero," by a large majority, in October 1997. The administration is strategically successful in formulating and implementing the vision and establishing the necessary institutional framework. Its road safety vision is to eventually have no one killed or seriously injured within the road transport system. On the institutional setting, the Swedish road safety administration stresses *shared responsibility*. The Swedish National Road Administration, as the central administrative agency, has been commissioned with the overall responsibility for road safety within the road transport system and shall monitor and actively promote developments within this area. As the road manager, it is also responsible for road safety on the state road network.

16.4.2 Rural–Urban Divide

The rural–urban divide in road safety has been recognized worldwide. In a road safety report on European countries, it was found that 50%–75% of the traffic collisions causing injuries happened in urban built-up areas (OECD 2002). Nonetheless, more than 60% of the fatalities in traffic collisions happened in the rural areas. The risk of fatality in collisions was much higher on roads in rural areas than in urban areas. Furthermore, there seems to be a distinctive risk-taking driving culture in the rural areas (Eiksund 2009).

In North America, Mueller et al. (1988) found that the rate of motor vehicle–pedestrian collisions was higher in urban areas, but the death rate in collisions was generally higher in the rural areas. The rural–urban divide was related to vehicle speed, availability of emergency care, age and sex distribution of the population, and proximity to definitive medical care. Moreover, the National Highway Traffic Safety Administration of the United States (2008) showed that road fatality rates were higher in the rural than urban areas from 1997 to 2006. In 2006, 56% of all fatal collisions in the country happened in rural areas, but only 23% of the population lived there. In addition, more rural drivers were found to have been drink-driving, speeding, and driving unrestrained than urban drivers. In Canada, Kmet and Macarthur (2006) found that the collision fatality and hospitalization rates among children and youths in Alberta were much higher in rural than urban areas. On average, rural children and youths in Alberta were five times more likely to lose their lives and three times more likely to be hospitalized in road collisions than their urban counterparts from 1997 to 2002. The reasons were related to the greater exposure to motor vehicle travel, more risky road environment, poorer accessibility to medical care, and less compliance with road safety regulations in rural areas.

Consistent with the earlier findings in Europe and North America, Li et al. (2008) found that the percentage of pre-hospital deaths in traffic collisions was higher in rural than urban Taiwan. The reason was mainly attributable to unrestrained driving and the delayed emergency medical system. In mainland China, Duan's study (2002) shows that traffic collisions in mainland China mainly occurred in rural areas, on highways with higher classifications, and on roads of mixed classifications. In addition, Loo et al. (2011) found that urban roads in China had alarmingly higher rates of collisions. They are situated in urban areas where traffic volume is high and road traffic conditions are particularly complicated. To improve road safety in urban China, emphasis should be put on urban and transport planning. Planning professionals need to assess the impact of urban development on safety problems. In addition, more frequent traffic inspection is needed. Regular speed checks, road blocks, and traffic patrols should be carried out to avoid violation of traffic regulations. Education on traffic rules and traffic rights should be promoted to raise the awareness on road safety among road users. Furthermore, as the urban traffic volume is high, the comprehensive introduction of segregated lanes for motor vehicles, bicycles, and pedestrians can serve the dual purpose of enhancing both efficiency and safety.

Nonetheless, rural collisions are more deadly in mainland China. In particular, the risk of fatality in collisions was notably higher on Expressways and Class One

Highways connecting cities and countryside. In rural places, road safety measures should target at reducing the high death rates in road collisions. These highways traversing the rural areas are relatively free from congestion, but medical services are less accessible than in urban China. Thus, facilities such as vehicle speed monitors and speed humps or bumps should be installed to prevent speeding. Moreover, the availability of medical and emergency rescue services should be improved in rural China so as to save more lives in case of road collisions. For instance, emergency rescue teams should be set up in rural regions where collision rates are high. Special transport, such as helicopters, should be made available in remote rural areas where road transport is poorly developed. Furthermore, more road safety measures to improve the visibility on roads should be implemented in rural China. For collisions happening at night, about 50% of them happened on roads without lighting, but roughly 70% of all deaths were killed in these collisions. Installing more street lights on roads should be a direct method. Nevertheless, funding and resources in rural China may be inadequate, and some of the roads in rural China are located in remote areas with no electricity. In the rural context, lighting facilities should be installed with high priority at hazardous hot zones. In a high-density urban context, dangerous road locations are likely to exhibit characteristics of hot spots, which are spatially concentrated at highly specific locations like intersections (Loo 2009). However, dangerous road locations in rural areas are more likely to present themselves as hot zones, which cover longer sections of roads with multiple risk factors like long slopes (Loo 2009). A systematic program to identify dangerous road locations using the appropriate methodology should take these spatial characteristics into consideration (Loo 2009). Also, warning signs should be placed at these locations in order to alert drivers. Furthermore, drivers should be educated to develop the habits of keeping their headlights on all the time at night and sound their sirens to alert others when they make a turn on roads without lighting at night. All in all, more measures to improve the visibility and safety should be launched on rural roads without lighting facilities. Last but not least, road safety measures in rural areas should target at promoting the safe use of tractors, trailers, and special vehicles. One should bear in mind that collisions involving tractors, trailers, and special vehicles had exceptionally high death rates. More directions and training on using these vehicles should be given. Talks and other publicity campaigns should be carried out to raise farmers' awareness on road safety. In the longer term, a licensing system could be introduced to the drivers of tractors, trailers, and special vehicles. Subsidies could also be given to farmers to maintain their tractors, trailers, and special vehicles and/or to replace old ones with safer ones. Moreover, these rural road safety improvement measures should not be restricted to inner provincial units only. Beyond the broad regional classifications, there was a negative relationship between the proportion of urban population and the death rate per 100 collisions at the provincial level. In other words, the rural road safety problems are prevalent in a large part of China. Such challenges should best be taken up by the central government with a national road safety strategy. Currently, a road safety strategy is still lacking in China (Duan 2002). In the future, a national road safety strategy can be developed.

16.5 STRATEGY IN STAGES

Based on Loo et al. (2007) and UN-ESCAP (2002), administrations with the least sophisticated (Phase I) road safety activities can be considered to be at the early stage of road safety development. Administrations with relatively more (Phase II) and the most sophisticated (Phase III) road safety activities belong to the intermediate and advanced stages, respectively. While the characteristics of administrations in different stages differ substantially, the following discussion will use administrations at the intermediate stage to illustrate how the road safety strategy can be implemented in stages and improvements be made via the short-, medium-, and long-term approaches.

16.5.1 SHORT-TERM APPROACH

Improvements in two road safety components require fewer additional resources and can be implemented relatively easily. They are vision and objectives. For administrations at the immediate stage of road safety development, the weakest component is usually the lack of a clear vision at the top government level. As such, the general public and different actors involved in road safety work do not share a common desirable vision or an overall direction. To formulate a road safety vision, not much additional resource or preparation time is required. Hence, improvement is feasible in the short term. Similarly, not much extra input of physical, human, and financial resources is required in setting strategic objectives that help to materialize the vision. These objectives serve dual purposes. On the one hand, they help to define the key road safety action areas. On the other hand, they allow relevant targets to be set accordingly.

16.5.2 MEDIUM-TERM APPROACH

In the medium term, improvements in three other road safety components can be planned. They are targets, institutional framework, and quantitative modeling. At present, road safety targets are not common in administrations at the immediate stage of road safety development. The setting of targets requires a longer planning horizon, because a thorough analysis has to be conducted to ensure that they are feasible and meaningful in contributing to the fulfillment of the road safety vision and objectives. In addition, having a clear and well-structured institutional framework can ensure higher efficiency in implementing the road safety strategy. For administrations at the late phase of the intermediate stage, there usually exist a number of road safety agencies, both public and private. Hence, the task is to better organize and coordinate their activities. This involves defining the roles and responsibilities of different agencies clearly. While not many additional resources are required (unless new agencies are to be established), the coordination will take time. Finally, administrations that have entered the intermediate stage for some time should process a comprehensive data system of road accidents. This lays solid grounds for modeling work. Quantitative modeling helps to analyze and interpret the accident statistics.

Extra resources need to be devoted, but the amount is much less when compared to the efforts of building and maintaining of the existing road accident databases. Thus, more efforts should be spent to make the best use of these databases to advise policy makers.

16.5.3 Long-Term Approach

Finally, improvements in the remaining four road safety components may require much longer time to be implemented. They are action plan, evaluation and monitoring, research and development, and funding. Before the formulation of a more cohesive and long-term action plan, a systematic review of the current effort is required. New measures, which may require additional resources, should be considered. Synergy among road safety measures should also be well thought out. At the advanced stage, the action plan is no longer a "priority" action plan, but it should be formulated as an indispensable component of the entire road safety strategy so that the plan directly works and contributes toward the fulfillment of the targets, objectives, and vision set. Hence, first, a long planning horizon is required. Second, the establishment of evaluation mechanisms also requires much preparation work. Evaluation mechanisms are typically either absent or very weak in administrations at the intermediate stage. To ensure impartiality, new divisions may be required to take charge of the evaluation work. Third, the strengthening of research and development takes time. As a start, funds need to be set up and allocated to finance research projects on understanding and improving road safety. After all, breakthroughs in the reduction of accidents can be made only with such advancements. Generally, research and development is an endeavor that possibly takes several years or even longer to materialize. Finally, it may be difficult for administrations at the intermediate stage to set aside dedicated funds, establish funding mechanisms, and diversify the sources of funds efficiently. To obtain stable funding from the government, legislations may be required. Taping decent and consistent sources of funding from the private sector also requires much discussion. There are possibilities of more active involvement of the insurance companies in the funding mechanisms. These factors make the improvement of the funding component a long-term process.

16.6 CONCLUSION

This book is about the spatial analysis of traffic collisions. First and foremost, a scientific spatial analysis of traffic collisions depends on a good-quality traffic collision database with high accuracy, precision, and reliability. Yet, the value of a spatial analysis will depend not only on advanced methodological knowledge to guide the correct specifications of the statistical models but also on sound theoretical understandings to inform the choice of variables and the interpretation of results. No doubt, these insights do not lie within any specific disciplinary boundary. The study of Loo et al. (2013) is an illustration of value and importance of multidisciplinary efforts toward sustained road safety benefits. The involvement and active participation of the community (e.g., through the setting up of safe communities of the World Health Organization), private sector (e.g., automobile manufacturers and insurance industries), and government (including the police and different administrative departments) are indispensable for the long-term improvements of road safety. In relation, the road safety strategy serves as the *magnet* for gathering expert knowledge, financial resources, and public support essential for reducing traffic collisions and their impact on society. Through a cycle of the formulation, implementation, monitoring, review, and updating of a road safety strategy, sustained efforts to improve road safety can be maintained despite changes in government leaders, economy, and other political and social conditions. Ultimately, traffic collisions lead to unnecessary human loss (many as premature deaths for people with good physical health and/or at a young age) and lifelong sufferings (notably to traffic injury victims, family members, and friends) that every society should aim to reduce.

REFERENCES

Aeron-Thomas, A., A. J. Downing, G. D. Jacobs, J. P. Fletcher, T. Selby, and D. T. Silcock. 2002. Review of road safety management practice: Final report. Global Road Safety Partnership, Geneva, Switzerland. http://www.grsproadsafety.org/ (accessed February 1, 2005).

Allsop, R. E., N. N. Sze, and S. C. Wong. 2011. An update on the association between setting quantified road safety targets and road fatality reduction. *Accident Analysis & Prevention* 43: 1279–1283.

ATC. 2002. The national road safety strategy 2001–2010. Australian Transport Council, Canberra, Australian Capital Territory, Australia. http://www.dotrs.gov.au/atc/strategy.pdf (accessed August 17, 2002).

CHP. 2002. 2000 Annual report of fatal and injury motor vehicle traffic collisions. California Highway Patrol, Sacramento, CA. http://www.chp.ca.gov/html/switrs2000.html (accessed August 20, 2002).

DfT. 2001. A road safety good practice guide. Department for Transport, London, U.K. http://www.roads.dft.gov.uk/roadsafety/goodpractice/ (accessed August 21, 2001).

DfT. 2002a. Road safety advisory panel papers. Department for Transport, London, U.K. http://www.roads.dft.gov.uk/roadsafety/strategy/rsap/index.htm (accessed August 23, 2002).

DfT. 2002b. Tomorrow's roads: Safer for everyone; The Government's road safety strategy and casualty reduction targets document for 2010. Department for Transport, London, U.K. http://www.roads.dft.gov.uk/roadsafety/strategy/tomorrow/ (accessed August 22, 2002).

DfT. 2004. Tomorrow's roads—Safer for everyone: The first three year review. Department for Transport, London, U.K. http://www.dft.gov.uk/stellent/groups/dft_rdsafety/documents/page/dft_rdsafety_028165.hcsp (accessed January 26, 2005).

DGPPC. 2001. White paper on traffic safety in Japan (Abridged edition). Directorate General for Policy Planning and Coordination, Cabinet Office, Tokyo, Japan. http://www8.cao.go.jp/koutu/taisaku/2001wp-e.pdf (accessed August 21, 2002).

DTLR. 2002. Road safety research compendia. Transport Research, Department for Transport, Local Government and the Regions, London, U.K. http://www.research.dtlr.gov.uk/transport/02.htm#2 (accessed August 23, 2002).

Duan, L. R. 2002. Road safety in China: Problems and strategies. In *Road Safety—Strategy and Implementation*, eds. S. C. Wong, W. T. Hung, and H. K. Lo, pp. 51–71. Shenzhen, China: China Public Security Publisher.

Eiksund, S. 2009. A geographical perspective on driving attitudes and behaviour among young adults in urban and rural Norway. *Safety Science* 47: 529–536.

Elvik, R. 1993. Quantified road safety targets: A useful tool for policy making? *Accident Analysis & Prevention* 25: 569–583.

Elvik, R. 2001. Quantified road safety targets: An assessment of evaluation methodology. TØI Report 539/2001. Institute of Transport Economics, Oslo, Norway.

Haddon Jr., W. 1970. On the escape of tigers: An ecologic note. *American Journal of Public Health and the Nation's Health* 60 (12): 2229–2234.

Haight, F. A. 1985. Road safety: A perspective and a new strategy. *Journal of Safety Research* 16: 91–98.

Halden, D. and G. Harland. 1997. Planning road safety activities in Scotland. *Highway & Transportation* 44: 18–22.

Kmet, L. and C. Macarthur. 2006. Urban–rural differences in motor vehicle crash fatality and hospitalization rates among children and youth. *Accident Analysis & Prevention* 38 (1): 122–127.

Li, M. D., J. L. Doong, K. K. Chang, T. H. Lu, and M. C. Jeng. 2008. Difference in urban and rural accident characteristics and medical service utilization for traffic fatalities in less-motorized societies. *Journal of Safety Research* 39: 623–630.

Loo, B. P. Y. 2009. The identification of hazardous road locations: A comparison of the black-site and hot zone methodologies in Hong Kong. *International Journal of Sustainable Transportation* 3 (3): 187–202.

Loo, B. P. Y., C. B. Chow, M. Leung, T. H. J. Kwong, S. F. A. Lai, and Y. H. Chau. 2013. Multi-disciplinary efforts toward sustained road safety benefits: Integrating place-based and people-based safety analyses. *Injury Prevention* 19: 58–63.

Loo, B. P. Y., W. S. Cheung, and S. Yao. 2011. The rural–urban divide in road safety: The case of China. *The Open Transportation Journal* 5: 9–20.

Loo, B. P. Y., W. T. Hung, H. K. Lo, and S. C. Wong. 2005. Road safety strategies: A comparative framework and case studies. *Transport Reviews* 25 (5): 613–639.

Loo, B. P. Y., S. C. Wong, W. T. Hung, and H. K. Lo. 2007. A review of the road safety strategy in Hong Kong. *Journal of Advanced Transportation* 41 (1): 3–37.

Mueller, B. A., F. P. Rivara, and A. B. Bergman. 1988. Urban–rural location and the risk of dying in a pedestrian-vehicle collision. *Journal of Trauma & Acute Care Surgery* 28 (1): 91–94.

National Highway Traffic Safety Administration. 2008. Traffic safety facts, 2006 data: Rural/urban comparison. NHTSA, U.S. Department of Transportation, Washington, DC.

NRSC. 2000. Road safety strategy 2010. National Road Safety Committee, Wellington, New Zealand. http://www.ltsa.govt.nz/publications/docs/2010Strategy.pdf (accessed August 17, 2002).

NTRSC (Road Safety Council of the Northern Territory). 2001. *The Northern Territory road safety strategy 2001–2003: Safer territory roads.* Darwin, Australia: Department of Transport and Works.

OECD. 2002. Safety on roads: What's the vision? Paris, France: Organisation for Economic Co-operation and Development.

OECD Scientific Expert Group. 1994. Targeted road safety programmes. Paris, France: Organisation for Economic Co-operation and Development.

OTS. 1998. Annual progress report—Federal fiscal year 1998. Office of Traffic Safety, Elk Grove, CA. http://www.ots.ca.gov/profile/files/98annual.pdf (accessed August 18, 2002).

Rumar, K. 2002. Road safety work: Problems, strategies and visions. In *Road Safety—Strategy and Implementation*, eds. S. C. Wong, W. T. Hung, and H. K. Lo, pp. 1–17. Shenzhen, China: China Public Security Publisher.

Runyan, C. W. 1998. Using the Haddon matrix: Introducing the third dimension. *Injury Prevention* 4 (4): 302–307.

UN-ESCAP. 2002. Road safety action plans and programmes. United Nations Economic and Social Commission for Asia and the Pacific, Bangkok, Thailand.

van Uden, J. H. A. and A. H. Heijkamp. 1995. Intrinsic road safety: A new approach? *Safety Science* 19: 245–252.
Wegman, F., J. Vanselm, and M. Herweijer. 1991. Evaluation of a stimulation plan for municipalities in the Netherlands. *Safety Science* 14: 61–73.
Wong, S. C., N. N. Sze, H. F. Yip, B. P. Y. Loo, W. T. Hung, and H. K. Lo. 2006. Association between setting quantified road safety targets and road fatality reduction. *Accident Analysis & Prevention* 38 (5): 997–1005.

Appendix: STATS19 Data Record Sheets

FIGURE A.1 STATS19 vehicle records. (Reprinted from UK Department for Transport, STATS19 road accident injury statistics—Report form, September 2004, http://webarchive.nationalarchives.gov.uk/20110503151558/http://dft.gov.uk/pgr/statistics/datatablespublications/accidents/casualtiesgbar/, accessed January 19, 2015. With permission from UK Image Library of The National Archives.)

Appendix: STATS19 Data Record Sheets

Sept. 2004

MG NSRF/A

ACCIDENT STATISTICS

*FATAL / SERIOUS / SLIGHT

Incident URN

Other ref.

1.3 ACCIDENT REFERENCE

1.9 **TIME** H H M M   DAY* Su M T W Th F S   1.7 **DATE** D D M M 2 0 Y Y

1st Road Class & No. or (Unclassified - UC) (Not Known - NK)

1st Road Name

Outside House No. or Name or Marker Post No.

at junction with/or   metres N S E W *of

2nd Road Class & No. or (Unclassified - UC) (Not Known - NK)

2nd Road Name

Town   Sector /Beat No.

County or Borough

Parish No. or Name   1.10 **Local Auth No.** (if known)

1.11 **Grid Reference**   E →   N ↑

REPORTING Name   Number
OFFICER BCU/Stn   1.2 Force   Tel Number

| 1.5 | Number of vehicles | |
| 1.6 | Number of casualties | |

1.14	ROAD TYPE	X
Roundabout		1
One way street		2
Dual carriageway		3
Single carriageway		6
Slip road		7
Unknown		9

| 1.15 | Speed Limit (Permanent) | |

1.16	JUNCTION DETAIL	X
Not at or within 20 metres of junction	00
Roundabout	01
Mini roundabout	02
T or staggered junction	03
Slip road	05
Crossroads	06
Multiple junction	07
Using private drive or entrance	08
Other junction	09

JUNCTION ACCIDENTS ONLY

1.17	JUNCTION CONTROL	X
Authorised person	1
Automatic traffic signal	2
Stop sign	3
Give way or uncontrolled	4

1.20a	PEDESTRIAN CROSSING- HUMAN CONTROL	X
None within 50 metres	0
Control by school crossing patrol	1
Control by other authorised person	2

1.20b	PEDESTRIAN CROSSING- PHYSICAL FACILITIES	X
No physical crossing facility within 50m	0
Zebra crossing	1
Pelican, puffin, toucan or similar non junction pedestrian light crossing	4
Pedestrian phase at traffic signal junction	5
Footbridge or subway	7
Central refuge—no other controls	8

1.22	WEATHER	X
Fine without high winds	1
Raining without high winds	2
Snowing without high winds	3
Fine with high winds	4
Raining with high winds	5
Snowing with high winds	6
Fog or mist—if hazard	7
Other	8
Unknown	9

1.23	ROAD SURFACE CONDITION	X
Dry	1
Wet/Damp	2
Snow	3
Frost/Ice	4
Flood (surface water over 3cm deep)	5

1.21	LIGHT CONDITIONS	X
Daylight: street lights present	1
Daylight: no street lighting	2
Daylight: street lighting unknown	3
Darkness: street lights present and lit	4
Darkness: street lights present but unlit	5
Darkness: no street lighting	6
Darkness: street lighting unknown	7

1.24	SPECIAL CONDITIONS AT SITE	X
None	0
Auto traffic signal out	1
Auto traffic signal partially defective	2
Permanent road signing or marking defective or obscured	3
Roadworks	4
Road surface defective	5
Oil or diesel	6
Mud	7

1.25	CARRIAGEWAY HAZARDS	X
None	0
Dislodged vehicle load in carriageway	1
Other object in carriageway	2
Involvement with previous accident	3
Pedestrian in carriageway - not injured	6
Any animal in carriageway (except ridden horse)	7

1.26	Did a police officer attend the scene and obtain the details for this report?	X
Yes	1
No	2

Subject to local directions, boxes with a grey background need not be completed if already recorded

* Circle as appropriate
UNCLASSIFIED

FIGURE A.2 Attendant circumstances. (Reprinted from UK Department for Transport, STATS19 road accident injury statistics—Report form, September 2004, http://webarchive.nationalarchives.gov.uk/20110503151558/http:/dft.gov.uk/pgr/statistics/datatablespublications/accidents/casualtiesgbar/, accessed January 19, 2015. With permission from UK Image Library of The National Archives.)

MG NSRF/C

Sept. 2004

2.8 DIRECTION OF VEHICLE TRAVEL

1. Using the Example shown complete the FROM and TO boxes for the vehicles concerned, indicating direction of travel FROM and TO
2. If PARKED enter '00'

Vehicle 001		Vehicle 002	
FROM	TO	FROM	TO

Vehicle 003		Vehicle 004	
FROM	TO	FROM	TO

EXAMPLE

FROM	TO
1	3

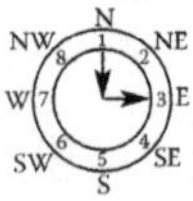

CASUALTY RECORD

3.4 VEHICLE REFERENCE NUMBER
Enter VEH No. which CASUALTY occupied (for pedestrians, code vehicle that struck them) e.g. 001,002 etc.

Casualty 001	0	Casualty 002	0
Casualty 003	0	Casualty 004	0
Casualty 005	0	Casualty 006	0

3.18 CASUALTY HOME POSTCODE
or Code: 1- Unknown
2- Non-UK Resident

Casualty 001	
Casualty 002	
Casualty 003	
Casualty 004	
Casualty 005	
Casualty 006	

3.7 SEX OF CASUALTY ✗

		CASUALTY 1	2	3	4	5	6
Male	1						
Female	2						

3.8 AGE OF CASUALTY (Estimate if necessary)
For children less than a year enter 00

Casualty 001		Casualty 002	
Casualty 003		Casualty 004	
Casualty 005		Casualty 006	

3.6 CASUALTY CLASS ✗

		1	2	3	4	5	6
Driver/Rider	1						
Veh./pillion Passenger	2						
Pedestrian	3						

3.9 SEVERITY OF CASUALTY ✗

		1	2	3	4	5	6
Fatal	1						
Serious	2						
Slight	3						

3.13 SCHOOL PUPIL CASUALTY ✗

		CASUALTY 1	2	3	4	5	6
School pupil on journey to or from school	1						
Other	0						

3.15 CAR PASSENGER (not driver) ✗

		1	2	3	4	5	6
Not a car passenger	0						
Front seat passenger	1						
Rear seat passenger	2						

3.16 BUS OR COACH PASSENGER ✗
(17 passenger seats or more)

		1	2	3	4	5	6
Not a bus or coach passenger	0						
Boarding	1						
Alighting	2						
Standing passenger	3						
Seated passenger	4						

PEDESTRIAN CASUALTIES ONLY

3.10 PEDESTRIAN LOCATION ✗

		CASUALTY 1	2	3	4	5	6
In carriageway, crossing on pedestrian crossing facility	01						
In carriageway, crossing within zig-zag lines at crossing approach	02						
In carriageway, crossing within zig-zag lines at crossing exit	03						
In carriageway, crossing elsewhere within 50m of pedestrian crossing	04						
In carriageway, crossing elsewhere	05						
On footway or verge	06						
On refuge, central island or central reservation	07						
In centre of carriageway, not on refuge, island or central reservation	08						
In carriageway, not crossing	09						
Unknown or other	10						

3.11 PEDESTRIAN MOVEMENT ✗

		CASUALTY 1	2	3	4	5	6
Crossing from driver's nearside	1						
Crossing from driver's nearside-masked by parked or stationary veh'	2						
Crossing from driver's offside	3						
Crossing from driver's offside-masked by parked or stationary veh'	4						
In carriageway, stationary-not crossing (standing or playing)	5						
In carriageway, stationary-not crossing (standing or playing), masked by parked or stationary veh'	6						
Walking along in carriageway-facing traffic	7						
Walking along in carriageway-back to traffic	8						
Unknown or other	9						

3.12 PEDESTRIAN DIRECTION ✗

		CASUALTY 1	2	3	4	5	6
Standing still	0						
North bound	1						
Northeast bound	2						
Eastbound	3						
Southeast bound	4						
Southbound	5						
Southwest bound	6						
Westbound	7						
Northwest bound	8						
Unknown	9						

3.19 PEDESTRIAN INJURED IN THE COURSE OF 'On The Road' WORK
Work actively carried out on public road (e.g. delivery services, road maintenance, postal delivery, traffic control etc.) ✗

		1	2	3	4	5	6
No	0						
Yes	1						
Not known	2						

LOCAL STATISTICS

Subject to local directions, boxes with a grey background need not be completed if already recorded

UNCLASSIFIED

FIGURE A.3 Casualty details. (Reprinted from UK Department for Transport, STATS19 road accident injury statistics—Report form, September 2004, http://webarchive.nationalarchives.gov.uk/20110503151558/http:/dft.gov.uk/pgr/statistics/datatablespublications/accidents/casualtiesgbar/, accessed January 19, 2015. With permission from UK Image Library of The National Archives.)

Index

A

Anti-drink-drive publicity campaigns, 232
ArcGIS software, 22, 30–31, 141
ArcObject module, 141
Arizona Local Government Safety Project (ALGSP) model, 163
Automobile Association (AA), 1, 220
Average annual daily traffic (AADT)
    observed collision rate, 164
    Poisson model, truck collision frequency, 188
    traffic volume index, 181
    ZIP model, 189

B

Basic spatial unit (BSUs)
    boundary problem, 148
    collision density, 152
    collision reporting system, 151
    HRLs, 169
    MAUP, 147
    network segmentation, 149
    traffic collisions, 148
Bicyclists
    improvement strategies, 262
    motor vehicle collision fatalities, 63–64
    no-car households and, 61
    parking provision, 117
    requirements of, 262
    risky method of transport, 262
    taxonomy, facilities, 262
"Bikeability" program, 276
Breath tests, 232
BSUs, *see* Basic spatial unit (BSUs)

C

Campaigns, road safety
    definition, 279
    drink-drive enforcement, 280
    enforcement level, 282
    individual campaigns, 280–281
    mass media advertising, 280
    RBT and speed cameras, 280
    on road collisions, effects of, 281–282
    scientific outcome-based evaluation, 281
    shock-based "fear appeals," 280
    small group, specific themes, 280

Child pedestrians
    age-related differences, 127
    antisocial, 62
    collision risk, 61
    driveway-related, 128
    educational measures, 273
    household resides, 62
    injury death rate, 60, 64
    land use, 63
    lower socioeconomic group families, 67
    motor-vehicle collisions, 125
    population-density, 62
    shopping sites, 63
    social classes, 64
    social exclusion, 62
    urban environments, risk factors, 130
Children and youth
    behavior changes, 273
    in classrooms, 274
    educational measures, 273
    ethnic minorities, 273
    objectives, 273–274
    probationary license, 276–277
    road and traffic environment, 273
    school education, cycle safety, 275–276
CI, *see* Confidence intervals (CI)
Collision–exposure relationship, 185
Collision frequency
    EB methods, 199
    HRLs, 164
    linear regression models, 186
    Poisson model, 34
    road safety strategy, 193
    statistical definitions, 164–165
    vehicular traffic exposure, 185
Collisions-in-networks
    distance in networks, 139
    geovalidation before collision analysis, 139–141
    nodes and links, 138
    1D events, 138
    spatial accuracy and precision, 138–139
Component Object Model (COM), 141
Confidence intervals (CI), 127, 164–165, 167
Critical number (CN), 163
Critical rate (CR), 163
"Cycling Proficiency Test," 276

Index

D

Darwin matrix for traffic calming, 255
Density functions
    DDS, 21
    frequency distribution, 22
    HRL, 22
    K-function method, 23
    location measurement, 22
    neighborhood, 21
    point density, 21
    reference density, 21
    spread and shape measurement, 22–23
Department for Transport (DfT), 120, 299–300
Deprivation, socioeconomic factors
    child injuries, 60
    collision risk, 65
    community sense, 61
    households, 63
    Index of Multiple Deprivation, 60
    injury mortality, 60
    investigations, 64
    lone parenthood, 65
    motor vehicle collision, 63
    neighborhoods, 60
    pedestrian/cyclist casualties, 60–61
    road traffic hazards, 60
    traffic injury, 64
Descriptive statistical analysis, 10
Disability-adjusted life years (DALYs), 40–41
Discrete density surface (DDS), 21
Distance-based methods
    Euclidian approach, 8
    GIS, 8
    journey time distance, 8
    network distance, 8
    SQL functions, 8
Drink-driving behaviors
    anti-drink-drive publicity campaigns, 232
    breath tests, 232
    collision risk, 231
    drug use, 232
    "exposed"/"not exposed," 99
    intensive publicity campaigns, 231
    media campaigns, 100
    penalties, 232–233
    post-collision management, 100
    randomized breath tests, 99
    and road collisions, 231
    teenage drivers, 99
Driveways
    Auckland study, 127–128
    awareness, media analysis, 126
    Chambers report in 2007, 126
    commercial parking, collision in, 125
    conventional police reporting, 123
    HASS/LASS database system, 125
    human factor, 126
    medical and second academic/analytical, 123
    public/community awareness, lack of, 127
    Queensland Ambulance Service and CARRS-Q, 125
    risks of, 123
    shared, 126
    Subaru Liberty range of station wagons, 128
    U.S. child fatalities, 123–124
    vehicles' visibility index, 126
Driving under the influence of drugs (DUID)
    adolescent marijuana use, 233–234
    cannabis, prevalence of, 233
    cognitive and experimental studies, 234
    fatal traffic accident, 233
    in Hong Kong, 234–235
    National Highway Transportation Safety Administration, 234
    traffic accidents, cause of, 233
Drug-driving
    dose–response relationship, 100
    issues, 100
    publication bias evidence, 100
DUID, *see* Driving under the influence of drugs (DUID)

E

EB methods, *see* Empirical Bayes (EB) methods
Elderly, road traffic collisions
    campaigns (*see* Campaigns, road safety)
    cognitive tests, 279
    collision rates examination, 279
    geodemographics, road users
        campaigns on, 284
        functions, 284
        MAST, 285–286
        "social marketing," 284
        Thames Valley Safer Roads Partnership, 284–285
    group sessions, 279
    intelligent transport system, 278
    older drivers, 278
    physical vulnerability, 277
    programs, 278–279
    publicity (*see* Publicity, road safety)
    self-selection bias, 279
    Western world issues, 278
Empirical Bayes (EB) methods
    collision frequency, 199
    effectiveness, 204
    helmet law enforcement, 203–204
    hierarchical Bayes methods, 201
    NB probability, 200
    Poisson probability, 200

Index

reference population, 199
spatial–temporal patterns, 200
statistical models, 199
Engineering safety; *see also* Location-specific treatments; Physical engineering; Vulnerable road users
  collision migration, 242
  collision situation and engineering remedies, 242–243
  complex and extensive, 241
  construction measures, 242
  control sites, selection of, 267–268
  countermeasure selection criteria, 242–243
  designs and management, 242
  driver and road environment, 241
  experimental design challenges, 265–266
  factors, 242
  insurance company claims, 265
  management measures
    accident risk, 257
    area-wide traffic calming, 259
    black spot treatment, 259
    braking distance maintenance, 258
    characteristics, 253
    collisions, continual reduction, 261
    Darwin matrix for traffic calming, 255–256
    deaths and serious injuries and shared responsibility, 250–251
    design choices, 253–254
    driver interventions, 249
    environmental street, 259
    institutional management functions, 251–252
    interventions, 252
    local communities, 254
    model, 251–252
    pedestrian streets, 260
    results, 252–253
    road infrastructure and maintenance, 253, 256
    safety management, 251
    sustainable road safety engineering, 260
    system-wide interventions, 250
    targeted results and institutional leaderships, 250
    traffic control, 258
    transport task, demands, 261
    urban and rural networks, 253
    urban play streets, 260
    vehicle speeds and traffic volumes, 255
    winter maintenance, roads, 258
    "Winter Service Plan," 257
  parameters, 264
  post-implementation monitoring, 264
  road design and equipment, 241
  road maintenance, 241
  statistical tests/procedures, designs and criteria, 266–267
  traffic control, 241
Environmental Systems Research Institute (ESRI)
  ArcGIS density measure, 31
  GIS-based spatial data validation system, 141
Euclidian/Manhattan techniques, 9
Exploratory spatial data analysis (ESDA), 22, 25–26
Exposure measure
  and collision frequency, 209–212
  HRLs, 209
  mobile phones, 210
  POP, STP and PPT, 210
  population-based methods, 210
  safety performance functions, 209
  vehicular traffic data, 209

G

Gatso camera, 225
Generalized linear model (GLM), 198
Geodemographics, 59
  campaigns on, 284
  census lead systems, 68
  commercially lead systems, 68
  communication programs, 69
  deprivation indicators, 67
  driver and casualty, 75
  functions, 284
  lead systems, 68
  in London, United Kingdom, 75–77
  MAST, 285–286
  mobility and constraints, 69
  Mosaic type, 75–76
  nonspatial analysis, 75
  overrepresentation, 68
  reduction strategies, 68
  residential layouts and housing types, 68
  risk exposure, 69
  road collision analysis, 67
  skepticism, 76
  "social marketing," 284
  sociocultural, 68
  and socioeconomic variables, 67
  Thames Valley Safer Roads Partnership, 284–285
  traffic injury risk, 67
Geographical information systems (GIS); *see also* Collisions-in-networks
  local engineering measures, 6
  Manhattan/Euclidian distance, 8
  postcodes and appended Mosaic type, 75
  research and operational need, 2
  spatial data validation system, 141–142

Geographically weighted regression (GWR)
  distance decay, 33
  expansion parameters, 32
  HRLs, 32
  kernel function, 33
  LISA, 32
  multiple deprivation, 33
  Poisson regression model, 33
  regression model, 34
  spatial autocorrelation techniques, 32
GIS, *see* Geographical information systems (GIS)
GIS-based spatial data validation system, 142–143
Global estimation, public health
  death rate, 41–47
  fatality rate, 42
  motor vehicles, 42
  road collision injuries, 41
Global positioning systems (GPS), 2, 140, 151, 208
Graduated License System (GLS), 277
Gross domestic product (GDP), 49, 53
GWR, *see* Geographically weighted regression (GWR)

H

Haddon matrix, 293
Hazardous road locations (HRLs), 209
  blacksites, 161
  BSUs, 162
  CN, 163
  collision prediction models, 164
  confidence intervals, 164
  CR, 164
  definition, 19
  empirical collision pattern, 164
  geovalidation, 169
  hot/black spots, 162, 165
  identification process, 174
  investigation/analysis process, 162
  network contiguity, 162
  nonspatial analysis, 162
  phases, 19–20
  PILs, 161
  risk reduction potential, 165
  RPs, 169
  simple density functions, 22
  space intervention, 163
  statistical definitions, 165
  traffic collisions, 161
Home Accident Surveillance System (HASS), 125
Hong Kong, road collision data
  district board coverage, 143
  geovalidation, 141–142
  GIS-based spatial data validation system, 142–143
  link-node system, 143
  road network database, 143–144
  TRADS database, 141
  traffic collisions in, 143–144
Hot zone, HRLs
  BSUs/RPs, 169
  collision statistics, 169
  link-attribute, 170–172
  model-building process, 170
  Monte Carlo simulations, 172
  spatial interdependency, 169
HRLs, *see* Hazardous road locations (HRLs)

I

International best practices, road safety
  administrations, evaluation of, 298–299
  Australia's road safety, 299
  California, targets setting, 299
  components, 299
  DfT, 299–300
  Great Britain (GB), 299
  Japan, 300
  New Zealand's, 300
  Sweden, 300
International Traffic Safety Data and Analysis (IRTAD)
  collision data, 47–48
  injury collisions, 48
  insurance data, 48
  Joint Transport Research, 47
  OECD, 42
  road fatalities, 47
  WHO, 47

J

Joint Transport Research, 47

K

Kernel density estimations (KDEs)
  ArcGIS, 30–31
  band 2, 30–31
  band 4, 30, 32
  bandwidth, 27–28
  cell density, 29
  circular neighborhood, 29
  clustering techniques, 28
  collision analysis, 28
  HRL techniques, 27
  interpolation technique, 27
  optimum bandwidth, 29–30
  risk levels, 28
  road safety, 27
  univariate data, 28
K-function method, 23

Index

Killed and seriously injured (KSIs)
  PIC and KSI prevented across Great Britain, 227–228
  "regression-to-mean" effect, 226

L

Land use planning
  accessibility effects, 113
  characteristics of areas, 114
  design stages, 111
  desires, wants, needs and possibilities of people, 112
  environment, livability and risks, 113
  factors, 112
  high-density areas, 115–116
  Millot's method, 114
  mobility impact, 112
  pedestrian, 116–117
  practices and "smart growth" land policies, 111
  properties, 114
  risks
    aggregate studies, 117
    area deprivation, 120
    car-based infrastructure, 119
    Chinese car penetration, 120
    DfT, 120
    disaggregate analysis, 117–118
    nonmotorized transport users, 119
    pedestrians/cycles/two-wheelers, 119
    rapid motorization, 118–119
    single-vehicle and multi-vehicle collisions, 121
    traffic congestion and road collisions, 118
    traffic flow, 117
    transportation management centers, 118
    types of trips, 118
  road network organization, 114
  safety effects and issues, 111–113
  single family housing, 116
  traditional areas, 115
  transport resistance, 112
  urban forms and development, 113–114
Leisure Accident Surveillance System (LASS), 125
Link-attributes, HRLs
  BSUs, 170
  collision frequency, 171–172
  GIS, 171
  Monte Carlo method, 170
  network connectivity, 170
  road networks, 172
  spatial relationships, 170
Local indicators of spatial association (LISA)
  global autocorrelation statistics, 26, 32
  local indices, 26
  maps, local spatial clusters, 154
  road collision analysis, 26–27

Location-specific treatments
  area-wide action, 244–245
  mass action, 244–245
  route action, 244
  single site, 244
London Accident Analysis Unit (LAAU), 1–2
Lost generation
  description, 286
  driver education and training programs, 286
  education, 287
  strategic targeting, 288

M

Market Analysis and Segmentation Tools (MAST)
  corporate public sector method, 285
  customer insight, 285
  online, 79
  "Safer Motorcycle Rider Campaign," 285
Millennium development goals (MDGs), 39–40
Mobile enforcement/mobile van camera, 225
Modifiable areal unit problem (MAUP)
  aggregation problem, 24
  boundary problem, 148
  BSUs, 147–148
  geovalidations, 148
  Poisson distribution, 149
  road safety research, 147
  scale problem, 148
  spatial distribution, 147
  statistical modeling, 148
  traffic collisions, 148
  2D point pattern, 147–148

N

National Highway Transportation Safety Administration, 234
Negative binomial methods
  AADT, 192
  collision frequency and risk, 191–192
  elasticity, 192
  empirical collision, 190
  maximum likelihood estimation, 191
  Poisson distribution, 190
  road safety, 190
  variance, 191
Neighbor analysis
  clustered pattern, 13
  collision prediction, 15
  decision tree clustering, 14
  dispersion patterns, 13
  edge effects, 15

k-mode clustering, 14
linear clustering effect, 16
minimum distance, 13
random distribution, 15
road safety literature, 12–13
spatial randomness, 15
VDM, 14
Network autocorrelation
benefit–cost method, 168
BSU, 161
collision frequency, 167–168, 173
communities, 173
decision makers, 174
down fluctuations, 166
EB methods, 168
GIS, 167–168
HRLs (see Hazardous road locations (HRLs))
injury and fatality, 168
investigation and treatment, 173
legislation and enforcement, 174
randomness, 166
ranking exercises, 166
road safety measurement, 169
score of priority, 174
Severity Index, 174
in site investigations, 166
spatial contiguity, 161
TL, 167
traffic collision patterns, 161, 173
2D kernel method, 161
Network collisions
aggregate analysis, 147
autocorrelation analysis (see Spatial autocorrelation analysis)
BSUs, 150–151
empirical collision pattern, 158
geographic coordinates, 151
geovalidation process, 151
GIS, 151
MAUP, 147–149
randomness, 147
road collision analysis, 147
Network segmentation
BSU, 149
dissolution, 149–150
GIS, 149–150
nodes, 149
traffic collisions, 149

O

Ordinary least squares (OLS) model, 199
Organisation for Economic Co-operation and Development (OECD)
International Transport Forum, 47
road network planning principles, 256

P

Pedestrian land use planning
area-wide schemes, 116
cycle planning, 117
cyclists, criteria, 116–117
integrated walking networks, 116
types of policies, 116
Pedestrians
average annual pedestrian injury rates, 66
casualty details, 10
and deprivation, 61–63
methodological flaws, 263
no-car households and, 60–61
pedestrian–motor vehicle conflicts, 263–264
"regression-to-mean," 263
risk and severity of, 263
separation countermeasures, 263
TRIS, search engine, 263
Physical engineering
low vs. high cost
collision reduction schemes in Oxfordshire, 246
consultation and implementation process, 248
HRL applications, 247–248
potential reductions (%), injury collision types, 246–247
safety countermeasures, 246, 248
roundabouts, 249
Poisson regressions, 179
AADT, 188
additive/multiplicative form, 187
Bernoulli experiment, 186
Chi-squared estimation, 187
collision frequency, 188
estimated parameters, 189–190
logarithm transformation, 187
probability density function, 188
resulting models, 187
single-vehicle collisions, 189
variance, 188
ZIP model, 188–189
Policy, socioeconomic factors
collision propensity, 80
community-based, 78, 80
customer insight, 78
evidence, 77
hazard management risk, 78
injury and death, 77–78
MAST online, 78–79
membership, 78
public health, 80
road safety policy, 77
sheer density, 81
social cohesion, 80
Population density (POP), 210

Index

Potential for collision reduction (PCR), 165, 168
Potential path tree (PPT) method, 210
Priority investigation locations (PILs), 161
Probationary license
  GLS evaluations, 277
  health and development, longitudinal study, 277
  learner's permit, 276
  minimum driving age, New Zealand, 277
  monthly collision rates, 276
  responsibility levels in learning, 276
  young drivers, 276
Public health issues
  accidents, 56
  broader development process, 50
  communities, 39, 49
  DALYs, 40–41
  data reliability, 51
  death and injury, 40
  decision-making process, 49
  financial resources, 48
  global estimation, 41–43
  hazards evidence, 55–56
  healthy lifestyles, 39
  hemorrhaging resources, 49
  humanitarian disaster, 39
  injury and death, 51, 53
  international road infrastructure, 54–55
  IRTAD, 43, 47–48
  MDGs, 40
  motor vehicle injuries, 39
  national resource planning, 49
  recording deficiencies, 51
  road collision costing, 53–54
  road risk, 50
  speed–fatality relationships, 54–55
  traffic fatalities, 40
  transport infrastructure, 39
  UN General Assembly, 40
  vehicle collisions, 49, 51–52
  vulnerability, 40
  WHO, 40, 51
Publicity, road safety
  aims and objectives, 282
  audience motivation, 283
  mass media campaigns, 282
  measurements, 283–284
  media selection, 283
  message content, 283
  target behavior and audience, 283
  types of data, 283

R

Random breath testing (RBT)
  description, 99
  drink-drive enforcement, 231, 280
  and speed cameras, 280

Red-light enforcement, 225–226
Reduction analysis techniques, 7
Reference density, 21
Reference points (RPs), 169
Research/policymaking analysis, 17
Risk-taking behaviors
  age and gender differences, 96–97
  behavioral adaptations, 89–90
  classification of papers, 86–87
  collision causation, 91–92
  "collision proneness," 89
  collision severity, 89
  community, regional and national levels, 87
  concepts of, 85–86
  culture and ethnicity, 98–99
  definition, 85
  drink driving, 99–100
  drug-driving, 100
  empirical research, 91
  excessive driving speed, 92
  good collision records, 87
  hot spot method, 90
  insurance business, "predicted loss" concept, 85
  low probability, negative outcome, 93
  managers, 91
  measuring, 94–96
  personality, role of, 93
  "risk compensation," 86, 88
  risk perception, 93
  risk thermostat, 91–92
  road use, risk consciousness, 88
  traffic environment, 89
  in transport systems, 86
  typology, 90–91
  WHO outlines, 93–94
Risk thermostat
  management process, 91–92
  risk compensation, 88
Road environment
  collision prediction models, 197–198, 201
  EB, 197, 199–200
  geometric features, 198
  GWR, 198–199
  helmet law enforcement, 203–204
  judgments, 204
  law enforcement activities, 202
  logistical regression, 198
  Maoming Transport Authority, 202
  mid-block location, 197–198
  overestimation, 202
  regression-to-mean problem, 197, 201
  research methodologies, 204
  safety improvement measures, 201
  site-specific safety, 197
Road safety administrations, 163, 173

Road safety education; *see also* Children and youth
  action plan, 296
  administration, 293
  changes, 294
  on children, 271
  components, 293–294
  Department for Transport, *Children's Road Traffic Safety*, 272
  driver education and training reviews, 271
  ethnic issues, 288–289
  evaluation and monitoring, 296
  funding, 297
  geographical variability
    international best practices, 298–300
    levels of details, 298
    rural–urban divide, 301–302
    scope and degree of sophistication, 298
  Haddon matrix, 293
  individual countries/administrations, 294
  initiatives, 294–295
  institutional framework, 297
  integrated, 271
  interventions, 271–272
  long-term approach, 304
  medium-term approach, 303–304
  objectives, 295
  post-license training, 272
  quantitative modeling, 297
  research and development, 296
  road collisions, reduction, 271
  short-term approach, 303
  targets, 295–296
  variability in delivery, 271
  vehicle control skills, 272
  vision, 295
Road Traffic Act 1934, 220, 224, 232–233
Road traffic collisions
  automobiles control, 3
  behavioral factors, 5, 7
  casualty and vehicle details, 10
  countermeasures, 5
  crash and incident, 2
  drivers and casualties, 10–12
  environmental risk factors, 7
  government agencies, 6
  intervention management, 8
  mobility management, 5
  quality data, 1
  robust research, 2
  safe collision, 1
  spatial heterogeneity, 17
  standardization, 7
  statistical return, 10
  temporal variations, 6
  transport system, 3
RPs, *see* Reference points (RPs)

S

Safety awareness planning; *see also* Driveways
  built environment, 122
  child pedestrian–vehicle collisions, 122
  crosswalk signs, 122
  framework, 120
  injury prevention and reduction, 120
  negative binomial spatial model, 122
  schools, 128–130
  university campuses, 122–123
School education, cycle safety
  "bikeability" program, 276
  geography, socio-economics and infrastructure, 275
  mandatory helmet laws, 275
  poor-quality environments and parental fear, 275
  "safe routes to school" programs, 275
Simple ranking (SR) method, 167
Socioeconomic factors
  census data, 70
  collision frequencies, 69
  community, 70
  database construction, 70, 72
  deprivation (*see* Deprivation, socioeconomic factors)
  empirical evidence, 59
  ethnicity, 65–66
  explanatory variables, 77
  exposure and inequality, 66–67
  geodemographics (*see* Geodemographics)
  geostatistics, 74
  health inequalities, 59
  Mosaic data, 70–71
  neighborhood, 59, 70
  omitted variable problem, 77
  policy and intervention
    collision propensity, 80
    community-based, 78, 80
    customer insight, 78
    evidence, 77
    hazard management risk, 78
    injury and death, 77–78
    MAST online, 78–79
    membership, 78
    public health, 80
    road safety policy, 77
    sheer density, 81
    social cohesion, 80
  qualitative data analysis, 73
  regression analysis, 73–74
  research methodology, 72
  risk-behavior, 77
  risk/likelihood, 72
  road collisions, 69

Index

statistics description, 73
typology analysis, 74–75
Space-time model
    collisions development, 207
    locations, 207
    road collision
        per population and per vehicle registered, 207–208
        per vehicle- and passenger-km, 208
        time-space measures, 208–209
    temporal importance, road collisions, 207
Space-time path (STP)
    groups of individuals, 209
    individual's life cycle and style, 209
    with 2D space, 208
Spatial analysis
    data quality and consistency, 135
    issues of, road collision data, 135
    nodes and arcs, 135
    road collision pattern and network, Hong Kong, 136–137
    2D and 1D space, random points, 135–136
Spatial autocorrelation analysis
    autocovariance, 152
    bandwidth, 155
    BSUs, 152
    collision density, 152
    density-based methods, 155
    dispersion and clustering, 152
    ESDA techniques, 25
    Gaussian function, 156
    GIS, 23
    global order effects, 25–26
    HRLs, 24
    KDE methods, 24, 155
    k intensity, 156
    LISA, 26–27, 154
    MAUP, 24
    minimum variance function, 156–157
    MISE, 156
    network proximity matrix, 153
    planar K-function, 157
    probability density function, 155
    quartic function, 156–157
    randomness, 152
    regression models, 155
    road safety analysis, 24
    RPs, 158
    SEM, 155
    socioeconomic variables, 25
    statistical significance test, 153–154
    traffic collisions, 152, 158
    2D space, 157
    Z-score method, 154
Spatial error model (SEM), 155
SPECS time-over-distance cameras, 226
Speed cameras
    additional collisions and injuries, 230
    "Cochrane Review," 227
    enforcement detection devices, 223
    frequency and severity, road accidents, 231
    Goldenbeld's dilemma classifications, 227, 229–230
    governments, local authorities and police, 226
    high collision frequency sites, 224
    joint ventures, 227
    management policy, 223
    PIC and KSI prevented across Great Britain, 227–228
    police speed enforcement tools, 223
    policing, 228–229
    "red routes" monitoring, 224
    research evaluations, 223
    risk compensation theory, 228
    safety cameras, 224
    telephone surveys, 224
    time stamps, 223
    types, 225–226
    UK National speed camera survey and reduction of injuries, 226–227
Speed enforcement
    characteristics, 215
    commission/omission errors, 215
    controversy, 221–222
    fixed-site engineering methods, 220
    future, 222–223
    high-risk collision sites, 221
    human intervention, 215
    penalty system, 216
    road collisions, 215
    speed limits, 220
    targeted and appropriate legislation, 216
    traffic, 215
Speeding; *see also* Speed enforcement
    AA, 220
    advantages and disadvantages, 217
    Australian Transport Safety Bureau, 219
    collision and casualty risk, 217
    excessive/inappropriate, 216
    high speeds and variations, 217
    low, middle and high bands, 217
    motorways traveling, 217
    regulation, 219
    Road Traffic Act 1934, 220
    road traffic collisions, 216
    setting of speed limits, factors, 219
    at slower speeds, 217
    spatial implications, 235
    speed cameras, 218
    trade off travel time *vs.* safety, drivers, 216–217
    UK 1865 Locomotive Act, 220
    U.S. federal and state studies, 218
    U.S. Transportation Research Board, 218

STATS19 vehicle records, 310–311
STP, *see* Space-time path (STP)

T

Thames Valley Safer Roads Partnership, 78, 284–285
Threshold levels (TL), 167
Topologically integrated geographic encoding and referencing (TIGER), 141
Traffic volume
  AADT, 179
  accidents, 184
  collision–exposure ratio, 180, 185
  collision frequency and risk, 179, 181
  crossing points, 184
  fatality numbers and rates, 181–183
  high-density environment, 181
  HRLs, 185
  linear regression model, 186
  negative binomial methods, 190–192
  Poisson regressions (*see* Poisson regressions)
  regression-to-mean problems, 193
  risk measures, 180
  road safety research, 179
  statistical models, 179, 181
  two-way junction, 184
  VKT/AADT, 180–181
Transportation Research Information Services (TRIS) database, 263
Transport Research Laboratory (TRL), 51
Transport system management, 3
Truvelo camera, 225
Two-dimensional (2D) space
  crime, 19
  density cluster functions, 20
  distance-weighting function, 34
  Euclidian distances, 34
  HRL, 19, 35
  point patterns, 16
  policy decision, 20
  quadrat methods, 20–21
  spatial heterogeneity, 19

U

UK 1865 Locomotive Act, 220
UK National Speed Camera Survey
  factors, 226
  KSIs, 226
  "regression-to-mean" effect, 226–227
University campuses
  limitations, 123
  pedestrian safety issues, 122
Urban development
  city's land use and infrastructure, 107
  "hot spots," pedestrian collisions, 107
  planning and land use environment, 108
  roads and local environmental conditions, 107
  socioeconomic changes, 108
  "suburbia" emergence, 108–109
  and transport decisions, 107
  victims, 109
Urban sprawl
  car-dependent communities, 109
  counties, 110
  definition, 109
  driving operations, 110
  index, census data, 110
  road fatalities and injuries, 109
  urban dwellers, 109
U.S. Transportation Research Board, 218

V

Value difference metric (VDM), 14
Variance methods, 11–12
Vehicle-kilometers traveled (VKT), 180–181, 186
Vulnerable road users
  bicyclists, 262
  pedestrians, 263–264
  road safety engineers, 261–262

W

"Winter Service Plan," 257

Z

Zero-inflated Poisson (ZIP) model, 188–189
Z-score method, 154